Technology, Peace and Security I
Technologie, Frieden und Sicherheit

Series Editor

Christian Reuter, Darmstadt, Germany

Technology plays a crucial role in various aspects of peace and security. This book series explores the intersection of computer science with peace and security studies. With a focus on cyber security and privacy, human-computer interaction as well as peace and conflict studies, it addresses topics such as peace informatics and technical peace research (cyber war, peace, arms control, and dual use), crisis informatics and information warfare (social media and collaborative technologies in conflict and crisis situations, misinformation and opinion manipulation), as well as usable safety, security, and privacy (resilient digital infrastructures, security and privacy enhancing technologies).

Technologie spielt eine entscheidende Rolle in verschiedenen Aspekten von Frieden und Sicherheit. Diese Buchreihe befasst sich mit Fragen an der Schnittstelle der Informatik mit der Friedens- und Sicherheitsforschung. Mit einem Fokus auf Cybersicherheit und Privatheit, Mensch-Computer-Interaktion sowie Friedens- und Konfliktforschung werden Themen wie Friedensinformatik und technische Friedensforschung (Cyberkrieg, Frieden, Rüstungskontrolle und Dual-Use), Kriseninformatik und Informationskrieg (soziale Medien und kollaborative Technologien in Konflikt- und Krisensituationen, Desinformation und Meinungsmanipulation) sowie nutzbare Sicherheit und Privatheit (resiliente digitale Infrastrukturen, Technologien zur Verbesserung der Sicherheit und des Datenschutzes) behandelt.

Markus Bayer

Deep Learning in Textual Low-Data Regimes for Cybersecurity

Springer Vieweg

Markus Bayer
Fachbereich Informatik
Technische Universität Darmstadt
Darmstadt, Germany

Dissertation approved by the Department of Computer Science of the Technical University of Darmstadt for the award of the academic degree Doctor rerum naturalium (Dr. rer. nat.) submitted by Markus Bayer, M. Sc. (born in Hanau)
First examiner: Prof. Dr. Dr. Christian Reuter
Second examiner: Prof. Dr. Lucie Flek
Date of submission: 14.08.2024
Date of disputation: 25.09.2024

ISSN 3004-9318 ISSN 3004-9326 (electronic)
Technology, Peace and Security I Technologie, Frieden und Sicherheit
ISBN 978-3-658-48777-5 ISBN 978-3-658-48778-2 (eBook)
https://doi.org/10.1007/978-3-658-48778-2

This Springer Vieweg imprint is published by the registered company Springer Fachmedien Wiesbaden GmbH, part of Springer Nature.
The registered company address is: Abraham-Lincoln-Str. 46, 65189 Wiesbaden, Germany

Foreword

Professionals in cybersecurity are overwhelmed by the rapidly increasing volume of publicly available data on cyber threats, complicating threat analysis for response teams. Initially, clustering techniques are explored to manage this data by grouping it into broad categories. Still, these methods need to be revised to provide the detailed analysis needed for precise threat identification and mitigation. This shortcoming highlights the necessity for more advanced approaches. Supervised machine learning offers a potential solution for predicting data relevance. Still, the dynamic nature of cyber threats limits the effectiveness of static classifiers, and training new classifiers for each incident is too labor-intensive and data-intensive.

This dissertation proposes a comprehensive solution utilizing low-data regime methods across different stages of the machine learning pipeline to enable effective training with minimal supervised data. Key strategies include: (1) Data Acquisition: Implementing an active learning strategy to optimize data labeling and enhance learning efficiency quickly. (2) Preprocessing: Developing data augmentation techniques to increase the diversity and volume of collected data. (3) Model Selection: Identifying an effective multi-level transfer learning strategy to leverage prior knowledge. And (4) Prediction: Enhancing model resilience and correcting errors through adversarial training.

These methods allow for training deep learning models for new cybersecurity incidents with minimal labeled data. Empirical evaluations show that using these methods in BERT-like models significantly improves performance over existing low-data regime techniques. While the primary focus is on cybersecurity, these methods are also applicable to other fields such as crisis informatics, business, and credibility assessment. The research emphasizes the growing importance of foundation models like GPT-4 and Llama 3 and the enduring relevance of the proposed techniques and thus makes an important contribution.

The studies included in this PhD thesis have been published as eight papers. In total, Markus has been involved in 16 scientific publications. His first publication began during the first semester of the PEASEC chair's existence in 2017 when Markus (a bachelor's student at that time) attended a seminar with us. Later, he did his bachelor's and master's theses in different areas, but luckily within our group, and then successfully continued for a PhD. Besides his research, he regularly provided us with tutorials and practical tips from the fast-developing area of deep learning, his area of expertise; he has actively been involved in our software development courses and further third-party acquisition. At the time of submitting his thesis he has already achieved scientific impact (about 850 citations), which is, in my opinion, excellent for a dissertation.

Markus Bayer has proven that he is capable of independent scientific work. Thus, in September 2024, his dissertation was accepted by the Department of Computer Science at the Technical University of Darmstadt for the degree of Dr. rer. nat. as the ninth PhD thesis in our research group PEASEC. I would like to see a further focus on topics of such high importance. Markus, thank you for your contribution and for allowing me to accompany you on your way to your PhD. I wish you all the best and every success for the future.

Prof. Dr. Dr. Christian Reuter
Professor for Science and Technology
for Peace and Security (PEASEC)
and Dean of the Department
of Computer Science at
Technical University of Darmstadt
Darmstadt, Germany

Acknowledgements

First and foremost, I would like to express my deepest gratitude to my direct supervisor, *Prof. Dr. Dr. Christian Reuter*. Your guidance has been a cornerstone of my academic journey, beginning even before my bachelor thesis and extending through this dissertation. The expertise and unwavering support you have provided have been invaluable. I am incredibly thankful for the countless hours you invested in reviewing my work, providing insightful feedback, and helping me navigate the complexities of my projects. I also extend my heartfelt thanks to my second referee, *Prof. Dr. Lucie Flek*, for your constructive advice and support. Additionally, I am grateful to *Prof. Dr. Sebastian Faust* and *Prof. Dr. Kristian Kersting*, who kindly join the doctoral committee.

I am deeply appreciative of the support and companionship of my research group, *Science and Technology for Peace and Security (PEASEC)* at the Department of Computer Science at the *Technical University of Darmstadt*. The collaborative spirit and intellectual exchange within our group have greatly enriched my research and broadened my perspectives. I would like to specifically thank *Dr. Marc André Kaufhold*, whose mentorship has been instrumental in shaping my academic path. Starting from my bachelor thesis and continuing through to this dissertation, your guidance and insights have profoundly influenced my growth and development. I would also like to acknowledge *Philipp Kühn* for his valuable discussions, assistance, and friendship. A special thanks goes to my student assisstant, *Tobias Frey*, for his support in preparing this dissertation.

My sincere thanks go to *CID GmbH* for their indispensable collaboration during my research, particularly in the areas of deep learning. I am profoundly grateful to *Dr. Björn Buchhold*, *Markus Neiczer*, *Marcel Keller*, *Maximilian Samsinger*, and *Daniel Frisinghelli* for their pivotal roles in shaping conceptual frameworks and enriching the discussions related to deep learning techniques. Their expertise and insights were crucial in navigating the complex challenges and significantly contributed to the success of this work.

Additionally, I would like to acknowledge all my co-authors, including *Dr. Thea Riebe*, *Daniel Hartung*, *Dr. Jörg Dallmeyer*, and *Ramin Shanehsaz* for their valuable contributions to this dissertation. Your insights and hard work have been instrumental. This dissertation was also made possible through the funding and support of the German Federal Ministry of Education and Research and the Hessian Ministry of Higher Education, Research, Science and the Arts within their joint support of the National Research Center for Applied Cybersecurity ATHENE, by the German Federal Ministry of Education and Research in the projects CYWARN (13N15407) and CYLENCE (13N16636), and the Lichtenberg high-performance computer at TU Darmstadt.

Finally, I want to express my deepest gratitude to my family and friends for their unwavering support and encouragement. To my parents, *Anna and Jürgen Bayer*, and my sister, *Jennifer Bayer-Maier*, thank you for your unconditional love and belief in me. To my friends, your companionship and understanding have been a constant source of strength. I am especially grateful to my girlfriend, *Samantha McGilley*, thank you for your patience, love, and for keeping me balanced and focused. Your presence in my life is a continuous source of great joy and motivation.

Sign language for "♡"

Markus Bayer

Abstract

In the field of cybersecurity, professionals face significant information overload as publicly disseminated data on cyber threats increases at an alarming rate. This overload complicates the task of analyzing cyber threat information, thereby posing a significant challenge for computer emergency response teams and other professionals. To address this challenge, we initially explore the utility of clustering techniques to manage the voluminous data by extracting coarse-grained groups. However, while clustering helps in understanding broad threat categories, it falls short in providing the fine-grained analysis necessary for precise threat identification and mitigation. This limitation underscores the need for more sophisticated, fine-grained analysis techniques. Although supervised machine learning is a viable solution for predicting precise data relevance and addressing this overload, the highly variable nature of cyber threats impedes the effectiveness of static classifiers for new incidents. Additionally, training new classifiers for each incident situation is too labor-intensive, requiring extensive expert-labeled data. This is especially pronounced in the field of deep learning, where the use of large datasets is essential for achieving unbiased generalization.

This dissertation presents a comprehensive solution by proposing and evaluating low-data regime methods across different stages of the supervised machine learning process to enable effective training with minimal labeled data: (1) Data Acquisition: We introduce an active learning strategy based on foundation models, such as GPT-4, to optimize data labeling and maximize learning efficacy. (2) Preprocessing: We employ and sensibly constrain GPT-2 and GPT-3 to function as data augmentation techniques, thereby improving the performance of pre-trained models by increasing diversity and volume of the collected data. (3) Model

Selection: We train a cybersecurity language model and pinpoint an effective multi-level transfer learning strategy, ensuring prior knowledge is leveraged. (4) Prediction: We develop a novel adversarial example generation method, grounded in explainable AI, to enhance model resilience and correct erroneous predictions.

These methods allow deep learning models to be trained for each new cybersecurity incident with a minimal number of labeled instances. Empirical evaluations demonstrate that these methods, when applied to BERT-like models, significantly outperform state-of-the-art low-data regime techniques, enabling effective and specialized collection of information in cyber threat situations. While the primary focus of this work is on enhancing cybersecurity information collection, the methods developed are broadly applicable across various domains, including those touched upon in this work, such as crisis informatics, business, and credibility assessment. This research also emphasizes the emergence of foundation models such as GPT-4 and Llama 3 and underscores the sustained importance of the proposed methods.

Author's Publications

In sum, 16 publications have been published in the context of the authors work. The following 8 publications are published as chapters in Part II of this thesis:

1. **Markus Bayer**, Marc-André Kaufhold, and Christian Reuter. "Information Overload in Crisis Management: Bilingual Evaluation of Embedding Models for Clustering Social Media Posts in Emergencies". In: *29th European Conference on Information Systems—Human Values Crisis in a Digitizing World, ECIS 2021, Marrakech, Morocco, 2020.* Ed. by Frantz Rowe, Redouane El Amrani, Moez Limayem, Sabine Matook, Christoph Rosenkranz, Edgar A. Whitley, and Ali El Quammah. 2021. URL: https://aisel.aisnet.org/ecis2021_rp/64 **CORE A (2018)**
2. **Markus Bayer**, Justin Lutz, and Christian Reuter. "ActiveLLM: Large Language Model-based Active Learning for Textual Few-Shot Scenarios". In: *Transactions of the Association for Computational Linguistics* (2025). https://doi.org/10.48550/ARXIV.2405.10808. arXiv: 2405.10808. URL: https://doi.org/10.48550/arXiv.2405.10808 **IF: 10.9 (2022)**
3. **Markus Bayer**, Marc-André Kaufhold, and Christian Reuter. "A Survey on Data Augmentation for Text Classification". In: *ACM Comput. Surv.* 55.7 (2023), 146:1–146:39. https://doi.org/10.1145/3544558 URL: https://doi.org/10.1145/3544558. **CORE A* (2020), IF: 16.6 (2022)**
4. **Markus Bayer**, Marc-André Kaufhold, Björn Buchhold, Marcel Keller, Jörg Dallmeyer, and Christian Reuter. "Data augmentation in natural language processing: a novel text generation approach for long and short text classifiers". In: *Int. J. Mach. Learn. Cybern.* 14.1 (2023), pp. 135–150. https://doi.org/10.1007/S13042-022-01553-3. URL: https://doi.org/10.1007/s13042-022-01553-3 **IF: 5.6 (2022)**

5. Marc-André Kaufhold, **Markus Bayer**, Daniel Hartung, and Christian Reuter. "Design and Evaluation of Deep Learning Models for Real-Time Credibility Assessment in Twitter". In: *Artificial Neural Networks and Machine Learning—ICANN 2021—30th International Conference on Artificial Neural Networks, Bratislava, Slovakia, September 14–17, 2021, Proceedings, Part V*. Ed. by Igor Farkas, Paolo Masulli, Sebastian Otte, and Stefan Wermter. Vol. 12895. Lecture Notes in Computer Science. Springer, 2021, pp. 396–408. https://doi.org/10.1007/978-3-030-86383-8_32. URL: https://doi.org/10.1007/978-3-030-86383-8_32 **CORE B (2020)**
6. **Markus Bayer**, Philipp Kuehn, Ramin Shanehsaz, and Christian Reuter. "CySecBERT: A Domain-Adapted Language Model for the Cybersecurity Domain". In: *ACM Trans. Priv. Secur.* 27.2 (Apr. 2024). ISSN: 2471–2566. https://doi.org/10.1145/3652594. URL: https://doi.org/10.1145/3652594 **CORE A (2020), IF: 2.3 (2022)**
7. **Markus Bayer**, Tobias Frey, and Christian Reuter. "Multi-level fine-tuning, data augmentation, and few-shot learning for specialized cyber threat intelligence". In: *Comput. Secur.* 134 (2023), p. 103430. https://doi.org/10.1016/J.COSE.2023.103430. URL: https://doi.org/10.1016/j.cose.2023.103430 **CORE B (2020), IF: 5.6 (2022)**
8. **Markus Bayer**, Markus Neiczer, Maximilian Samsinger, Björn Buchhold, and Christian Reuter. "XAI-Attack: Utilizing Explainable AI to Find Incorrectly Learned Patterns for Black-Box Adversarial Example Creation". In: *2024 Joint International Conference on Computational Linguistics, Language Resources and Evaluation (LREC-COLING 2024)*. LREC-COLING '24. 2024 **COLING: CORE A (2021)**

The following 8 papers are not included in the thesis, although their findings are supplementary to it:

9. Marc-André Kaufhold, Thea Riebe, **Markus Bayer**, and Christian Reuter. "'We Do Not Have the Capacity to Monitor All Media': A Design Case Study on Cyber Situational Awareness in Computer Emergency Response Teams". In: *Proceedings of the CHI Conference on Human Factors in Computing Systems, CHI 2024, Honolulu, HI, USA, May 11–16, 2024*. Ed. by Florian 'Floyd' Mueller, Penny Kyburz, Julie R. Williamson, Corina Sas, Max L. Wilson, Phoebe O. Toups Dugas, and Irina Shklovski. ACM, 2024, 580:1–580:16. https://doi.org/10.1145/3613904.3642368. URL: https://doi.org/10.1145/3613904.3642368 **Best Paper Award, CORE A* (2023)**

10. Philipp Kuehn, Mike Schmidt, **Markus Bayer**, and Christian Reuter. "ThreatCrawl: A BERT-based Focused Crawler for the Cybersecurity Domain". In: *CoRR* abs/2304.11960 (2023). https://doi.org/10.48550/ARXIV.2304.11960. arXiv: 2304.11960. URL: https://doi.org/10.48550/arXiv.2304.11960
11. Marc-André Kaufhold, **Markus Bayer**, Julian Bäumler, Christian Reuter, Stefan Stieglitz, Ali Sercan Basyurt, Milad Mirbabaie, Christoph Fuchss, and Kaan Eyilmez. "CYLENCE: Strategies and Tools for Cross-Media Reporting, Detection, and Treatment of Cyberbullying and Hatespeech in Law Enforcement Agencies". In: *Mensch und Computer 2023—Workshopband, Rapperswil, Switzerland, September 3–6, 2023.* Gesellschaft für Informatik e. V., 2023. https://doi.org/10.18420/MUC2023-MCI-WS01-211. URL: https://doi.org/10.18420/muc2023-mci-ws01-211
12. Thea Riebe, Tristan Wirth, **Markus Bayer**, Philipp Kuehn, Marc-André Kaufhold, Volker Knauthe, Stefan Guthe, and Christian Reuter. "CySecAlert: An Alert Generation System for Cyber Security Events Using Open Source Intelligence Data". In: *Information and Communications Security—23rd International Conference, ICICS 2021, Chongqing, China, November 19–21, 2021, Proceedings, Part I.* Ed. by Debin Gao, Qi Li, Xiaohong Guan, and Xiaofeng Liao. Vol. 12918. Lecture Notes in Computer Science. Springer, 2021, pp. 429–446. https://doi.org/10.1007/978-3-030-86890-1_24. URL: https://doi.org/10.1007/978-3-030-86890-1_24 **CORE B (2021)**
13. Marc-André Kaufhold, Jennifer Fromm, Thea Riebe, Milad Mirbabaie, Philipp Kühn, Ali Sercan Basyurt, **Markus Bayer**, Marc Stöttinger, Kaan Eyilmez, Reinhard Möller, Christoph Fuchß, Stefan Stieglitz, and Christian Reuter. "CYWARN: Strategy and Technology Development for Cross-Platform Cyber Situational Awareness and Actor-Specific Cyber Threat Communication". In: *Mensch und Computer 2021—Workshopband, Ingolstadt, Germany, September 5–8, 2021.* Ed. by Carolin Wienrich, Philipp Wintersberger, and Benjamin Weyers. Gesellschaft für Informatik e. V., 2021. https://doi.org/10.18420/MUC2021-MCI-WS08-263. URL: https://doi.org/10.18420/muc2021-mci-ws08-263
14. Philipp Kuehn, **Markus Bayer**, Marc Wendelborn, and Christian Reuter. "OVANA: An Approach to Analyze and Improve the Information Quality of Vulnerability Databases". In: *ARES 2021: The 16th International Conference on Availability, Reliability and Security, Vienna, Austria, August 17–20, 2021.* Ed. by Delphine Reinhardt and Tilo Müller. ACM, 2021, 22:1–22:11. https://doi.org/10.18420/muc2021-mci-ws08-263. URL: https://doi.org/10.1145/3465481.3465744 **CORE B (2021)**

15. Marc-André Kaufhold, **Markus Bayer**, and Christian Reuter. "Rapid relevance classification of social media posts in disasters and emergencies: A system and evaluation featuring active, incremental and online learning". In: *Inf. Process. Manag.* 57.1 (2020). https://doi.org/10.1016/j.ipm.2019.102132. URL: https://doi.org/10.1016/j.ipm.2019.102132 **CORE A (2020)**
16. Christian Reuter, Wolfgang Schneider, Daniel Eberz, **Markus Bayer**, Daniel Hartung, and Cemal Kaygusuz. "Resiliente Digitalisierung der kritischen Infrastruktur Landwirtschaft—mobil, dezentral, ausfallsicher". In: *Mensch und Computer 2018—Workshopband, Dresden, Germany, September 2–5, 2018*. Ed. by Raimund Dachselt and Gerhard Weber. Gesellschaft für Informatik e. V., 2018. https://doi.org/10.18420/MUC2018-WS12-0330. URL: https://doi.org/10.18420/muc2018-ws12-0330

Contents

List of Figures

List of Tables

Part I
Synopsis

1 Introduction

Deep learning, a cornerstone of modern computational solutions, is revolutionizing fields ranging from decision-making [251] and data analysis [286] to data categorization [234]. Cybersecurity, in particular, is one area that benefits from the capabilities of natural language processing (NLP) with deep learning. Integrating it into cybersecurity practices has become progressively essential in detecting, preventing, and mitigating sophisticated cyber threats and vulnerabilities prevalent today [273, 501]. As cyberattacks become increasingly intricate, pervasive, and dynamic, traditional rule-based or signature-based detection methods are proving insufficient [206, 500]. Deep learning, in contrast, shines with its capacity for automatic feature extraction, enabling substantial performance gains [167]. Yet, access to vast computing resources and extensive data is imperative for deep learning to exhibit its full potential; however, extensive data is limited in low-data regimes.

> In this thesis, a *low-data regime* in machine learning refers to scenarios specific to textual data where only a limited amount of labeled data is available for training models, irrespective of the volume of unlabeled or unsupervised data.

1.1 Motivation

Deep learning, despite its transformative impact on problems with abundant training data, has limited success in low-data regimes, thus constraining its applicability for emerging, niche, or label-intensive tasks [341]. This dissertation demonstrates that clustering can provide insights even without explicit supervised training.

M. Bayer, *Deep Learning in Textual Low-Data Regimes for Cybersecurity*, Technology, Peace and Security I Technologie, Frieden und Sicherheit, https://doi.org/10.1007/978-3-658-48778-2_1

However, this yields broad results without providing the fine-grained information often necessary. Recent advances have strived to reduce deep learning's reliance on large datasets [47, 212, 431] and promoted meta-learning, which involves the development of learning strategies that enable new tasks to be learned more quickly and efficiently based on past experiences [452]. There is a growing need to consolidate these advances and push the frontier, enhancing deep learning's efficacy in low-data regimes, especially for cybersecurity. This dissertation navigates the challenges of deep learning, in particular BERT-like transformer models in low-data regimes within the four main phases of the supervised machine learning process—Acquisition, Preprocessing, Model Selection, and Prediction—adapted and simplified from the frameworks proposed by Osisanwo et al. [317] and Verma et al. [454] (see Fig. 1.1).

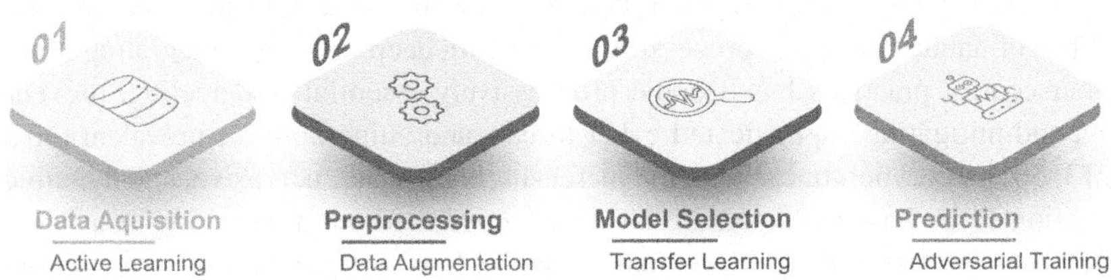

Fig. 1.1 Cornerstones of the supervised machine learning process from data acquisition to prediction in the real world, enriched by the main topics of this thesis (four points below)

Acquisition Phase: The first phase of the machine learning process involves obtaining labeled data from which a model can be trained. Active learning is a promising concept for reducing the number of labels for effective classification [198]. To complement the labeling process, it prioritizes specific instances over random sampling [393]. For instance, uncertainty sampling uses rapidly trained machine learning models to pinpoint instances with uncertain predictions [242], forming the decision boundary much faster and, by that, reducing labeling effort. However, current methods are unable to apply this to pre-trained transformer models because a mismatch between the active learning model and the final model architecture can lead to poor results [387, 397, 529]. In scenarios without this mismatch, they remain impractical due to long training times [517]. Furthermore, active learning strategies are not well-studied for low-data regimes, as they often face a cold-start problem. This refers to their inability to select meaningful instances without sufficient labeled data [68]. We investigate the potential of instruction-tuned language models, such as GPT-4 [315], as an active learning component in addressing the challenges above.

Preprocessing Phase: Once the labeled data has been collected, it can optionally be preprocessed, for example, with data augmentation strategies. Originating from computer vision, data augmentation refers to the artificial creation of new training instances, often from, but not limited to, existing data. By artificially enhancing the diversity of training data, these methods help to mitigate overfitting and data scarcity, enabling more effective model training even in low-data regimes [151]. However, this poses a significant challenge in the textual domain due to the intricacies of natural language transformations [209, 481]. Nonetheless, with the surge in research potential and sophisticated transformation techniques, the field has rapidly grown. Therefore, the field needs a holistic overview of textual data augmentation methods for classification, especially concerning pre-trained models. The advent of transfer learning has made models invariant to simple transformations and, by that, has rendered many augmentation techniques obsolete [29, 265], such as synonym-based [187, 214, 275], embedding-based [250, 365, 473], or simple noise-inducing [33, 99, 481] methods. This underscores the need for more sophisticated methods that can enrich contemporary models. While large language models are almost ideal in this case, given their ability to generate instances with novel linguistic patterns, at the time of this work, state-of-the-art approaches do not constrain them in a meaningful way [467] or are only designed for single-sentence instances [18].

Model Selection: Once the training data has been compiled, it is important to choose a suitable model. The well-known technique of transfer learning is particularly effective in scenarios with little data, showing significant performance improvements in many tasks [89, 339]. The underlying rationale is to pre-train models on abundant data and subsequently fine-tune them for specific tasks where data is sparse [346]. In this way, the model is already meaningfully initialized in the text domain, possibly modeling natural language better, leading to a beneficial state for the actual task. In particular, the field of cybersecurity would benefit significantly from a domain-adapted pre-trained model, as is the case in other fields such as biomedicine [238], law [59] and crisis informatics [229]. While CyBERT [347] constitutes a first model baseline, a substantially longer training of such a model with much more cybersecurity data would enhance existing systems and offer a basis for further research in this sector, ultimately leading to better cybersecurity tools. This can be sensibly combined with few-shot learning techniques, i.e., specific architecture selections or techniques to enable generalization from a small number of examples. Based on this, the dissertation explores the viability of a multi-level transfer learning technique for low-data regimes and further few-shot learning to enable specialized cyber threat information collection. While works from Riebe et al. [363] and Sceller et al. [381]

extract coarse-grained cybersecurity information, security professionals demand a more fine-grained technique that is flexibly adaptable to differing needs.

Prediction Phase: While deep learning systems demonstrate strong classification capabilities in the final task, they are vulnerable to small input manipulations ("adversarial attacks/examples") [17, 99], especially pronounced when trained in low-data regimes [130]. In recent research, works like BERT-Attack [247], BAE [124], Textfooler [184], or SMART [181] have tried to solve this problem with adversarial training, i.e. utilizing adversarial examples for model re-training. However, this often proves ineffective due to insufficient or very specific robustness impacts and evaluation methods that fail to capture anticipated bias reductions [126, 466]. Therefore, we tackle the question how to create adversarial examples that achieve a significant learning impact with adversarial training and how to assess this in a meaningful way.

At its core, this thesis seeks to amalgamate the methods discussed above into a cohesive framework tailored for cybersecurity professionals and computer emergency response teams (CERTs). CERTs are specialized teams responsible for both preventive and reactive measures related to computer security [49, 483]. One of their key activities involves collecting, analyzing, and disseminating information about emerging or existing cyber threats [196]—a process known as cyber threat intelligence (CTI) [279]. Given the constant influx in potentially relevant information content and the varying quality and credibility of information, executing the collection of CTI data is becoming increasingly challenging [246, 424]. CERTs report that they often do not have automated tools to manage and collect information from the various sources [196], leading to potential oversights in information and, ultimately, wrong decisions being made. The application of machine learning in the form of relevance classification could be the key component to stem information overload, as other fields dealing with information overload, such as crisis informatics, also show [198]. However, the cybersecurity domain faces the problem of high dynamics and individual relevance definitions, so classifiers need to generalize to new situations [85]. Meaning that it is almost infeasible to have only one classifier for the detection of relevant content, as new cyber situations often differ greatly from previous ones [22], e.g., in terms of individual requirements, the names of specific cyber threats, different attack methods, and affected products and functions. To make matters worse, these situations are time-sensitive, and cybersecurity teams are not able to label thousands of posts and articles every time a new incident is about to happen, resulting in a low-data regime for every new situation. Therefore, there is an emerging need for a machine learning framework that can address these challenges.

1.2 Aims and Research Question

This thesis aims to develop a comprehensive deep learning framework tailored to BERT-like transformer models for textual low-data regimes, with the goal of optimizing classifier performance in cybersecurity, specifically for the relevance classification of CTI information for CERTs and professionals.
The principal research question guiding this thesis is:

Main Research Question
How can the performance of deep learning classifiers for cyber threat intelligence texts be improved in low-data regimes?

To address this central question, the thesis will investigate several sub-questions corresponding to each step of the supervised machine learning process:

Data Acquisition: Active Learning

RQ1 **How can active learning in low-data regimes be enhanced by instruction-tuned large language models, enabling the resulting pre-trained models to gain high classification performance with reduced annotation effort?**
This segment proposes a novel strategy in active learning to lessen the manual effort in data labeling without compromising the classification performance of the resulting models. For this, instruction-tuned large language models, like GPT-4, are incorporated into the process.

| Chapter 7

Preprocessing: Data Augmentation

RQ2 **How can the advanced generative capabilities of large language models be leveraged and constrained for textual data augmentation to enhance classification performance, particularly in low-data regimes?**
This part overviews current data augmentation techniques in the textual data context. Based on this, sophisticated techniques based on large language models such as GPT-2 and GPT-3 are developed to increase classification performance in low-data regimes.

| Chapter 8, Chapter 9, Chapter 12

Model Selection: Transfer Learning

RQ3 **How can transfer learning methods be developed and applied to mitigate the dependency on extensive data requirements in deep learning applications?**
This part investigates the implementation and optimization of multi-level and domain-dependent transfer learning techniques to achieve higher classification performance with a focus on low-data regimes.

| Chapter 10, Chapter 11, Chapter 12

Prediction: Adversarial Training

RQ4 **How can textual adversarial examples be effectively formulated and evaluated in adversarial training to achieve significant learning effects?**
This segment focuses on generating textual adversarial examples to reinforce model resilience, which is especially important for models trained in low-data regimes [130]. It emphasizes misclassified examples, which can lead to stronger learning effects and higher performance if corrected. In addition, an evaluation method is designed which assesses the robustness improvements more effectively than current methods.

| Chapter 13

In addition to demonstrating how each method enhances its respective field of research and quantitatively assessing the resultant performance gains, we engage in a critical discourse on the relevance of these methods, particularly considering the advent of foundation models that appear to mitigate numerous challenges associated with low-data scenarios. In this synopsis, we use the term ‘foundation models’ to refer to large, pre-trained models such as GPT-3 [48] and Llama [446], which are capable of generating high-quality text and performing a wide range of tasks across various domains. This categorization helps to distinguish them from models like BERT [89] and GPT-2 [340], which, while also large language models pre-trained on extensive datasets, are primarily optimized for understanding and processing language with a focus on specific tasks such as classification or entity recognition, rather than general-purpose task-solving. As we explore the broader implications of these advances, it becomes imperative to assess whether the methods developed for low-data regimes hold their ground or even complement these powerful foundation models:

RQ5 **How do the techniques developed for enhancing deep learning in low-data regimes maintain their relevance in the era of foundation models?**

| Chapter 4

By addressing all research questions, this thesis aims not only to advance the scientific understanding and practical applications of deep learning for CTI information classification but also offers valuable insights and methods that can be generalized to other domains grappling with challenges in low-data scenarios.

1.3 Content and Structure of the Thesis

The remainder of this synopsis is structured as follows. Chapter 2 presents the research design of this dissertation and provides a brief overview of the research field and projects in which this dissertation is placed in, as well as the research methodology. Chapter 3 summarises the main findings of the studies of this dissertation. In Chap. 4, the relevance of the proposed methods in the context of foundation models is discussed and empirically assessed. The synopsis concludes with Chap. 5, in which the implications of this work are derived, the answers to the research questions are presented, and the limitations and future work of the dissertation are discussed.

Part II of this dissertation is then structured as follows. The general theme and research questions of this thesis are motivated in Chap. 6 where a framework for clustering instances to reduce information overload is shown. While the method works as intended, the real world application requires a specialized system that is able to retrieve more fine-grained and specialized information. The following studies address this requirement and are oriented towards the machine learning process in Fig. 1.1. The first work is placed within the data acquisition phase and explores how to implement a supporting active learning component based on foundation models (Chap. 7). Then data augmentation is introduced in the preprocessing phase, where at first a survey of augmentation approaches for text classification is given (Chap. 8), and then a novel data augmentation method is developed (Chap. 9). In the model selection phase, a classifier is transfer learned to perform credibility classification (Chap. 10). Furthermore, a state-of-the-art cybersecurity language model is trained (Chap. 11), which is then used in the following few-shot learning approach for identifying fine-grained information about cybersecurity threats (Chap. 12). Finally, in the prediction phase, a novel technique for creating adversarial examples with explainable AI is demonstrated, which helps to mitigate the incorrect behavior of the models used (Chap. 13).

Chapter 6 (Information Overload in Crisis Management) shows the principal problem of information overload in public data. To propose a fast solution for getting a comprehensive perspective on the various topics discussed in user-generated data, several clustering schemes are proposed and evaluated. Special emphasis is placed on embedding methods and the different languages that are communicated. This chapter has been published in the *European Conference on Information Systems* [30].

Contributions and Highlights

- A clustering system for reducing information overload in social media streams
- Comparison of embedding models across two languages and the proposal of a novel embedding evaluation task
- Proposal of a method for automatic cluster labeling

Author Statement

This paper constitutes a joint work of Markus Bayer, Marc-André Kaufhold, and Christian Reuter. As corresponding and leading author, Markus Bayer led the overall research design, management, and writing process of the paper. The authors conducted a literature review led by Markus Bayer, with additions and revisions by Marc-André Kaufhold. The research methodology, implementation and experimentation was conceptualized and conducted by Markus Bayer, with review and support from Marc-André Kaufhold. The results of the paper were interpreted by Markus Bayer, whereas the implications and contributions were written equally by Markus Bayer and Marc-André Kaufhold. Christian Reuter was the general advisor of this work and contributed with continuous feedback during all phases of the paper writing process.

CRediT author statement[1]:

Markus Bayer: Conceptualization, Methodology, Software, Formal analysis, Investigation, Data Curation, Writing—Original Draft, Writing—Review & Editing, Visualization, Project administration, Funding acquisition. **Marc-André Kaufhold:** Validation, Writing—Original Draft, Writing—Review & Editing, Funding acquisition.

[1] The contributor role taxonomy (CRediT) proposed by Brand et al. [45] provides a taxonomy for transparently describing scholarly contributions in a structured way [12].

Christian Reuter: Writing—Review & Editing, Resources, Supervision, Funding acquisition.

Chapter 7 **(ActiveLLM: Large Language Model-based Active Learning for Textual Few-Shot Scenarios)** demonstrates a novel active learning method for BERT classifiers. By using foundation models as active learning components that do not need to be trained, the method overcomes the long latency and cold-start problems of active learning strategies. In low-data regimes the method significantly outperforms other active learning strategies and state-of-the-art few-shot learning methods. This chapter has been published in the *Transactions of the Association for Computational Linguistics* [32].

Contributions and Highlights

- A novel active learning method based on foundation models like GPT-4 and Llama 3
- Overcoming the cold-start and model-mismatch problem of active learning strategies
- Comprehensive evaluations with various foundation models, other active learning methods, and few-shot learning

Author Statement
This paper constitutes a joint work of Markus Bayer, Justin Lutz and Christian Reuter. As corresponding and leading author, Markus Bayer performed the overall research design, management, and writing of the paper. Markus Bayer conceptualized and conducted the literature review, research methodology, implementation, experimentation, and interpretation. Justin Lutz assisted Markus Bayer in revising the paper and data presentation. Christian Reuter was the general advisor of this work and contributed with continuous feedback during all phases of the paper writing process.

CRediT author statement:

This paper constitutes **Markus Bayer:** Conceptualization, Methodology, Software, Formal analysis, Investigation, Data Curation, Writing—Original Draft, Writing—Review & Editing, Visualization, Project administration, Funding acquisition. **Justin Lutz:** Validation, Visualization, Writing—Review & Editing. **Christian Reuter:** Resources, Supervision, Funding acquisition.

Chapter 8 **(A Survey on Data Augmentation for Text Classification)** gives an overview of data augmentation methods that can be used for text classification. The basics of data augmentation, as well as its goals and applications, are explained. While the augmentation methods are described in detail, the state of the art is clearly shown, and further research perspectives, especially in the light of generative language models, are identified. This chapter has been published in the *Computing Surveys Journal* [29].

Contributions and Highlights

- Overview of goals and applications of data augmentation
- Comprehensive survey on data augmentation in text classification
- Data-structure-driven taxonomy and method-oriented categorization
- State-of-the-art review with comparisons and in-depth details of the methods
- Identification of future research perspectives

Author Statement
This paper constitutes a joint work of Markus Bayer, Marc-André Kaufhold, and Christian Reuter. As corresponding and leading author, Markus Bayer led the overall research design, management, and writing process of the paper. The research methodology was conceptualized and conducted by Markus Bayer, with review and support from Marc-André Kaufhold. The analysis and interpretation of the paper was performed by Markus Bayer. Marc-André Kaufhold assisted Markus Bayer in writing the paper. Christian Reuter was the general advisor of this work and contributed with continuous feedback during all phases of the paper writing process.

CRediT author statement:

Markus Bayer: Conceptualization, Methodology, Formal analysis, Investigation, Data Curation, Writing—Original Draft, Writing—Review & Editing, Visualization, Project administration, Funding acquisition. **Marc-André Kaufhold:** Validation, Writing—Original Draft, Writing—Review & Editing, Funding acquisition. **Christian Reuter:** Resources, Supervision, Funding acquisition.

Chapter 9 **(Text Generation Approach for Data Augmentation)** shows a novel data augmentation method for natural text. The developed synthetic data generation method achieves large performance improvements over the baseline and a common data augmentation method by combining the large language model GPT-2 with a specialized label conditioning technique and a filtering mechanism with the human in the loop. The study of this chapter has been published in the *International Journal of Machine Learning and Cybernetics* [28].

Contributions and Highlights

- Novel data augmentation method based on GPT-2
- Overcoming the challenge of data augmentation to benefit pre-trained models by creating unseen training data relevant to the task
- Comprehensive evaluations with domain-specific and low-data regime insights

Author Statement
This paper constitutes a joint work of Markus Bayer, Marc-André Kaufhold, Björn Buchhold, Marcel Keller, Jörg Dallmeyer, and Christian Reuter. As corresponding and leading author, Markus Bayer led the overall research design, management, and writing process of the paper. The authors conducted a literature review led by Markus Bayer, with additions and revisions by Marc-André Kaufhold, Björn Buchhold, and Marcel Keller. The research methodology, implementation and experimentation was conceptualized and conducted by Markus Bayer, with review and support from Marc-André Kaufhold, Björn Buchhold and Marcel Keller. The results of the paper were interpreted by Markus Bayer, whereas the implications and contributions were written equally by Markus Bayer and Marc-André Kaufhold. Christian Reuter and Jörg Dallmeyer were the general advisors of this work and contributed with continuous feedback during all phases of the paper writing process.

CRediT author statement:

Markus Bayer: Conceptualization, Methodology, Software, Formal analysis, Investigation, Data Curation, Writing—Original Draft, Writing—Review & Editing, Visualization, Project administration, Funding acquisition. **Marc-André Kaufhold:** Validation, Writing—Original Draft, Writing—Review & Editing, Funding acquisition. **Björn Buchhold:** Validation, Writing—Review & Editing. **Marcel Keller:** Validation, Writing—Review & Editing. **Jörg Dallmeyer:** Resources, Supervision. **Christian Reuter:** Resources, Supervision, Funding acquisition.

Chapter 10 **(Design and Evaluation of Deep Learning Models for Real-Time Credibility Assessment in Twitter)** contains a study regarding transfer learning of the BERT model for assessing credibility of Twitter posts in real-time. The study includes a data collection process and a comparison of several frameworks suitable for the credibility classification task. The pretrained language model BERT achieves the best results as it already contains knowledge which is essential for credibility classification. This chapter has been published in the *International Conference on Artificial Neural Networks* [194].

Contributions and Highlights

- Analysis of transfer learning for credibility classification
- A novel unified credibility dataset
- Insights into real-time credibility classification

Author Statement

This paper constitutes a joint work of Markus Bayer, Marc-André Kaufhold, Daniel Hartung, and Christian Reuter. Marc-André Kaufhold and Markus Bayer led the management and writing process of the paper. The authors conducted a literature review led by Daniel Hartung, with additions and revisions by Marc-André Kaufhold and Markus Bayer. The scientific quality of the research methodology, implementation and experimentation, which were conducted by Daniel Hartung, were supervised and validated by Marc-André Kaufhold and Markus Bayer. Markus Bayer contextualized the work within current machine learning research by including related studies and both interpreting and rewriting the results to address existing research gaps. The findings and contributions were written equally by Marc-André Kaufhold, Markus Bayer and Daniel Hartung. Christian Reuter was the general advisor of this work and contributed with continuous feedback during all phases of the paper writing process.

CRediT author statement:

Marc-André Kaufhold: Conceptualization, Validation, Writing—Original Draft, Writing—Review & Editing, Visualization, Project administration, Funding acquisition. **Markus Bayer:** Conceptualization, Validation, Investigation, Writing—Original Draft, Writing—Review & Editing, Visualization, Project administration, Funding acquisition. **Daniel Hartung:** Conceptualization, Methodology, Software, Formal analysis, Investigation, Data Curation, Writing—Initial Draft, Visualization. **Christian Reuter:** Resources, Supervision, Funding acquisition.

Chapter 11 (CySecBERT: A Domain-Adapted Language Model for the Cybersecurity Domain) presents a novel cybersecurity language model. This work contains a sensibly chosen cybersecurity dataset consisting of blogs, arXiv-paper, Twitter posts, and vulnerability database entries. The pretrained language model BERT is domain-adapted on this dataset. While we show that the novel model outperforms the state of the art in various cybersecurity tasks, we also place particular emphasis on preventing catastrophic forgetting. This chapter has been published in the *ACM Transactions on Privacy and Security* [31].

Contributions and Highlights

- A state-of-the-art cybersecurity language model based on BERT
- A large cybersecurity dataset consisting of blogs, arXiv-paper, Twitter posts, and vulnerability database entries
- Experiments for hyperparameter tuning in the light of catastrophic forgetting
- State-of-the-art comparisons for 15 different tasks

Author Statement
This paper constitutes a joint work of Markus Bayer, Philipp Kühn, Ramin Shanehsaz, and Christian Reuter. Markus Bayer and Philipp Kühn led the overall research design, management, and writing process of the paper. The research methodology and experiments were initially designed and executed by Ramin Shanehsaz and validated by Markus Bayer and Philipp Kühn. Philipp Kühn carried out the data collection and filtering. Markus Bayer refined and extended the research methodology and experiments, and performed the analysis and interpretation of the results. Markus Bayer drafted the paper with contributions from Philipp Kühn. Both were involved in incorporating necessary revisions. Christian Reuter was the general advisor of this work and contributed with continuous feedback during all phases of the paper writing process.

CRediT author statement:

Markus Bayer: Conceptualization, Methodology, Software, Validation, Formal analysis, Investigation, Data Curation, Writing—Original Draft, Writing—Review & Editing, Visualization, Project administration, Funding acquisition. **Philipp Kühn:** Conceptualization, Methodology, Software, Validation, Data Curation, Writing—Original Draft, Writing—Review & Editing, Visualization, Project administration, Funding acquisition. Investigation **Ramin Shanehsaz:** Writing—Initial Draft, Software, Investigation, Visualization. **Christian Reuter:** Resources, Supervision, Funding acquisition.

Chapter 12 **(Multi-Level Fine-Tuning, Data Augmentation, and Few-Shot Learning for Specialized Cyber Threat Intelligence)** proposes a novel pipeline for fine-grained and potentially individualized collection of cybersecurity information. The pipeline consists of a novel transfer learning technique based on multi-level fine-tuning, a new GPT-3 data augmentation technique, and state-of-the-art few-shot learning adapted for cybersecurity. Furthermore, this study presents a specialized CTI dataset annotated by three experts. The framework, enabling the rapid training of new, high-performance classifiers, achieves state-of-the-art results for solving this task. This chapter has been published in *Computers & Security* [27].

Contributions and Highlights

- Novel pipeline combining transfer learning, data augmentation, and few-shot learning
- New transfer learning technique based on multi-level fine-tuning
- Novel data augmentation technique based on GPT-3
- New specialized CTI dataset annotated by three experts and based on the 2021 Microsoft Exchange Server data breach

Author Statement
This paper constitutes a joint work of Markus Bayer, Tobias Frey, and Christian Reuter. As corresponding and leading author, Markus Bayer performed the overall research design, management, and writing of the paper. Markus Bayer conceptualized and conducted the literature review, research methodology, implementation, experimentation, and interpretation. Tobias Frey assisted with the literature review and implementation, and carried out the experiments. Christian Reuter was the general advisor of this work and contributed with continuous feedback during all phases of the paper writing process.

CRediT author statement:

Markus Bayer: Conceptualization, Methodology, Software, Validation, Formal analysis, Investigation, Data Curation, Writing—Original Draft, Writing—Review & Editing, Visualization, Project administration, Funding acquisition. **Tobias Frey:** Investigation, Data Curation, Software, Writing—Review & Editing **Christian Reuter:** Resources, Supervision, Funding acquisition.

Chapter 13 (**XAI-Attack: XAI for Adversarial Example Creation**) introduces a novel adversarial example creation method that identifies indicators of wrong predictions. For this, explainable AI methods are incorporated into adversarial example research. Furthermore, the study highlights the need for more accurate robustness evaluations and proposes a novel out-of-distribution evaluation for this purpose. Evaluations based on this setting, human annotations, and transferring attacks to other models show state-of-the-art performance of the adversarial example creation method. This chapter has been published in the *Proceedings of the 2024 Joint International Conference on Computational Linguistics, Language Resources and Evaluation* [32].

Contributions and Highlights

- A novel method for creating adversarial examples based on identifying indicators of wrong predictions
- Proposal to utilize XAI methods as importance functions for adversarial example creation
- A novel out-of-distribution evaluation setting for adversarial training that enables a more accurate assessment of robustness

Author Statement
This paper constitutes a joint work of Markus Bayer, Markus Neiczer, Maximilian Samsinger, Björn Buchhold, and Christian Reuter. As corresponding and leading author, Markus Bayer performed the overall research design, management, and writing of the paper. Markus Bayer conceptualized and conducted the literature review, research methodology, implementation, experimentation, and interpretation, with review and support from Markus Neiczer, Maximilian Samsinger, and Björn Buchhold. Christian Reuter was the general advisor of this work and contributed with continuous feedback during all phases of the paper writing process.

CRediT author statement:

Markus Bayer: Conceptualization, Methodology, Software, Validation, Formal analysis, Investigation, Data Curation, Writing—Original Draft, Writing—Review & Editing, Visualization, Project administration, Funding acquisition. **Markus Neiczer:** Validation, Writing—Review & Editing. **Maximilian Samsinger:** Validation, Writing—Review & Editing. **Björn Buchhold:** Validation, Writing—Review & Editing. **Christian Reuter:** Resources, Supervision, Funding acquisition.

2 Research Design

This chapter presents the research design of the dissertation. First, the research field of deep learning in NLP and CTI is introduced. In the following sections, the context in which the research is conducted is shown, and the general research approach and its methodology are presented.

2.1 Research Field and Foundations

This work aims to contribute to the research field of machine learning, more specifically deep learning in NLP, by exploring, developing, and evaluating techniques that are beneficial for textual low-data regimes. Furthermore, this aim is being realized in the field of cybersecurity, specifically in CTI research. In combination, deep learning methods are employed to condense vast amounts of cybersecurity data into relevant content.

The field of deep learning focuses on the development and adaptation of deep neural networks to achieve various goals, such as performance improvements [463], robustness [405], explainability [361], privacy [41], novel application contexts [219], and many more. In addition to novel architectures, different techniques are being developed within the deep learning process, such as optimization techniques, learning strategies, labeling techniques, or data manipulations [94, 318]. To assess whether the previously defined goal is achieved, the typical evaluation is empirical testing. Commonly, a comparison of the model with and without the technique (i.e., baseline) or against a model with another related technique (e.g., with the state of the art) is performed. For BERT-like models, this is typically done with extrinsic tasks, where a classifier is trained on the latent representation output of the encoder-only model. However, if the model itself, and not just the practical application, is of interest, it may be useful also to evaluate intrinsically, describing that

M. Bayer, *Deep Learning in Textual Low-Data Regimes for Cybersecurity*, Technology, Peace and Security I Technologie, Frieden und Sicherheit, https://doi.org/10.1007/978-3-658-48778-2_2

the representations are being evaluated, for example, in word similarity tasks. The evaluation metrics vary according to the goal defined previously. When aiming to improve performance or reduce variance, metrics such as accuracy or the F1 score are often used. Accuracy refers to the proportion of true results among the total number of cases examined, making it a straightforward measure for evaluating the correctness of a model's predictions. On the other hand, the F1 score is a metric that considers both precision (the proportion of true positive results among all positive predictions) and recall (the proportion of true positive results among all actual positives), providing a balance between the model's ability to correctly label positives without incorrectly labeling negatives. This is particularly helpful in situations with imbalanced class distributions [364]. Time improvement is trivially measured by the model's training and/or testing time, although which is more informative depends on the context. A work that wants to increase the speed of the training process must specify the training time, and a work that wants to increase the speed of classification in the application context (inference) must specify the speed of evaluation of the test examples or new instances. If a research paper aims to emphasize or address privacy, explainability, or adversarial examples, it may be more design-orientated, ultimately demonstrating results through examples or implementations. However, to measure, for example, adversarial robustness, a model may be tested with different adversarial combinations. Many works in deep learning also contain ablation studies, which are a systematic approach to understanding the contribution of each component or configuration of a technique by removing parts of it. Besides contributing the code or resulting models, works in machine learning can also include the creation and publication of datasets. This is of particular benefit to the research community, as training data is most important for machine learning models [25]. Like the list of research objectives, the list of research methods for validating, verifying or reproducing the objectives is also extensive.

While the deep learning research in this thesis can be used for other text classification problems, it is mainly applied for information collection in CTI and cybersecurity. Cybersecurity research also has a wide range of directions that include, for example, cryptography, network security, software and hardware security, data analytics, malware analysis, privacy technologies, and AI and IOT security. Studies on CTI are related to data analytics and software/hardware security but can also branch out in other directions. A great research cluster in this area uses machine learning models to reduce the amount of data from open sources, such as social media [363, 381]. These models are used, for example, to extract specific or the most important information. In CTI research, it is also common to analyze datasets and databases, for example, to assess the quality of information [219], to gain insights from the past

[368] or to identify trends [486]. As previously stated, this thesis is concerned with the extraction of specific CTI information in order to alleviate information overload.

2.2 Research Context

This dissertation is a central part of the CYWARN project and is part of the dedicated project within the Lichtenberg High Performance Computer, which enables the necessary computing resources. Furthermore, this work contributed to the CYLENCE project and the research areas of ATHENE.

CYWARN (13N15407) is a research project funded by the German Federal Ministry of Education and Research (BMBF). Its aim is to support CERTs with new strategies and technologies for gathering, analyzing and communicating the cyber situation. A demonstrator is created that enables the automated collection of cybersecurity sources as well as data evaluation with credibility analysis and information prioritization. I have been fully involved in the CYWARN project since it started in October 2020.

The follow-up project CYLENCE (13N16636), also funded by the BMBF, aims to develop strategies and mechanisms against cyberbullying and hate speech. The findings of this dissertation make a significant contribution to the advancement of detection methods for this project, which are also integrated into a demonstrator for further analysis by investigation and prosecution authorities. I have been involved in the CYLENCE project since it started in August 2023.

The Lichtenberg High Performance Computer Darmstadt has been active since 2012 and provides computing resources to German universities and public research institutions. The high-performance computer is in its second expansion stage (Lichtenberg II), which offers various NVIDIA accelerators for machine learning. As part of the CYWARN project and specifically for this dissertation, a four-year project enabled the use of these computer resources since January 2021. I am the project manager and main user.

The National Research Center for Applied Cybersecurity ATHENE is funded by the BMBF and the Hessian Ministry of Higher Education, Research, Science and the Arts, with the participation of the Technical University of Darmstadt, Fraunhofer SIT, Fraunhofer IGD and Darmstadt University of Applied Sciences. It is a cooperation model of university and non-university research that promotes security solutions for the economy, society, and the state. ATHENE is a cybersecurity and privacy research centre with the aim of improving cybersecurity and data protection. I have been involved with ATHENE and the research areas Secure Urban Infrastructures (SecUrban) and User-centered Security and Privacy (UCSP) since 2021 and 2022, respectively. My contributions are mainly directed at the User-Centered Technology Design for Cyber Situational Awareness (CYAWARE) project, in which I design usable security technologies enhancing the cyber situational awareness of CERTs, businesses, industry, government, and society. I am the project representative of CYAWARE since its beginning in January 2023.

2.3 Research Methodology

This dissertation encompasses a wide array of studies, each contributing to different phases of the machine learning process, from data acquisition and preprocessing to model selection and prediction. The research methodology of this dissertation consists of literature reviews to identify research gaps, followed by the design and development of tailored techniques and empirical evaluations. The specifics of the methodology are delineated in the following paragraphs in chronological order corresponding to the chapters of Part II and the supervised machine learning process, as illustrated in Fig. 2.1. Table 2.1 provides a synthesized overview of the research methodology.

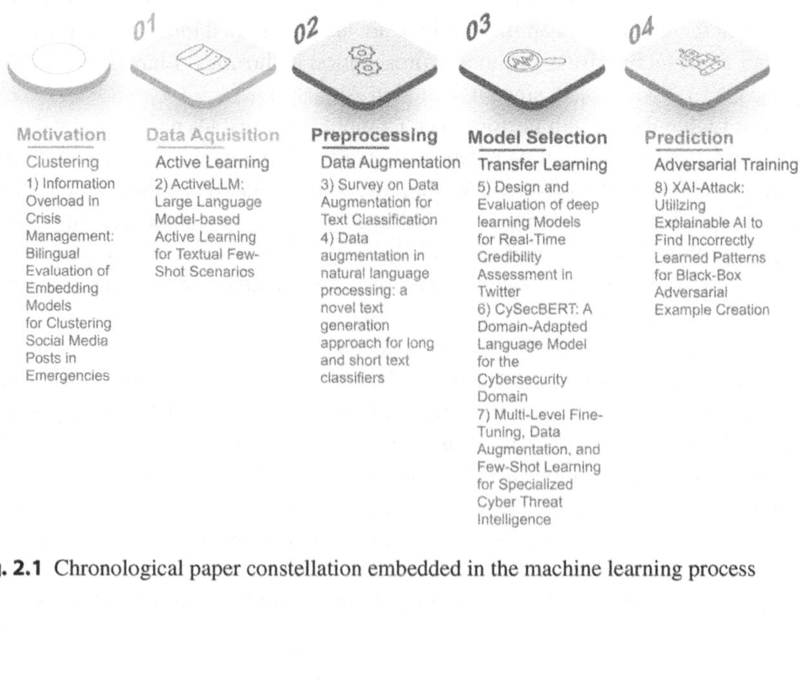

Fig. 2.1 Chronological paper constellation embedded in the machine learning process

Table 2.1 Overview of the research methodology in the various chapters of the dissertation

	Background & Review	Technique	Evaluation
6	1. Techniques to mitigate information overload in crises 2. Clustering, embedding models, and grouping possibilities in crises	NLP pipeline with different embedding schemes for clustering social media posts	Internal clustering metrics and clustering time
7	Active learning for transformer models	Active learning with foundation models	Classification performance
8	Data augmentation for text classification	1. Extensive literature survey 2. Data-structure- driven taxonomy and method-oriented categorization 3. In-depth description of methods 4. Identification of future research potentials	Comprehensive results comparison and state-of-the- art synthesis
9	Data augmentation for NLP	Novel data augmentation method for long and short text data	Classification performance and generated data quality
10	Machine learning approaches for assessing credibility	Fine-tuning BERT for novel application context	Classification performance and classification time
11	BERT models adapted for different domains	Novel cybersecurity domain-adapted BERT language model	Classification performance of domain-dependent extrinsic and intrinsic, and general tasks
12	1. Transfer learning 2. Data augmentation for text classification 3. Few-shot learning 4. CTI	1. Novel transfer learning method 2. Novel data augmentation method for text classification 3. State-of-the-art few-shot learning	Classification performance and generated data quality
13	Textual adversarial examples	1. Adversarial example generator 2. Robustness evaluation setting	1. Robustness against attacks 2. Human evaluation 3. Classification performance 4. Transferability to other models

2.3.1 Motivation

The research of this dissertation begins with an examination of information overload in social media using the example of crisis situations. First, a literature review is conducted on techniques to combat information overload and on different ways of grouping posts in crisis situations. Based on the information overload problem and the challenge of developing a system that does not depend on the emergency workers' effort, an NLP pipeline is proposed for clustering the data without requiring labeled input. Empirical evaluations are then used to determine the best embeddings and fastest methods for crisis datasets in German and English. These evaluations are based on internal evaluation metrics that do not require ground truth labels, which we preferred over external evaluations because of the different grouping possibilities identified in the literature review. The study also proposes a technique for automatically assigning labels to the clusters to quickly identify the information contained within a particular cluster. The details of this study are presented in Chap. 6. This study has a motivational value for the whole dissertation as the extracted clusters are too coarse-grained to gather specific information, which is particularly pronounced in relation to CTI.

2.3.2 Data Acquisition

As previously demonstrated in the Introduction (see Fig. 1.1), acquiring labeled data is the initial step in the supervised machine learning process. The following study is the first in a series of techniques designed to help reduce the amount of labeled data needed for fine-grained classification of CTI information. It proposes to extend the labeling process with an active learning component. Initially, the study reviews active learning methods for transformer models. In light of the mixed results observed in model-mismatch scenarios and the otherwise impractical long waiting times, as well as the general cold-start problem, the active learning method ActiveLLM is developed, which utilizes the zero-shot capabilities of foundation models. This method is then evaluated on low-data regime variants of standard NLP tasks and the specialized CTI dataset of Chap. 12 in terms of classification performance using a BERT classifier as the final model. Initial experiments are aimed at identifying the most appropriate prompt design for the foundation model. This is followed by a comparison of different foundation models and a comparison of ActiveLLM with random sampling, other active learning methods, and state-of-the-art few-shot learning. In the final experiments, ActiveLLM is used to overcome the cold-start problem of other active learning strategies. The specifics of this method, including the literature review and evaluation, are described in Chap. 7.

2.3.3 Preprocessing

Regarding deep learning in NLP, the subsequent preprocessing phase of the machine learning process is typically limited to tokenization. Yet, in low-data regimes, data augmentation can considerably expand and even enhance the data from the previous stage. The first of the two data augmentation works is an extensive literature review on data augmentation for text classification, addressing the need for an overview of the diverse research directions and potential future prospects (see Chap. 8). It follows a method-oriented categorization and proposes a data-structure-driven taxonomy (see Fig. 2.2). As well as explaining and comparing the data augmentation techniques in detail, it also contextualizes them and compares their results, leading to a synthesis of the state of the art and reflections on the future potential of this research. The findings of this survey are particularly important for the next study, in which a new method of data augmentation is proposed (highlighted in Fig. 2.2). The developed method focuses on the use of larger language models, as the

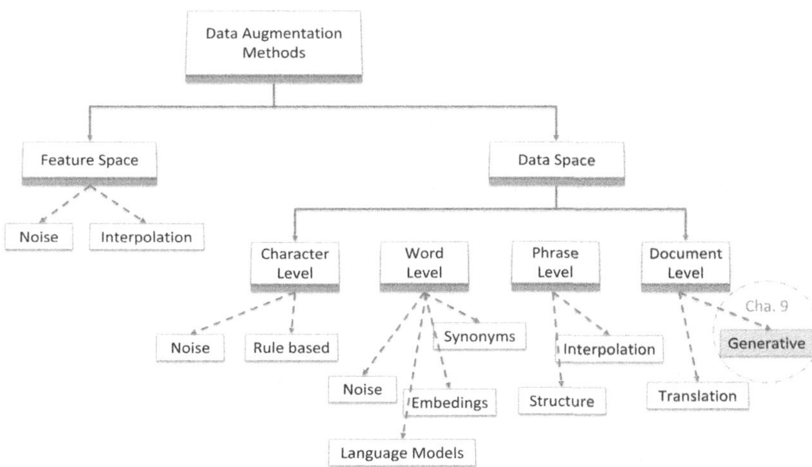

Fig. 2.2 Taxonomy of textual data augmentation methods from Chap. 8. Highlighting the generative methods of the document level, in which the study of the Chap. 9 is categorized

literature review identified that they allow more sophisticated text transformations with novel linguistic patterns. The method is also designed to meet the challenge of transforming long texts, which is particularly rare in the research field but prevalent in various domains, such as in CTI, exemplified by cybersecurity blogs or websites. The performance is evaluated using 11 different datasets, including sentiment analysis, news classification, and crisis informatics, against a baseline and state-of-the-art method. In addition, the study includes ablation studies to verify that all parts of the method contribute to the final improvement, as well as experiments with different amounts of augmentation data. We also review the generated data from a qualitative perspective. The details of the method, all evaluation results, and the data inspection are presented in Chap. 9.

2.3.4 Model Selection

Choosing an appropriate model and approach to transfer knowledge is vital when dealing with low-data regimes. The next three papers explore the different aspects of transfer learning.

Since the credibility of posts and news is also important for CTI, the first study uses transfer learning to develop a deep learning credibility classifier. The work abstracts from the concept of cybersecurity due to the absence of any pertinent data and focuses instead on the broader subject of credibility in social media, where there are some openly available datasets. The study covers an overview of existing machine learning approaches for assessing the credibility of Twitter posts. A new large dataset is then created by selecting smaller existing datasets, unifying the instances, and merging them. As models for credibility assessment need to cover a wide range of topics while accounting for the nuances of natural language, this paper explores the concept of transfer learning of the pre-trained model BERT. In the evaluation, this transfer learning is compared to conventional deep learning methods with regard to classification performance and real-time suitability. The details of this study are presented in Chap. 10.

While this constitutes a first step for credibility assessment in cybersecurity, there is still a need for domain-specific data. This demand can be improved by the developments of the next paper, which deals with the creation of a language model specifically for the field of cybersecurity (see Chap. 11). The rationale for constructing such a model is to gain an understanding of the field and minimize the need for extensive data for adaptation processes, i. e. fine-tuning. In the study, the model is trained to form a base model for cybersecurity-specific texts by further training the already pre-trained model BERT. The design principle is to replicate and iterate over related works to produce the most valuable and novel model possible for the cybersecurity domain. The body of research is extended with design considerations and experiments on catastrophic forgetting, a phenomenon describing that a further trained model forgets its previous knowledge. The resulting CySecBERT model is compared with state-of-the-art models and methods using various extrinsic and intrinsic cybersecurity tasks, including word similarity testing, clustering, classification, and sequence tagging. Additional experiments are used to ensure that the final model is still capable of performing well in general tasks.

In the last paper of the model selection phase, several new low-data regime methods are proposed. These techniques are used to gather specialized CTI information with very little data, as current methods can only collect coarse-grained information, lacking the flexibility to adapt to different needs. The study begins with a brief literature review on transfer learning, data augmentation, few-shot learning, and CTI. A pipeline is then proposed, consisting of a combination of a new transfer learning concept with multi-level fine-tuning, a new iteration of the data augmentation technique in Chap. 9, and state-of-the-art few-shot learning. The study also addresses a major issue in cybersecurity research, namely the lack of up-to-date datasets for particular areas of concern [380], by publishing a labeled dataset in which specialized CTI information is captured. The evaluation compares the classification performance of the newly proposed pipeline with a baseline, a state-of-the-art few-shot method, and a best-case scenario involving more than 50 times the amount of data. The evaluation also includes ablation studies on the different parts of the entire pipeline and an inspection of the augmented data. The details of this study are presented in Chap. 12.

2.3.5 Prediction

Our last research is positioned within the prediction phase of the machine learning process. At this point, when a model is fully trained, several deep learning studies have pinpointed a robustness issue for out-of-distribution instances [99, 124, 126]. Specifically, minor, intentionally altered data instances, termed adversarial examples or adversarial attacks, can cause the model to make incorrect predictions. Therefore, the last study deals with the re-training of classifiers to improve robustness and, ultimately, classification performance. This paper includes a literature review of adversarial example generation and adversarial training methods in NLP. Current methods often suffer from only small or very specific robustness gains, partly due to ineffective evaluation methods. Therefore, our method, XAI-Attack, is designed to combine explainable AI and adversarial example generation to find incorrectly learned patterns in wrong-classified examples. Furthermore, a novel out-of-distribution evaluation setting is proposed to better assess the robustness of models trained with adversarial training. In this setting, XAI-Attack is evaluated against a baseline and several state-of-the-art methods on the adversarial GLUE benchmark by Wang et al. [466]. In addition, the adversarial generation method is also evaluated within the standard adversarial training experiment on the normal GLUE benchmark. The generated adversarial examples are inspected for quality and validated through a human evaluation process in which two annotators label normal and adversarial data to find inconsistencies. The final experiment investigates whether the adversarial examples for one model are transferable to other models. The details of this study are presented in Chap. 13.

Findings 3

The findings of this work are diverse, with some pertaining to the application context of this dissertation, CTI, and others to machine learning research. We demonstrate the viability and effectiveness of these findings through rigorous experimentation and analysis. Our findings not only shed light on the nuances of these domains but also provide actionable insights and methods that can be applied to enhance the performance and robustness of machine learning models.

3.1 Motivation: Clustering

As discussed in the previous chapter (see Chap. 2) and illustrated in Fig. 2.1, this dissertation initially addresses the topic of clustering to combat information overload. Chapter 6 reveals that crisis informatics follows the direction of general NLP research with the increasing use of embedding methods such as Word2Vec [283]. However, by the time of this study, most relevance detection systems are based on supervised classifiers [3, 142, 171], which, parallel to CTI situations, do not capture the dynamics of crisis situations and are often too costly and time-consuming if trained for each new incident. Therefore, we introduce a clustering process comprising NLP preprocessing, embedding creation, and clustering algorithms. This process aims to mitigate social media information overload during crises. We place particular emphasis on embeddings, given the significance of the resulting latent representations to the clusters formed. We assess ten different embedding models. These include crisis [8, 9, 169] Twitter [128] domain-dependent Word2Vec models, the state-of-the-art sentence encoding model Sentence-BERT [356], and the language-specific FastText model [38, 186], all tested on both an English and a German dataset. The clustering results unambiguously demonstrate that selecting the appropriate embedding strategy is important for meaningful clustering. Our novel

M. Bayer, *Deep Learning in Textual Low-Data Regimes for Cybersecurity*, Technology, Peace and Security I Technologie, Frieden und Sicherheit, https://doi.org/10.1007/978-3-658-48778-2_3

conclusions contrast with prior studies on domain-dependent crisis embedding models [204, 245], which, in their solely extrinsic (downstream) evaluations, indicate that these models surpass other pre-trained embeddings. Indeed, domain-specific models prove advantageous when they are perfectly aligned with a given domain. However, if there is a language shift to German, we find that a language-centric model is more appropriate (see Table 3.1). Our work not only reveals novel insights into the crisis domain, but also proposes an intrinsic embedding evaluation method. This method involves comparing clustering outcomes directly from the embeddings, offering considerable value to NLP research. Finally, in this study, we recommend an automatic labeling technique that constructs template tweets that mirror humanitarian categories. This allows clusters to be assigned and labeled accordingly, providing direct insight into the clusters.

Table 3.1 Summarized and simplified clustering results of embedding models on an English and a German crisis dataset. 52M and 364M indicate the number of instances in millions on which the model was trained. For more details, see Sect. 6.4 of Chap. 6

Rank	English	German
1.	Crisis Word2Vec 40M [169]	FastText [38]
2.	Crisis Word2Vec 364M [8]	GloVe [330]
3.	FastText [38]	Crisis Word2Vec 364M [8]

Although this study does not directly address the CTI problem space, it shares many parallels with the crisis domain. Therefore, implications for both domains can be derived. A cyber threat incident can be viewed as a crisis situation with equally high dynamics regarding past incidents. For example, different crisis scenarios, actors, and locations are equivalent to different threat scenarios, actors, and affected products, functions, and paths. Furthermore, crisis informatics deals with gaining situational awareness in social media by finding relevant information depending on different requirements. This is equal to the specialized CTI problem. The CTI clustering work by Kuehn et al. [220] and further practical results from the CYWARN project indicate that clustering can be equally useful for cybersecurity events. However, these results also highlight the necessity for more specialized and individualized information collection for CERTs and professionals. Therefore, a rapidly adaptable supervised machine learning framework is essential, motivating the following chapters of this dissertation.

3.2 Data Acquisition: Active Learning

In Chap. 7, we design the active learning component ActiveLLM that effectively selects instances for labeling using foundation models. Since foundation models do not require training data to make informed selections, our method overcomes the cold-start problem, making it particularly suitable for low-data regimes. In our low-data regime experiments with BERT classifiers as final models, we first determine the optimal prompt configuration and then compare the performance of various foundation models for actively selecting instances, including GPT-4 [315], GPT-4o, Llama 3 [264], and Mistral Large [6]. While ActiveLLM with different foundation models performs significantly better than random sampling in most cases, the most consistent results are obtained with GPT-4, even reaching improvements of up to 17 percentage points in accuracy. The subsequent experiments show that ActiveLLM not only significantly outperforms random sampling and other active learning strategies in low-data regimes, but is also superior to the state-of-the-art few-shot learning techniques ADAPET and PERFECT [271] (see Fig. 3.1). In the final experiment, we demonstrate that our method can be combined with other active learning strategies to mitigate their cold-start problems. ActiveLLM can be applied directly through various chat interfaces, receiving feedback within seconds. This makes it much more practical than previous active learning strategies, which require several minutes of waiting even for efficient methods [517]. Finally, our experiments indicate that using

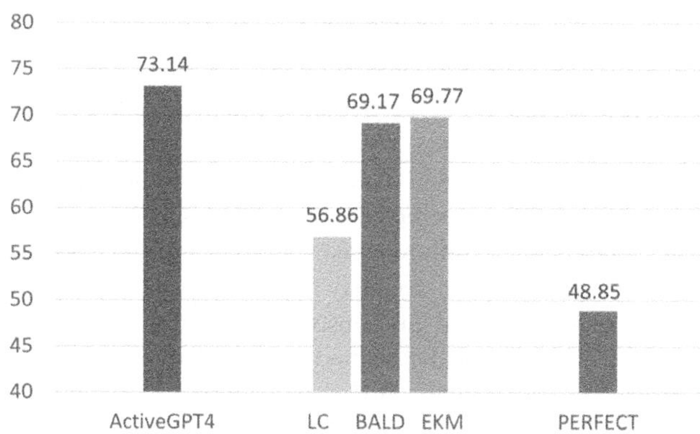

Fig. 3.1 Accuracy comparison of ActiveLLM, the few-shot method PERFECT, and common active learning strategies: Least Confidence (LC), BALD, and Embedding KMeans (EKM). Evaluated on a few-shot scenario (32 training instances) of the AGNews dataset (with BERT classifier). For more details on this experiment, see Sect. 7.4 of Chap. 7

much larger models for the active learning process decouples the dependency of the active learning and successor model (model-mismatch problem).

3.3 Preprocessing: Data Augmentation

As the use of data augmentation for text data continues to grow, there is a need for a consolidated overview in the research field. In Chap. 8, we address this gap by examining over 100 data augmentation techniques, categorizing them into 12 distinct groups (see Fig. 2.2). Additionally, we contextualize these methods by contrasting their design and performance. In doing so, we distill the most promising strategies for varied applications, thereby projecting future research trajectories. In addition to recognizing the demand for improved evaluation metrics, better comprehension of when and why augmentations are effective, and enhanced usability, a key discovery is the need for advanced techniques capable of introducing novel textual content. As also reported by Longpre, Wang, and DuBois [265], this is essential for improving the performance of pre-trained transformer models. Advances in NLP, particularly transfer learning, have rendered many simple data augmentation methods obsolete, as the pre-trained models inherently account for these data transformations. Therefore, we pinpoint future research avenues. Among these, the integration of large language models seems especially promising, given their potential to incorporate their vast pre-trained knowledge. Stemming from this insight, Chap. 9 proposes a data augmentation technique using the language model GPT-2. This study has an emphasis on short as well as long texts, as prior data augmentation with GPT-2 [18] was only able to augment single-sentence instances. The method consists of sensibly fine-tuning and prompting the GPT-2 model, coupled with document embedding filtering based on Sentence-BERT [356] for removing faulty augmentations. We enhance the diversity of the filtering process by incorporating minimal human supervision to select a threshold above which augmentations are discarded. We evaluate our method on 11 different tasks with the pre-trained ULMFit model [156]. The results show significant accuracy improvements of up to 3.56 and 15.53 percentage points in the real world and constructed low-data regimes. The ablation studies indicate that each part of the method contributes to the final improvement and that more augmentation is better. However, we also notice that the method can degrade performance when the task is too broadly defined, where GPT-2 is not able to infer the right context. The results on the constructed low-data regimes of the SST-2 task (100, 300, 500, and 700 instances) compared with the baseline and the data augmentation method EDA [481] are shown in Fig. 3.2. A revised version of this approach with GPT-3 is incorporated into the framework of Chap. 12, where it is applied to a CTI few-shot task with only 32 training instances. Due to the much larger size and context window of GPT-3, no fine-tuning is required, and the training data is incorporated directly into the prompt. The evaluation of this study indicates that this data augmentation approach improves classification performance even when combined with sophisticated few-shot techniques and when compared to popular data augmentation methods, such as EDA and round-trip translation. Further findings of this study are illustrated in the model selection stage, which is the subsequent step in the supervised machine learning process.

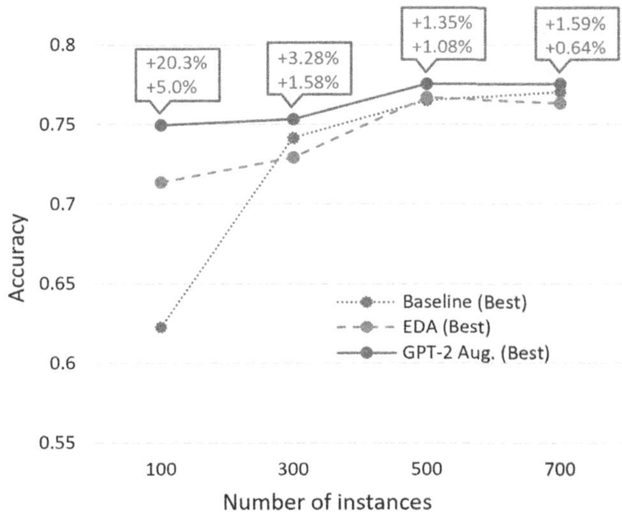

Fig. 3.2 Accuracy of the GPT-2 data augmentation method, EDA [481] and no augmentation (baseline) on different subsets of SST-2 with the ULMFit model. The number of instances of each subset is shown on the x-axis. The plot shows the best results from five runs each. For more details on this experiment, see Sect. 9.4 of Chap. 9

3.4 Model Selection: Transfer Learning

In our preliminary study on model selection and transfer learning (refer to Chap. 10), we evaluate the pre-trained BERT model for credibility assessment. As explained earlier, due to the absence of cybersecurity-specific credibility data, we focus on the general concept of credibility assessment and develop a credibility dataset by amalgamating existing works. The full dataset combines five Twitter credibility datasets, resulting in 5,225 instances with 2,619 credible instances. Our evaluations on this dataset demonstrate the dominance of the pre-trained transformer model over conventional multi-layer perceptrons and recurrent neural networks in terms of classification accuracy. However, in terms of real-time applicability, BERT models are orders of magnitude slower, while recurrent neural networks offer the best compromise. Still, it can be concluded that the BERT model classifications are sufficiently fast for a multitude of applications, and this will improve further as computational technologies advance. Consequently, this thesis focuses on transformer models.

The performance of BERT in a specific domain can even be improved by further training on domain-specific data, as demonstrated by BioBERT [238] for biomedical data or SciBERT [34] for scientific data. As of the time of this study, there is no cybersecurity-aligned BERT model, with the exception of the very lightweight CyBERT model trained by Ranade et al. [347]. In Chap. 11 we introduce the thoroughly trained CySecBERT model. In this chapter, we contribute an open cybersecurity dataset consisting of 4.3 million documents from different papers, NVD data, Twitter posts, and blog entries. Based on this data, we domain-adapt the BERT model through further pre-training. The resulting CySecBERT model is trained diligently to prevent catastrophic forgetting, ensuring the model's initial knowledge remains intact. By evaluating seven fully trained models with different learning rates, epochs, and dataset sizes, we conclude that additional training should avoid extreme updates, e.g., with excessively high learning rates. Conversely, it should also not be too minimalistic, e.g., with a too-small dataset or too few epochs compared to the original BERT training process. The results of two intrinsic evaluation tasks (word similarity and clustering) demonstrate that CySecBERT enhances the representation space of domain-specific words in comparison to the BERT baseline and CyBERT by Ranade et al. [347]. Moreover, in the extrinsic cybersecurity evaluation of five downstream tasks, including classification and sequence tagging, our model once again demonstrates superior performance compared to both. The summarized results are shown in Table 3.2.

The CySecBERT model is also integral to Chap. 12, which explores this dissertation's application objective of collecting specialized and individualized CTI information for cybersecurity. Our first contribution in this realm is a unique dataset that encompasses Twitter posts related to the 2021 Microsoft Exchange Server data breach. This dataset is binary-labeled by three cybersecurity experts based on its relevance to CTI. It comprises a full-shot training and development set (with 1800/600 instances), a few-shot training and development set (32/32 instances), and a testing set (with 601 instances). To address the challenge posed by this dataset, we present an innovative framework based on GPT-3 data augmentation, multi-level fine-tuning, and cybersecurity-adapted few-shot learning. As detailed above, the

Table 3.2 Summarized results of the CySecBERT experiments. Extrinsic tasks: Full-shot CTI dataset from Chap. 12 (F1), CySecAlert dataset by Riebe et al. [363] (F1), and software versions (SV), software names (SN), and attack complexities (AC) of NVD descriptions from the OVANA dataset [219] (F1). Intrinsic tasks: Cybersecurity word similarity task (F1) and Log4j clustering task (silhouette score). For more details on these experiments, see Sect. 11.4 of Chap. 11

Name	CTI	CySecAlert	SV	SN	AC	Word Sim.	Clu.
BERT	0.8599	0.8779	0.9247	0.8837	0.3323	0.4382	0.130
CyBERT [347]	0.8766	0.8647	0.9298	0.8834	0.3336	0.4861	0.113
CySecBERT	**0.8869**	**0.8883**	**0.9302**	**0.8871**	**0.3472**	**0.6382**	**0.172**

data augmentation technique is a further development of the GPT-2 technique from Chap. 9. It employs the considerably larger GPT-3 model, without fine-tuning but distinct prompting and the embedding filtering method. As a pioneering effort, we introduce the concept of multi-level fine-tuning. Here, we train a neural network progressively towards the specific task (as depicted in Fig. 3.3). Training BERT can be viewed as the initial phase where the model acquires basic language knowledge.

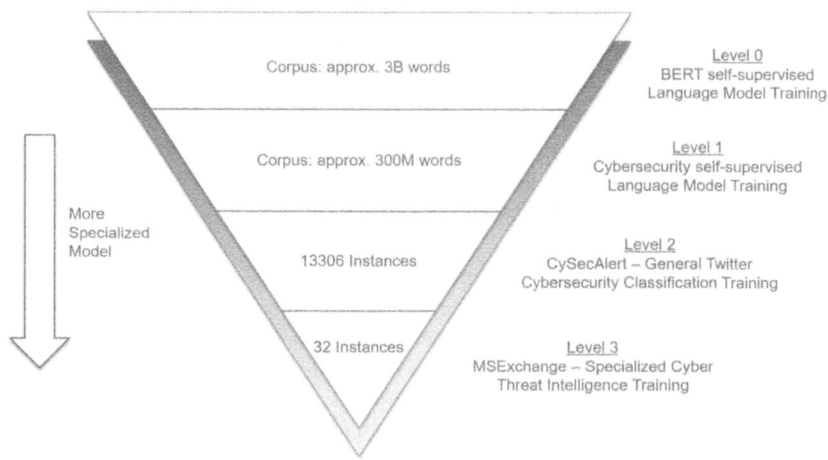

Fig. 3.3 Multi-level fine-tuning process that shows the model becoming more specialized as it is guided to the actual task

Subsequent training on cybersecurity data produces CySecBERT, which enables the model to recognize cybersecurity-specific terms and phrases. Following this, we utilize the dataset from Riebe et al. [363] to acquaint the model with cybersecurity-relevant Twitter posts. Afterward, we further train the model using the final few-shot dataset. The preceding training sessions play a pivotal role at this stage, allowing for effective classification even with very few instances. As the final step in this framework, we adapt the few-shot learning strategy ADAPET to cater to specialized CTI tasks. Using the few-shot set from the aforementioned dataset, we outperform the current state-of-the-art ADAPET strategy by approximately 13 percentage points in accuracy (see Table 3.3). Ablation studies on our method pipeline indicate that each component of the framework impacts the overall result, with the most notable enhancements achieved through multi-level fine-tuning.

Table 3.3 Shortened evaluation results of the methods of Chap. 12. Comparing BERT with 1800 instances (best case), BERT with 32 instances (baseline), BERT with 32 instances and ADAPET (state-of-the-art few-shot learning), and CySecBERT with 32 instances and the methods of Chap. 12. The values in brackets show the standard deviation. For more details on this experiment, see Sect. 12.4 of Chap. 12

Method	Train Size	Model	Accuracy
Best Case	1800	BERT	85.36 (0.07)
Baseline	32	BERT	49.65 (1.90)
ADAPET	32	BERT	65.89 (1.35)
Chap. 12	32	CySecBERT	79.13 (0.56)

3.5 Prediction: Adversarial Training

To address robustness issues of trained models for out-of-distribution instances in the form of adversarial examples or even attacks, we introduce XAI-Attack (see Fig. 3.4) in Chap. 13. The method generates adversarial examples for BERT-like models that can be utilized to re-train the model, making it more resistant to such manipulations (adversarial training). In contrast to current methods, we analyze incorrect predictions from which incorrect behavior, such as spurious correlations or shortcut rules [126], can be derived. This is implemented by extracting words that are most responsible for the incorrect predictions. To this end, we propose to use explainable AI methods as importance functions, which not only allows for more sophisticated importance scores than previous methods, but also enables applicability in a full black-box setting, i.e., no model internals or soft labels are required for generation. Validated by human evaluation, our method generates adversarial examples that maintain a high label-preserving quality. This can also be observed in the examples presented in Table 3.4. Moreover, we conceptualize a novel adversarial training experiment to address a prevalent issue in current adversarial training. Through standard adversarial training experiments on the original datasets, we observe that test sets often share biases with training and development

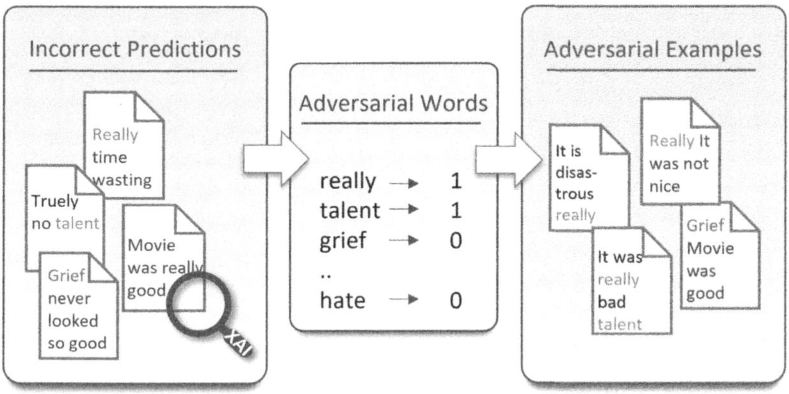

Fig. 3.4 Overview of XAI-Attack: 1. Incorrect predictions are analyzed by explainable AI to extract adversarial words with their potential label changes. 2. The adversarial words are optionally filtered. 3. The remaining adversarial words are inserted into correct predictions, from which the resulting incorrect instances are adversarial examples

Table 3.4 XAI-Attack adversarial examples from different datasets (inserted adversarial words shown in blue)

Task	Instance	Label	Prediction
QNLI	Question: What issue has been plaguing the civil disobedience movement. Sentence: *q* It has been argued that the term "civil disobedience" has always suffered from ambiguity and in modern times, become utterly debased.	Not Entailment	Entailment
SST2	*enjoy* not only unfunny , but downright repellent .	Negative	Positive
SST2	*better* good old-fashioned slash-and-hack is back !	Positive	Negative
QQP	Question1: What is abstract class and methods? Question2: *Java* What is abstract class and methods in java?	Not Duplicate	Duplicate

sets for a given task. This overlap may result in an adversarially trained model, stripped of this bias, receiving a diminished test score. Hence, in our suggested evaluation setting, we employ the adversarial GLUE dataset [466] to gauge the model's performance, serving as a proxy for its robustness. Our results demonstrate that our proposed adversarial example method enhances robustness by up to 23 percentage points in accuracy for adversarial tasks. Additionally, we show that the adversarial examples crafted for one transformer model (distilBERT) can often be applied to other transformer models, such as BERT and RoBERTa.

Discussion

4

As presented in the preceding chapter, the methods developed in this dissertation significantly enhance text classification in low-data regimes. However, as foundation models such as GPT-4 or Llama exhibit strong few- or zero-shot capabilities, it is necessary to examine how they compare to the methods developed in this dissertation. This chapter initially identifies the strengths and limitations of these foundation models in the context of the low-data regime methods of this dissertation. It then considers the potential benefits of applying low-data regime methods to foundation models. The last part assesses their empirical relevance by performing experiments on the CTI dataset of Chap. 12 with the dissertation's full pipeline and the foundation models GPT-3.5 and GPT-4.

4.1 The Advent of Foundation Models

One significant discovery from the development of foundation models is that many challenges in deep learning research can be overcome with increased data and computing capacity, aligning with "The Bitter Lesson" described by Sutton [429]. The bitter lesson states that general-purpose methods that scale effectively with increased computation will eventually outperform those that rely on human knowledge and manually encoded discoveries. Sutton [429] advocates for approaches that utilize computational power to enable systems to learn and discover independently. As some methods of this dissertation arguably fall under what Sutton [429] describes as manually encoded discoveries, we believe that the proposed pipeline in its current state will eventually be outperformed by foundation models in terms of performance. However, this does not diminish the relevance of these methods, as performance is only one of many critical factors in application-specific scenarios. In addition, part

M. Bayer, *Deep Learning in Textual Low-Data Regimes for Cybersecurity*, Technology, Peace and Security I Technologie, Frieden und Sicherheit, https://doi.org/10.1007/978-3-658-48778-2_4

of the continued relevance can be attributed to the rapid adoption of emerging foundation models within the developed methods, as seen in Chaps. 7, 9 and 12, without relying on them excessively. In the following, we will discuss when these foundation models are good substitutes and when the framework used in this dissertation should be preferred. In detail, this section aims to discuss the weaknesses and advantages of foundation models, the utility of our approaches for future studies, and their potential benefits.

A comprehensive history of foundation models is beyond the scope of this thesis, so we will directly discuss the strengths and weaknesses of them. The reader is referred to the survey of Zhao et al. [532] for a detailed survey.

4.1.1 Strengths

The debate among leading figures about whether foundation models possess true reasoning capabilities or merely act as sophisticated statistical word predictors—mockingly termed "stochastic parrots" [35]—is well-known (see the quotes below). This dissertation, however, focuses on the undeniable successes of foundation models, which render many advanced machine learning methods and research objectives obsolete, significantly impacting the content of this work. The following points highlight the strengths of the foundation models compared to the methods developed in this dissertation.

> *"[..] Emily Bender and Timnit Gebru have called these systems "stochastic parrots," which in my view is a little unkind—to parrots– but also vividly captures something real: a lot of what we are seeing now is a kind of unreliable mimicry" [..]*
>
> — Gary Marcus, What was 60 Minutes thinking, in that interview with Geoff Hinton?[1]

> *"[..] I agree with @geoffreyhinton that LLM have *some* level of understanding and that it is misleading to say they are 'just statistics.'" [..]*
>
> — Yann LeCun[2]

Performance: First and foremost, current foundation models achieve state-of-the-art results across a broad spectrum of tasks, often without the need for any few-shot examples, in contrast to the pipeline of this dissertation. In fact, it is challenging

---

[1] https://garymarcus.substack.com/p/what-was-60-minutes-thinking-in-that

[2] https://x.com/ylecun/status/1667947166764023808

to identify text-based tasks where these models do not perform competently (an example will be discussed later). For instance, foundation models have surpassed human performance in several areas:

- Medical: GPT-4 exceeds the passing score of the United States Medical Licensing Examination by over 20 points [308].—GPT-4 also has the potential to outperform specialists in ophthalmology [21].
- Law: GPT-4 achieves a score within the top 10% of examinees on a simulated bar exam [315].
- Code: Based on the Gemini foundation model, AlphaCode 2 ranks between the 'Expert' and 'Candidate Master' categories on Codeforces and performs better than 85% of participants [236].
- Business: GPT-4 handles two-thirds of Klarna's customer service chats, doing the work of about 700 full-time employees with the same level of customer satisfaction[3].

As previously stated, since the performance of foundation models seems to be improving, we expect that if they have not yet surpassed the pipeline discussed in this dissertation, they will do so in the future.

Ease of use: Abstracting from the years of research that went into this dissertation (as certainly, more research went into GPT), the pipeline developed still necessitate a multifaceted combination of machine learning disciplines and considerable expertise. For specialized CTI classification, only the labeling of data is required, as our framework can handle the rest. This effort is approximately equivalent to that required to use GPT for this task. However, for other cybersecurity tasks, while the overarching methods of the dissertation remain relevant, certain adjustments are necessary: The data augmentation approaches require prompt redesign, and for transfer learning, a new general dataset might be required along with a redesigned few-shot learning template. In contrast, applying GPT involves merely prompt redesign. This distinction becomes even more pronounced in domains where the CySecBERT model is inapplicable, necessitating the development of a new BERT-like model for those domains. In addition, the accessibility of most foundation models through API services offered by hosting companies abstracts hardware requirements, resolves many deployment issues, and facilitates model interchangeability.

---

[3] https://www.klarna.com/international/press/klarna-ai-assistant-handles-two-thirds-of-customer-service-chats-in-its-first-month/

Progress: The community surrounding foundation models is extensive and highly active, fostering a plethora of models, research, development environments[4], fine-tuning methods (e.g., LoRA [158], QLoRA [88]), prompt engineering techniques [64], quantization strategies [497], and more. The current demand and pace of foundation model research suggest that progress is unlikely to decelerate in the near future. Conversely, smaller and BERT-like models receive less attention from the community, resulting in fewer innovations.

Explainability: Because of the general generative capabilities of foundation models, they can be asked to provide explanations or uncertainty estimates for their decisions [329]. Although ongoing research continues to investigate the reliability of self-explanations provided by these models, such capabilities represent a significant advancement toward enhancing trust in foundation models [470]. The classifiers developed in this dissertation, however, are limited to predicting classes and providing unreliable confidence scores [136]. Although explainable AI methods such as LIME [361] or SHAP [267] can be used, these also have various limitations, particularly with regard to their informative value [294].

4.1.2 Weaknesses

In this section, we aim to outline the limitations associated with the use of foundation models. Our primary focus is on very large language models with at least 50 billion parameters. Smaller foundation models like Llama 3 (7 billion) [264] and Phi-3 (3 billion) [1] showcase impressive capabilities and do not exhibit some of the weaknesses detailed below. Nonetheless, their performance currently falls short, when considering that at the time of writing, models of this small size perform worse than GPT-3.5, which will be shown later in Sect. 4.3 to be inferior to the methods of this dissertation.

Dependency: A major drawback of these models is their inefficiency. For instance, the GPT-3 model, which has 175 billion parameters, requires approximately 350 GB of memory [47], making it only feasible for very large corporations and institutions to manage its storage requirements. Storing a model like GPT-3 on personal clusters or in the cloud is impractical for the tasks described in this dissertation, leading to a reliance on companies that offer these models via an API. This reliance introduces challenges such as opacity in the processing pipeline, potential unavailability,

[4] https://github.com/ggerganov/llama.cpp

and lack of flexibility. Furthermore, API usage incurs significant continuing costs, unlike the deployment of a BERT model, which requires a higher initial investment for the GPU but then comparably lower continuous operating expenses (power, cooling, maintenance). For tasks such as collecting relevant CTI information, many predictions need to be made over a long period of time, which is currently more cost-efficient when not relying on an API.

Sustainability: The cost efficiency also translates into lower energy consumption and carbon emissions for BERT-like models. Although estimates vary with newer hardware, the training of BERT-like models is responsible for approximately 650 kg of CO2 emissions [422], while GPT-3's training emitted around 550,000 kg of CO2 [328]. While this only addresses the training of those models, it is also reflected in deployment. As said before, GPT-3 needs about 350 GB of storage space, whereas the models utilized for inference in this dissertation consume merely 4–12 GB of memory. All numbers should viewed as estimates that vary based on numerous factors, including the specific methods used for calculation, the source of energy, and the efficiency of the infrastructure. However, even with better hardware and parallelization through API scheduling, CO2 emissions are likely to be higher in the foreseeable future.

Privacy: Utilizing an API also introduces data protection concerns since external providers handle the data processing. Many scenarios may prohibit the sending of data to external servers. For instance, if CERTs wish to use closed-source data, processing it on external servers may pose significant issues. While this dissertation's methods sometimes utilize foundation models via an API for training purposes (not for inference), only a minimal amount of data is transmitted, resulting in fewer data protection concerns, costs, and CO2 emissions.

Ethical concerns: The general reliance on non-transparent companies for foundation models raises several oligopolistic and ethical issues. If the development and deployment of foundation models are confined to a few companies that are capable of sustaining the necessary computational power, this could create a reinforcing cycle where only these companies can afford to invest in even larger models. Such a scenario could enable these entities to act as gatekeepers, controlling access and usage of their products. While this is a general market problem, it is exacerbated for AI systems, which have the potential to significantly impact information dissemination, decision-making, and automation in society. Without proper regulation, these companies could potentially encode biases, censor content, or suppress dissenting opinions, as has already been observed with some of the current models [450].

Hallucinations: A significant issue with current large language models is their tendency to generate plausible yet false information, a phenomenon known as "hallucination," i.e., text that is not accurate, relevant, or grounded in reality [528]. Ultimately, it may be difficult for the user to determine whether the result is accurate or a hallucination. This issue could undermine trust in these models and potentially impede progress. However, this limitation is less problematic within the scope of this dissertation, as the model's primary function is to predict a class rather than generate content. It can be a problem, for example, when the foundation model is asked to explain its answer.

Limited Dynamic Program Synthesis: Foundation models can memorize and retrieve vast amounts of reasoning patterns, also called program templates, which can be seen as instruction manuals that the model follows to solve tasks. François Chollet illustrates this with the example of the Caesar cipher, a complex algorithm that many foundation models can confidently apply to any text using common shift keys like 3 or 5 [73]. These models have memorized the program template from various examples on the Internet. However, Chollet explains that when using an uncommon, arbitrary key like 9, the models fail because they lack a generalized program and cannot dynamically synthesize it [73]. This limitation is also why, despite their extensive knowledge, foundation models struggle with tasks requiring calculations [382] or reasoning with abstractions. A good example is the Abstraction and Reasoning Corpus (ARC) [72], where the current best algorithms can solve only 31% of the problems, compared to humans, who can solve about 80% [228]. The limited dynamic program synthesis may apply to the specialized CTI task in this dissertation, where an individual definition for relevant cybersecurity content can be seen as a novel key in the Caesar cipher example. However, we also believe that foundation models large enough to have memorized an abstract program template for cybersecurity relevancy classification should be able to master this task when provided with a precise description of the definition. This rationale could also explain the difference between the GPT-3.5 and GPT-4 models in the CTI task of the empirical comparisons in Sect. 4.3, assuming that the data was not leaked to GPT-4.

Benchmark Contamination: The suitability of current dataset benchmarks to assess the actual capabilities of foundation models is questionable, as numerous studies have reported [87, 261, 536]. As also hypothesized for the following evaluation of GPT-4 in Sect. 4.3, the confidentiality of test sets in common benchmarks cannot be guaranteed as they might be part of the training data. Therefore, testing and comparing those models is very challenging. This issue partly stems from the non-disclosure of the research behind large models like GPT-4 and their training datasets.

As stated earlier, some weaknesses, such as privacy and ethical concerns, may soon be addressed as smaller models like Llama and Alpaca demonstrate high performance and present the opportunity to run on local hardware[446]. However, these models still require substantial computational resources. They are at least ten times larger than the BERT-base model, with the smallest foundation models having about one billion parameters. For results comparable to GPT-3.5, models are currently at least 60 times the size of BERT; the smallest model currently achieving this is Starling-LM-7B-beta, as indicated by the LMSYS Chatbot Arena Leaderboard[5]. Notably, these smaller yet capable models are often trained on high-quality data, sometimes generated by larger foundation models [299, 433]. This approach to data augmentation has already been identified as a promising research direction in Chap. 8 and even explored with GPT-2 in Chap. 9 and with GPT-3 in Chap. 12, highlighting that the dissertation's methods might even be relevant for training foundation models.

4.2 Low-Data Regime Methods for Foundation Models

The previously mentioned strengths and weaknesses of foundation models demonstrate that their suitability as an alternative to the methods presented in this dissertation depends on the specific circumstances of the task. Nonetheless, the methods presented in this dissertation could also play a significant role in the training procedures of foundation models. The remarkable success of current foundation models largely relies on reinforcement learning from human feedback (RLHF), a strategy that incorporates techniques to fine-tune language models based on human preferences, corrections, and feedback, necessitating human-labeled data [319]. Therefore, the argument that labeling is labor-intensive and potentially inefficient applies to foundation models as well. Moreover, research on foundation models suggests that advancements require more high-quality data [53, 532].

For instance, the ActiveLLM method could guide the labeling process for RLHF towards more diverse topics and crucial training instances. After data collection, it can be further enhanced using the augmentation strategies developed in this work. The use of other foundation models to enrich training data, as demonstrated by the foundation models Alpaca [433] or Orca [299], aligns with the approaches discussed in Chaps. 8 and 9. It can also be used to reduce bias and align foundation models with human expectations [478]. Additionally, transfer learning methods and few-shot learning, as recommended in Chap. 11 and 12, remain invaluable for

[5] https://chat.lmsys.org (accessed on the 19.06.2024)

optimizing foundation model performance in specific domains underrepresented in their training data. A foundation model that is tailored to a specific domain, in a similar way to CySecBERT, or even multiple layers of fine-tuning, may result in enhanced performance in those domains. Finally, the significance of adversarial examples has arguably increased with the advent of foundation models, which can be easily misled, circumventing the safety measures implemented by researchers. While the method of Chap. 13 remains relevant, the feasibility of generating adversarial examples in the same manner for foundation models needs further investigation.

4.3 Empirical Relevance of the Designed Approaches

To analyze the empirical relevance of the designed approaches in the context of foundation models, such as GPT-3.5 [47], we conduct further experimentation with the CTI dataset proposed in Chap. 12, which best represents the application goal of this dissertation. This dataset models the task of CERT teams collecting relevant and specialized CTI during cybersecurity incidents—a task that is inherently a few-shot learning problem, as professionals do not have the time to compile massive datasets during incidents that meet their relevancy criteria.

4.3.1 Experimentation Settings

This evaluation uses version 2 of the CTI dataset[6], which also contains 32 instances for few-shot learning, 1,800 instances for the normal training set, 600 instances for the development set, and 601 instances for the test set, all with an approximately equal binary class distribution. First, the settings of Chap. 12 are replicated, by conducting a full-shot experiment with BERT-base, as well as few-shot experiments with BERT-base, ADAPET, and the multi-level fine-tuning, data augmentation, and few-shot learning procedures from Chap. 12. The experiments are then expanded to include the active learning method ActiveLLM from Chap. 7, and finally, the adversarial training method from Chap. 13, thus encompassing the full pipeline. In this experiment context, the adversarial training method requires some additional hold-out data. We randomly select 32 instances from the development set of the dataset and use them to generate adversarial examples (the use of additional data is marked in the results).

[6] Published here: https://github.com/PEASEC/msexchange-server-cti-dataset

To determine the empirical relevance of the dissertation methods, they are compared with the GPT-3.5-turbo and GPT-4 language models. To do this, the GPT language models must be given a classification prompt from which the classification predictions can be obtained. The prompt template starts with a role allocation and a description of the task: "*In the following, you act as a labeler who must binary label whether the instances are relevant to Cyber Threat Intelligence and to cybersecurity experts during a cyber threat event (1=relevant, 0=irrelevant). You will receive the following annotated instances for reference.*" The few-shot instances are then concatenated in the format *instance ->1/0*. The prompt concludes with "*Please annotate the following Tweet only with 1 (= relevant) or 0 (= irrelevant):*" followed by the instance to be predicted. In all our experiments, we test GPT-3.5-turbo with and without (zero-shot) chain of thought (CoT) prompting, a technique to enhance the output capabilities of instruction-tuned language models by instructing the model to answer step by step [213]. For this, the prompt ends with "*Please annotate the following Tweet by first explaining your thought process and then annotating it with 1 (= relevant) or 0 (= irrelevant). The output format should be Explanation: [your explanation] Label: 1 or 0*" followed by the instance to be predicted. We also attempt to perform non-zero-shot CoT [74, 479, 530], i.e. we include explanations with each presented few-shot example. In order to avoid the need for a human in the loop to create these label explanations, ChatGPT is used to generate them.

In our goal to recreate the most realistic scenario, we hypothesize that cybersecurity professionals encountering a new incident might have to create labeling guidelines for defining relevance. Currently these labeling guidelines are just discarded as it is not trivially possible to incorporate them directly into encoder transformer models like BERT. However, for decoder transformer models with sufficient context size capabilities, these guidelines can be incorporated into the prompt. For the experiments, we must point out that the labeling guidelines of the CTI dataset (provided in Appendix A.4 of the Electronic Supplementary Material) also contain some data examples to describe and illustrate certain labeling decisions (which is marked in the results). In our final evaluation run, we employ the optimal GPT-3.5 setting (few-shot examples, guidelines, and no CoT) with the most advanced foundation model at the time, GPT-4 [315].

The script and output of the GPT experiments are freely available[7].

[7] The experiments were conducted on the 11.07.2023—https://github.com/PEASEC/GPT-CTI-Evaluation

4.3.2 Results and Interpretation

The results of this experiment are presented in Table 4.1. The full-shot experiment with 1800 training instances demonstrates the potential classification performance of a BERT classifier. When applying the same classifier model BERT-base to the few-shot experiment with 32 instances, we observe a drop of more than 37 percentage points in accuracy. Even the state-of-the-art few-shot learning method ADAPET [431] fails to achieve high performance, with 66.22% accuracy. The approach in Chap. 12 with multi-level fine-tuning, GPT-3 data augmentation, and few-shot learning significantly improves baseline (+27.28 percentage points) and state-of-the-art (+12.81 percentage points) results. The incorporation of the active learning method ActiveLLM from Chap. 7 results in an additional slight improvement in accuracy, with a score of 79.30%. The final phase of this dissertation's framework is the integration of XAI-Attack described in Chap. 13, which brings together the entire pipeline. Adversarial training with XAI-Attack reduces performance by 1.76 percentage points in accuracy, which is to be expected as the method corrects shortcut learnings and spurious correlations visible in training as well as development, and test sets. Therefore, in Chap. 13 we propose an out-of-distribution evaluation setting to better measure robustness. Nevertheless, an accuracy score of 77.54% for the full pipeline is still significantly better than ADAPET (+11.32 percentage points). Since XAI-Attack requires additional data for comparative analysis, it has been separated from the experiments of the true low-data regime.

When GPT-3.5 is used in the true low-data regime with few-shot instances, it also improves over the baseline and state of the art (ADAPET), but is still significantly inferior to our approach (–13.41 percentage points). We also include GPT-3.5 without the few-shot learning instances, but with the labeling guidelines (also considered a low-data regime experiment). Interestingly, while this is still significantly worse than our method, it is an improvement over just using the few-shot instances in the prompt. The reason for this may be that it is easier to understand a particular concept of relevance through a comprehensive description of its differences, rather than relying solely on examples from which to derive the aforementioned distinctions. Even when incorporating both the few-shot examples and the guidelines into the prompt for GPT-3.5, the performance is still lower than that of our proposed framework. This shows that despite providing more information and some prompt engineering, GPT-3.5 still cannot match the performance of our method. This may be due to the specialized task setting that the MSExchange dataset presents, containing very specific cybersecurity terms that a language model like CySecBERT, integral to our approach, might better recognize. Moreover, as explained in Sect. 4.1.2, foundation models have the ability to memorize and fetch vast amounts of program templates, but cannot synthesize new ones on-the-fly (limited dynamic program synthesis) [73]. A model like GPT-3.5 might not have memorized an abstract program template for cybersecurity relevance classification. Therefore, it confuses this specific task with program templates of certain cybersecurity information relevance definitions it has encountered in the training data, leading to incorrect decisions for a novel definition.

Table 4.1 Few-shot experiments of the methods of this dissertation compared to GPT-3.5-turbo and GPT-4 in different configurations. The performance of the BERT models are measured as the average accuracy of 5 runs[8]. Different constellations of full training (Full Tr.), presented few-shot examples (FS), presented labeling guidelines (GL), and chain-of-thought prompting (CoT) are marked. ✓ + indicates that non-zero-shot CoT with explanations from ChatGPT is used. Abbreviations: Few-shot learning (FSL), data augmentation (DA), and multi-level fine-tuning (MLF)

	Full Tr.	FS	GL	CoT	Acc.
Full-Shot					
BERT-base	✓	✗	✗	✗	89.02
Low Data Regime					
BERT-base	✗	✓	✗	✗	51.75
↳ FSL (ADAPET)	✗	✓	✗	✗	66.22
↳ DA, MLF, & FSL (Chap. 12)	✗	✓	✗	✗	79.03
↳ ActiveLLM, DA, MLF, & FSL	✗	✓	✗	✗	79.30
GPT-3.5-turbo	✗	✓	✗	✗	65.89
GPT-3.5-turbo	✗	✓	✗	✓	61.90
GPT-3.5-turbo	✗	✗	✓	✗	69.71
GPT-3.5-turbo	✗	✗	✓	✓	65.22
Additional Human Data					
BERT-base ↳ ActiveLLM, DA, MLF, FSL, & XAI-Attack	✗	✓	✗	✗	77.54
GPT-3.5-turbo	✗	✓	✓	✗	75.70
GPT-3.5-turbo	✗	✓	✓	✓	63.73
GPT-3.5-turbo	✗	✓	✓	✓ +	70.05
GPT-4	✗	✓	✓	✗	90.18

[8] These results differ slightly from those in Chap. 12 as we used version 2 of the dataset.

Although CoT prompting is touted as a major improvement for instruction-tuned language models [480, 531], it did not enhance performance in this experiment. In our entire evaluation, despite varying temperatures and testing zero-shot and normal CoT, we observed poorer results with it. This may be attributed to the integration of the CoT component in the prompt, as studies indicate that the phrasing of prompts can significantly affect performance [471]. In general, while prompt engineering may offer some improvement, it is unlikely to bridge the substantial performance gap to our method.

In our last experiment setting, we test GPT-4 to gauge the potential of these large foundation models (including few-shot examples and guidelines), as this model possesses considerably more knowledge, including more specific terms and concepts. With an accuracy of 90.18%, GPT-4 even surpasses the BERT-base full-shot classifier trained with 1800 instances. Although we have to take into account the additional information provided by the guidelines in this setting (compared to the true few-shot learning task), the high performance is still surprising. This is such an improvement that we have to consider the possibility that the published dataset might have been in the training data of GPT-4. Unfortunately, the training data for GPT-4 has not been made public. Therefore, it is impossible to ascertain if the MS Exchange CTI dataset was used in the model's training, which could account for the high values.

In conclusion, our approach with active learning, data augmentation, multi-level fine-tuning, few-shot learning, and adversarial training is still relevant performance-wise. The use of GPT-3.5 may be very powerful for general few-shot learning techniques, but is significantly worse than our method even with more task-related information. While GPT-4's results exceed those of our method, it is uncertain whether the MSExchange dataset was included in the model's training data.

Conclusion 5

This dissertation explores deep learning in textual low-data regimes within the context of gathering specialized and individualized CTI information. CERTs are often overwhelmed by the vast amount of potentially relevant open data during cybersecurity incidents. While clustering alone is insufficient for extracting fine-grained, individualized information, supervised machine learning is similarly limited by the requirement of substantial training data and the highly dynamic nature of these situations. Aiming to enhance the supervised machine learning process—including data acquisition, preprocessing, model selection, and prediction—this dissertation proposes several state-of-the-art methods to address emerging cyber threats by effectively training new highly specialized deep learning classifiers with minimal data.

The following sections synthesize the research findings by answering the research questions, present the theoretical and practical contributions derived from the dissertation, propose limitations and future work, and finally, present ethical and societal considerations.

5.1 Synthesis of Research Findings

To our knowledge, this dissertation is the first to extensively explore deep learning in low-data regimes specifically for cybersecurity research. It introduces state-of-the-art techniques in active learning, data augmentation, transfer learning, and adversarial training, presenting a novel framework that integrates these concepts to overcome the limitations of low-data scenarios. Despite the widespread research in machine learning for cybersecurity, only a limited number of studies tackle the dynamic nature and constant domain shifts encountered in textual data. The prevailing literature on cybersecurity often focuses on network data for intrusion

M. Bayer, *Deep Learning in Textual Low-Data Regimes for Cybersecurity*, Technology, Peace and Security I Technologie, Frieden und Sicherheit,
https://doi.org/10.1007/978-3-658-48778-2_5

detection methods [309, 537], are not fully aligned with the latest advancements in deep learning research [96, 121, 189, 506], and/or generally address only aspects of managing low data, such as through data augmentation [248, 354]. Notably, there are few recent works that use deep learning for information gathering in the CTI domain [215, 222, 436] but to our knowledge, none explore textual low-data regimes in conjunction with current deep learning techniques for this.

This work is motivated by the use of unsupervised clustering techniques, which can segregate data into coarse-grained groups. Within the field of crisis informatics, the work of Chap. 6, along with Alam, Ofli, and Imran [10], is pioneering in its approach to clustering based on textual information, as opposed to spatial or temporal dimensions [266, 333, 376]. In contrast to the work of Alam, Ofli, and Imran[10], our technique features more consistent design choices and rigorous evaluations using state-of-the-art embedding techniques on English and German datasets. Given the general inadequacy of coarse-grained cluster information for many CTI applications and the extensive training data requirements of supervised classifiers, which are constrained by the dynamics of CTI situations, this dissertation is guided by the following research questions concerning low-data regimes.

RQ1 **How can active learning in low-data regimes be enhanced by instruction-tuned large language models, enabling the resulting pre-trained models to gain high classification performance with reduced annotation effort?**

This research question arises from the challenges of model-mismatch and cold-start issues in active learning for pre-trained models. Research has shown that a mismatch between the active learning model and the final model often results in equal or worse performance [387, 397, 529]. However, using pre-trained models, such as BERT, for both active learning and the final model is often impractical due to the long waiting times during annotation, even when considering efficient variants [517]. ActiveLLM of Chap. 7 addresses this challenge by utilizing foundation models, which, due to their zero-shot capabilities, do not require training during the process and can provide feedback within seconds, as only a small amount of unlabeled data needs to be used. Furthermore, most active learning strategies cannot work effectively with insufficient data because they face the cold-start problem [68]. The works of Ein-Dor et al.[100] and Grießhaber, Maucher, and Vu [134] explore active learning with less data, but ActiveLLM is the first to address active learning for true few-shot scenarios with pre-trained models. Leveraging the zero-shot capabilities of foundation models overcomes the cold-start problem and can also be used to mitigate the cold-start problem in other active learning strategies. Our active learning method even outperforms the state-of-the-art few-shot learning methods PERFECT [271] and ADAPET [431]. In addition, as shown in the evaluations of the full pipeline of this dissertation (Sect. 4.3), ActiveLLM can further improve results when used in combination with multi-level fine-tuning, GPT-3 data augmentation, and ADAPET [431] few-shot learning.

Therefore, designing active learning with foundation models overcomes the challenges of current research methods and enables high classification performance in low-data regimes, even when combined with methods derived from subsequent research questions.

RQ2 **How can the advanced generative capabilities of large language models be leveraged and constrained for textual data augmentation to enhance classification performance, particularly in low-data regimes?**

Contrary to other studies from a similar timeframe [113, 402] that also identified the research gap of a missing overview of the various textual data augmentation methods, our work of Chap. 8 provides a comprehensive overview of data augmentation methods for text classification. We offer the first holistic taxonomy of text data augmentation and an in-depth description of the approaches, setting these methods in direct comparison to establish the current state of the art and suggest future research directions. A major insight is that pre-trained language models have rendered many simple data augmentation strategies obsolete, as these models are inherently invariant to those transformations. This enabled us to develop our own data augmentation strategy in Chap. 9 that leverages the GPT-2 model. Due to its generative capabilities and considerable size, it allows for a sophisticated data augmentation method that produces novel linguistic patterns necessary for augmenting pre-trained models. The two related works at the time of this study using this model, are only applicable to single-sentence instances [18], or do not constrain the model sufficiently to ensure label preservation [467]. In contrast, we develop the data augmentation method to be able to augment long textual instances and constrain it with a human-in-the-loop embedding filtering method. The method significantly improves classification quality over the baseline and state-of-the-art data augmentation in various domains, including low-data regimes. An adaptation of this method is proposed in Chap. 12, where the GPT-2 model is replaced by the much larger GPT-3 model and the training instances are incorporated directly into the prompt. This is similar to the GPT-3 data augmentation method by Yoo et al. [512] but also includes the embedding filtering method of Chap. 9. In the CTI few-shot learning task with only 32 training instances, this allows classification performance improvements even when sophisticated few-shot techniques are applied. It also outperforms the popular round-trip translation and EDA [481] data augmentation strategies.

Thus, by harnessing the generative capabilities of large language models like GPT-2 or GPT-3 and meaningfully constraining them with document embeddings, significant improvements in various domains and low-data regimes are achieved. While this study already applies the concept of transfer learning, the following studies are proposed to further advance it.

RQ3 **How can transfer learning methods be developed and applied to mitigate the dependency on extensive data requirements in deep learning applications?**

The first study addressing this research question (Chap. 10) follows the general trend of deep learning in NLP by exploring BERT models in the context of social media credibility assessment. Due to the lack of cybersecurity-specific datasets available at the time of this study, we abstract the task to the broader concept of general credibility assessment in social media. Although some works have considered the use of deep learning [260, 371, 490], to the best of our knowledge, we were among the first to apply transfer learning with a BERT model for this task. Our study demonstrates that BERT models significantly outperform recurrent neural networks and multilayer perceptrons in this task, although they are less time-efficient for predicting instances. Nevertheless, BERT is sufficiently efficient, justifying the dissertation's focus on these models.

The advancements of pre-trained models in NLP also highlight that domain-adapted pre-trained models can bring notable improvements in relevant domains [34, 238], resulting in even lower data requirements. Recognizing the absence of a refined base model for cybersecurity, apart from the lightweight, cybersecurity-aligned CyBERT [347], we introduce a thoroughly trained BERT model for cybersecurity in Chap. 11. In contrast to CyBERT [347], our CySecBERT model is trained substantially longer and utilizes a much larger corpus (CyBERT: 17,000 documents and one epoch versus our CySecBERT: 4,300,000 documents and 30 epochs). The results of both extrinsic and intrinsic evaluations clearly illustrate that CySecBERT improves the performance of downstream tasks such as classification and sequence tagging, and refines the representation space for cybersecurity-related terms. Additionally, this study extends research by being the first to control for catastrophic forgetting in domain-adaptive pre-training. Following general advice from related works [89, 426], the full training of seven CySecBERT models indicates that avoiding overly aggressive updates and overly minimalistic training results in less catastrophic forgetting, ultimately leading to a better model for cybersecurity. With our study and the openly accessible model and dataset, we are paving the way for improved cybersecurity transfer learning, enhancing both research and application.

This is directly exemplified in the study of Chap. 12, where the data requirements of this research question are central. In this study, the results of the transfer learning phase, including CySecBERT, and an updated version of the data augmentation strategy from Chap. 9 are incorporated into multi-level fine-tuning with a few-shot learning technique. This study is the first to address the collection of specialized/individualized CTI information during cybersecurity incidents, in contrast to the coarse-grained cybersecurity information collection from Riebe et al. [363] or Sceller et al. [381]. Recognizing the importance of collecting very specific CTI information based on the individual needs of professionals, we also publish an annotated dataset to encourage research. For the data augmentation method, as described above, we use the GPT-3 model instead of the GPT-2 data augmentation method in Chap. 9. The biggest improvement, however, is achieved with the novel multi-level fine-tuning technique which, to our knowledge, has not been studied before. Combined with the cybersecurity-adapted few-shot learning ADAPET [431] method, the results with 32 training instances demonstrate a substantial improvement over the baseline and the state-of-the-art, even approaching the full training results with 1800 training instances.

Enhancing the transfer learning procedure from BERT models with a domain-adapted cybersecurity model and multi-level fine-tuning significantly reduces data requirements, making these models highly capable of extracting information in CTI situations. However, models trained with less data are especially brittle and can be easily fooled [130], which motivates the next research question.

RQ4 **How can textual adversarial examples be effectively formulated and evaluated in adversarial training to achieve significant learning effects?**

This research question addresses the gap in state-of-the-art adversarial example generation and training methods, which often achieve only very specific robustness gains due to their less generalizable design and ineffective evaluation methods [126, 466]. With XAI-Attack in Chap. 13 we pioneer a method that creates adversarial examples based on patterns of incorrectly classified instances, thereby increasing the general learning effect when used in adversarial training. To identify these patterns, we employ explainable AI methods, which also facilitate more sophisticated importance scores with a broader function space and fewer constraints on the model than the simple importance functions currently used (e.g., in [247], [124], [184]). As our study also shows, that common datasets contain biases in the training set that echo in the development and testing sets, we design a novel out-of-distribution evaluation setting based on the adversarial GLUE dataset [466]. Evaluations of XAI-Attack in this and other settings confirm that it achieves significant learning effects, outperforming state-of-the-art adversarial example generation and training methods (BAE [124], BERT-Attack [247], SMART [181]) and showing baseline improvements of up to 23 percentage points in accuracy.

Hence, adversarial examples based on incorrect classifier patterns can achieve generalization and, thus, considerable learning effects. A novel out-of-distribution evaluation setting based on the adversarial GLUE dataset [466] can effectively capture this.

While this research question is the last to directly address the developed methods of this dissertation, it remains to be questioned whether these methods will retain their relevance in light of foundation models.

RQ5 **How do the techniques developed for enhancing deep learning in low-data regimes maintain their relevance in the era of foundation models?**

To empirically answer this research question, we compare the methods of this dissertation against foundation models on the CTI dataset in Sect. 4.3. While the few-shot results of this experiment show that the low-data regime methods are significantly better than the GPT-3.5 model, GPT-4 even surpasses the accuracy performance of a BERT model with the full training set of 1800 instances. However, due to the exceptional accuracy, it can be speculated that GPT-4 may have seen the testing set data with all labels beforehand, as the whole dataset is openly available on GitHub. Even if we assume that the GPT-4 results are valid, this dissertation discusses different weaknesses of these large foundation models that favor smaller models with specific methods like those in this dissertation. These include dependency on large companies providing these models, along with ethical and privacy concerns, high costs (for resources or APIs), and ecological aspects. On the other hand, foundation models are very easy to use and may even be able to explain their decisions. While there are many aspects that highlight the relevance of the methods in this dissertation for now and the foreseeable future, the extreme progress on foundation models and the hardware suggests that they, as well as most current machine learning research, will eventually fall under the bitter lesson of AI.

"The biggest lesson that can be read from 70 years of AI research is that general methods that leverage computation are ultimately the most effective, and by a large margin." [..]

—Rich Sutton, The Bitter Lesson

Nevertheless, cybersecurity research is important today and cannot be postponed to the future. In addition, methods such as data augmentation for synthetic data, active learning for RLHF, transfer learning methods for fine-tuning, and adversarial training for making large models robust may still be needed, seen less as competition against and more as an improvement for foundation models.

Finally, the main research question of this dissertation can be answered:

Main Research Question
How can the performance of deep learning classifiers for cyber threat intelligence texts be improved in low-data regimes?

By providing a component for each phase in the supervised machine learning process, introduced at the beginning of this dissertation, the quality of textual cybersecurity classifiers can indeed be significantly improved. The active learning procedure, ActiveLLM, aims to assist the data acquisition process by sensibly selecting instances with a high learning impact, even in few-shot scenarios. After the data is labeled, it can be preprocessed using a data augmentation strategy based on GPT-2 or GPT-3, while being constrained by an embedding filtering method. Then, in the model selection phase, the multi-level fine-tuning procedure, incorporating the novel CySecBERT model, fosters better adaptation in few-shot scenarios, as the model has a superior general understanding of the topic. Finally, since learned models, especially those from few-shot scenarios, tend to exhibit vulnerabilities in the form of adversarial examples, XAI-Attack can be used to increase robustness.

The full framework is tested on the CTI dataset in Sect. 4.3, where the potential of this work can be grasped. Besides XAI-Attack, every method contributes to increasing the score, with a baseline improvement of up to 27.55 percentage points in accuracy. We show that the methods of the framework surpass state-of-the-art few-shot learning methods, such as ADAPET, PERFECT, and even the foundation model GPT-3.5. A decrease in performance from the adversarial training method was expected, as we show in the corresponding chapter (Chap. 13) that XAI-Attack can identify and mitigate biases in the training data that are also apparent in the testing data. Even with this decrease, it can be concluded that the framework significantly improves classifier quality for cybersecurity texts. Beyond addressing the primary research question, this dissertation demonstrates that the developed methods address broader machine learning challenges, underscoring their universal applicability across various domains.

5.2 Theoretical and Practical Contributions

Each study in this dissertation has its own theoretical and practical contributions, as shown in Table 5.1, details of which can be found in the respective chapters. When examining this dissertation from a broader perspective, the following contributions can be derived.

Table 5.1 Contributions of the chapters in Part II. The types of contributions are differentiated into artifacts (A), datasets (D), theoretical contributions (T), and empirical contributions (E)

Topic	Cha.	Contribution	Type
Motivation	6	A system for reducing information overload in social media streams	A
		Comparison of embedding models and advice on the implementation of them	E
		Proposal of an intrinsic embedding evaluation task	T
Active Learning	7	Innovative active learning method with foundation models	A
		Proposal to overcome cold-start and model- mismatch problems	T
Data Augmentation	8	Method-driven textual data augmentation taxonomy	T
		Holistic state-of-the-art comparison of data augmentation in text classification	E
		Identification of future research perspectives of textual data augmentation	T
	9	Novel long and short text data augmentation method based on the GPT-2 language model	A
		Textual data augmentation that proves beneficial for pre-trained models	E
		Insights into the application of data augmentation for low-data regimes	E
	12	Novel data augmentation method based on the GPT-3 language model	A
		New insights on data augmentation with strong few-shot learning	E

(Continued)

Table 5.1 (Continued)

Topic	Cha.	Contribution	Type
Transfer Learning	10	Insights into the real-time capability of pre- trained models for credibility assessment	E
	11	Novel credibility dataset	D
		Novel, state-of-the-art language model for the cybersecurity domain	A
	12	Encountering catastrophic forgetting with learning rate, dataset size and number of epochs	E
		A cybersecurity dataset containing relevant cybersecurity sources for the language model training process	D
		New few-shot learning technique based on multi-level fine-tuning	A
		A specialized CTI few-shot dataset for further research purposes	D
Adversarial Training	13	Black-box adversarial example generation method by incorporating explainable AI and focusing on misclassified examples	A
		Novel out-of-distribution evaluation setting for better assessment of robustness	T
		Novel insights of learned biases of transformer models and how to mitigate them	E

5.2.1 Deep Learning

As practical contributions, this dissertation introduces sophisticated artifacts and recommendations amalgamated into a framework for creating performant deep learning models, particularly suitable for textual low-data regimes. These methods, along with the application code and trained models, are made publicly available. They can be used both in combination and isolation, offering versatile solutions for various problems. For instance, a classifier can be trained solely with XAI-Attack or through ActiveLLM without employing other methods. Notably, the data augmentation methods, CySecBERT, ActiveLLM, and XAI-Attack are applicable beyond low-data contexts. Furthermore, in the studies of Chaps. 6, 9, and 10, we evaluate or discuss the time consumption and resources required, ensuring practical applicability of these approaches. This dissertation also empirically demonstrates that all methods significantly improve the performance of pre-trained models such as BERT, ensuring the benefit for practical scenarios. Further evaluations and discussions in light of foundation models show that the developed methods are still advantageous in many respects and may also be useful in training or in combination with foundation models. In general, with the developed methods and recommendations, practitioners can overcome the need for much labor-intensive labeling and create performant and robust classifiers.

Part of the theoretical contributions includes strategies and recommendations for improving deep learning in textual low-data scenarios at various points in the design process. They advance the field by addressing unmet challenges and developing methods that advance the state of the art, accompanied by comprehensive evaluations and analyses. Beginning with research overviews in different chapters, we analyze the current state of the art, identify promising directions, and highlight sensible research gaps. Our approaches are designed to fill these gaps, enhancing the field across various metrics while also analyzing shortcomings and limitations that pave the way for future research. However, as already mentioned, these methods not only deal with low-data regimes but also improve the state of the art in their respective disciplines. In addition to sharing our code and models, we release datasets and evaluation techniques essential for furthering research in these areas, thereby empowering academic progress. In studies like Chap. 8 and Chap. 13, we critically discuss current evaluation methods and propose novel settings and perspectives to overcome shortcomings. Finally, each proposed system is thoroughly interpreted and discussed to derive implications for current research and to highlight potential limitations.

5.2.2 Cyber Threat Intelligence

Our main practical contribution in the area of CTI is the deep learning framework that enables specialized and individualized information gathering despite the highly dynamic nature of the field. This dissertation first identifies the challenges faced by practitioners that hinder the application of deep learning in the field of CTI. These challenges are then addressed by providing the code and resources necessary to rapidly develop high-performance classifiers with every new cyber threat incident. Among these resources, the cybersecurity community can particularly benefit from the rigorously trained CySecBERT model. Experts can use our resources directly or adapt them, e.g., by fine-tuning the CySecBERT model for specific tasks or by using the framework of Chap. 12 to train an individualized CTI classifier with little data. The methods we develop are tested on data from the CTI and general cybersecurity domains to ensure domain-oriented applicability. Furthermore, we demonstrate that while foundation models such as GPT-3.5 are capable of performing well for collecting specialized CTI information, there are several reasons to prefer the methods of this dissertation. However, our studies also identify limitations and future work that needs to be done so that practitioners can best benefit from the developed methods.

The proposed methods are poised to advance the field of machine learning and indirectly impact cybersecurity research and foster adaptation processes in research by proposing theoretical contributions. This dissertation reveals a misalignment between current CTI research and practical needs despite the existence of sophisticated deep learning strategies. We emphasize the need for models that can identify fine-grained cybersecurity information tailored to specific needs, rather than making general predictions. By integrating cybersecurity with deep learning research, we offer models and strategies, such as the CySecBERT model and GPT-3 prompting guidelines, for straightforward insights and adaptations. Furthermore, we publish datasets, such as the cybersecurity corpus used to train CySecBERT or the annotated MSExchange dataset from Chap. 12, to facilitate improved cybersecurity research based on our findings and to align CTI research with the latest advances in machine learning.

5.3 Limitations and Future Work

The broader limitations of this dissertation are delineated across the main domains: deep learning, cybersecurity, and interdisciplinary research.

5.3.1 Deep Learning

In our study, we intentionally refrained from defining a specific threshold for what constitutes a low-data regime, as this can vary significantly across different tasks. Since it is not possible to determine what constitutes "enough" data for a deep learning model—given its dependence on problem complexity, data quality, and other factors—it is also not feasible to establish a clear boundary for the amount of data that qualifies as a low-data regime. Binary classification few-shot scenario evaluations are often conducted with 32 training data instances [431, 449]. However, we believe it is more reasonable to assess each task on its own merits. Still, for the active learning and data augmentation steps, it is valuable to inspect when performance saturates or even degrades, as they can generate infinitely many instances. Although we conducted these evaluations in the active learning and data augmentation studies, more comprehensive assessments with different datasets would be insightful for practical applications. Further studies on this topic could provide insights into how much synthetic data can be generated without saturation, as well as how much real data is needed for the methods to work effectively without substantial distribution shifts.

Another limitation is that this dissertation does not focus on the computational or time resources required, nor on the environmental impact of the methods. Practically, it is necessary to determine, for example, whether to employ the data augmentation process with GPT-3 or if GPT-2 might suffice. Similarly, the multi-level fine-tuning process consumes significant computing resources, and from a financial perspective, labeling more data might be more cost-effective in scenarios where the different levels of the fine-tuning process cannot be reused. Additionally, while the active learning method is an efficient way to achieve good performance in terms of labeling costs, from an environmental standpoint, direct labeling might be preferable to using a foundation model in the active learning method. Furthermore, only a small part of our study addresses the time resources required for deploying the developed methods. While the studies on clustering with embedding models (Chap. 6) and credibility assessment with BERT (Chap. 10) have dedicated metrics for the evaluation of time, the studies on the GPT-2 data augmentation method (Chap. 9) and ActiveLLM (Chap. 7) only briefly discuss time constraints. For instance, XAI-Attack (Chap. 13) can require substantial time resources depending on the hold-out data, which was not considered in the respective chapter. Waiting for a method, and ultimately for the resulting model, can have serious consequences when deployed in time-critical environments like cybersecurity incidents. Generally, we encourage further studies on the practicality of the developed approaches in terms of computational and time resources, as well as environmental impact.

5.3.2 Cybersecurity

While this dissertation focuses on the practical requirements and challenges of cybersecurity professionals and CERTs, it does not report on the interaction with those professionals. The methods are currently only designed to improve classification performance metrics, not to enhance usability or user experience. This, however, is a very important aspect. While the methods presented in this dissertation might save on labeling costs, they might be too complicated for non-deep learning experts to execute, potentially leading to higher deployment costs. Furthermore, when aiming to train a new classifier with each new incident, the methods might need to be extended with a labeling interface and then automatically executed. Additionally, while not necessary, some methods might require parameter tuning to achieve the expected results, which also incurs the cost of expert knowledge. Therefore, we look forward to studies that explore and enhance the human interaction between the methods proposed in this dissertation and cybersecurity professionals, such as CERTs.

While we aim to establish the task of collecting specialized and individualized CTI information within the general field of cybersecurity, for instance, with the annotated dataset, there are some challenges that come with this novelty. We have laid the foundation for the continual training of CTI classifiers with every new incident, but we could only test our framework with one dataset and based on our constraints, such as social media data. Therefore, our view on additional situations is limited. Similarly, as the field has not been thoroughly explored and we only have contact with certain CERT teams, the overarching challenges are only partly visible for now. While we believe that the topic of CTI will continue to grow, we hope that more studies will follow to explore the specialized and individualized collection of CTI information.

5.3.3 Interdisciplinary Research

Given the interdisciplinary nature of this dissertation, it was not possible to cover every facet of each field comprehensively. For instance, within the realm of cybersecurity, our research focused on information gathering in CTI without looking at network or closed-source data. In the field of deep learning, our attention was primarily on classification of textual data and transformer models, excluding other task settings and architectures.

The interdisciplinary approach of this dissertation also reveals itself on a smaller scale. For example, a thesis solely dedicated to adversarial training might offer a more in-depth discussion on the robustness of such training, contain more rigorous analyses of the generated data, and evaluate additional out-of-distribution tasks. Therefore, we recognize the value of research that is fully immersed in its discipline, as it is as crucial as interdisciplinary work.

Furthermore, the interdisciplinarity of this dissertation comes at the cost of general applicability in other domains. While many deep learning methods, like ActiveLLM, XAI-Attack, and the GPT-2 data augmentation method, are evaluated for various domains, methods like multi-level fine-tuning and insights into catastrophic forgetting in CySecBERT are only evaluated in the application domain. Therefore, it remains to be shown whether these studies also apply to other domains.

5.4 Ethical and Societal Considerations

In our work, ethics has been a primary concern, and we have continuously reevaluated our methods to ensure responsible behavior. However, research on topics such as CTI information gathering and adversarial examples inherently involves risks of misuse. For instance, a system capable of identifying specialized CTI information can be used to gather insights on how to exploit vulnerabilities or deploy attacks on vulnerable machines. Similarly, our adversarial example generator, XAI-Attack, designed to enhance the robustness of models, can also be misused to attack deployed deep learning classifiers. Additionally, research on deep learning classifiers in low-data regimes can enable malicious actors to train models for unethical purposes with relative ease.

We also acknowledge that we have not explicitly focused on analyzing social biases that may result from or be inherent in our methods. For example, if the CySecBERT model is used for classification or GPT models are employed as data augmentation methods, the results could exhibit biases. While this might not significantly impact most applications, in some cases, such biases could lead to severe consequences, including discrimination.

As authors, we wish to express our concerns regarding the potential misuse of our work. Nevertheless, we advocate for an open-source mentality, recognizing the significant benefits these technologies can offer. We emphasize that the responsibility for ethical use lies with the users of these methods, in line with ongoing discussions within the open-source community.

Part II
Publications

Information Overload in Crisis Management: Bilingual Evaluation of Embedding Models for Clustering Social Media Posts in Emergencies

6

Abstract

Past studies in the domains of information systems have analyzed the potential and barriers of social media in emergencies. While information disseminated in social media can lead to valuable insights, emergency services and researchers face the challenge of information overload as data quickly exceeds the manageable amount. We propose an embedding-based clustering approach and a method for the automated labeling of clusters. Given that the clustering quality is highly dependent on embeddings, we evaluate 19 embedding models with respect to time, internal cluster quality, and language invariance. The results show that it may be sensible to use embedding models that were already trained on other crisis datasets. However, one must ensure that the training data generalizes enough so that the clustering can adapt to new situations. Confirming this, we found out that some embedding models were not able to perform as well on a German dataset as on an English dataset.

Original Publication *Markus Bayer, Marc-André Kaufhold, and Christian Reuter. "Information Overload in Crisis Management: Bilingual Evaluation of Embedding Models for Clustering Social Media Posts in Emergencies".* In: 29th European Conference on Information Systems—Human Values Crisis in a DigitizingWorld, ECIS 2021, Marrakech, Morocco, 2020. *Ed. by Frantz Rowe, Redouane El Amrani, Moez Limayem, Sabine Matook, Christoph Rosenkranz, Edgar A. Whitley, and Ali El Quammah. 2021.* URL:https://aisel.aisnet.org/ecis2021_rp/64[1]

[1] Minor typographical errors in the original publication have been corrected for this dissertation.

M. Bayer, *Deep Learning in Textual Low-Data Regimes for Cybersecurity*, Technology, Peace and Security I Technologie, Frieden und Sicherheit, https://doi.org/10.1007/978-3-658-48778-2_6

Material

Experimental Code: https://github.com/PEASEC/Evaluation-of-Embedding-Models-for-Clustering-Social-Media-Posts-in-Emergencies

6.1 Introduction

In the past 20 years, social media was not only used in everyday life but also during almost every major natural and man-made crisis, including the 2001 September 11 attacks, 2012 Hurricane Sandy, 2013 European floods, or the COVID-19 pandemic, to gather and spread disaster-related information [289, 322, 357]. This user-generated content comprises multimedia files (e.g., audio, photo, video) and textual information (e.g., situational updates, public mood, specific information) that has the potential to increase situational awareness and improve crisis response for crisis volunteers, emergency personnel, and other involved persons [161, 313]. However, emergency services face issues of information quality and overload under the time-critical constraints of large-scale emergencies, which requires an efficient and effective way to structure the incoming volumes of social big data [167, 312]. To address this challenge, many algorithms, frameworks, methods, and tools arose from the research areas of machine learning [10, 167], information systems [101, 116], social media analytics [108, 420], and crisis informatics [143, 321].

To tackle information overload, researchers came up with supervised machine learning classifiers to estimate the relevance of postings to the situation [2, 142] or to categorize the postings into groups of different information types [54, 171, 304]. Despite the value of such supervised approaches, the gathering and labeling of case-specific training data are costly and highly time-consuming, which is particularly problematic in disaster situations [198]. Little research, however, has focused on unsupervised techniques such as clustering, which utilize similarity measures to identify patterns in the data to form groups and do not need any training based on labeled data [503]. To find similarities between social media messages, it is necessary to convert them into vectors, which ideally share a similar contextual meaning. The contextual conversion can be produced by embedding models, such as the Word2vec model [283].

In our literature review, we identified a variety of potentials for research. First, researchers have brought up both general and domain-dependent embedding models [10, 128]. However, there is a lack of knowledge on which ones perform better in emergencies. Furthermore, current clustering approaches are primarily learned on

English data [244]. This calls for a cross-language evaluation of different embedding models. Since disasters are often characterized by time-critical constraints, performant clustering algorithms and embedding models are required to allow an almost real-time application. To further increase the interpretability and value of the built clusters in a disaster situation, a short description or a label for each would be desirable [10]. In summary, the goal of this work is to establish an efficient and effective methodology for clustering social media posts in disaster situations so that emergency personnel and other actors can gain a quick overview of the gathered data. The main aspect is to evaluate different state-of-the-art word and document embedding methods with respect to clustering and disaster situations. Thus, we seek to answer the following research questions: *To what degree are domain-dependent embeddings helpful for clustering the dynamic data in emergencies (RQ1)? Which embeddings are more invariant with respect to the language of the data (RQ2)? Which embedding methods are suitable for the time-critical analysis of Twitter data in emergency situations (RQ3)?*

In order to answer these questions, the paper is structured as follows: First, we present related work on the foundations and techniques regarding information overload and clustering before outlining the research gap (Sect. 6.2). Based on these foundations, we present the method and implementation of embedding models and the clustering approach (Sect. 6.3). Thereafter, we describe the selected datasets, evaluation criteria, and results of the evaluation (Sect. 6.4). In summary, we evaluate 19 methods for creating the document embeddings on two different datasets and use k-means for clustering, from which the produced groups are evaluated with internal evaluation methods. Furthermore, we discuss the prospect of automatically labeling the clusters (Sect. 6.5). The paper finishes with a discussion of the results and implications, the conclusion, and an outlook (Sect. 6.6).

6.2 Related Work

In information systems, social media analytics is defined as "the process of social media data collection, analysis, and interpretation in terms of actors, entities, and relations" [419]. It aims to combine, extend, and adapt methods and tools for the analysis of social media data [108, 420]. When applied to the domain of crisis informatics, it is often combined with interfaces for real-time analytics and machine learning algorithms [167, 314]. This section presents techniques to mitigate information overload in crises and proposes clustering as a solution to reduce the amount of data presented to emergency personnel.

6.2.1 Foundations and Techniques to Mitigate Information Overload in Crises

When tens of thousands of social media messages are generated during large-scale emergencies [358], authorities have to deal with the issue of information or social media overload [233]. Amongst others, the concept is examined in management information systems and can be caused by personal factors, information characteristics, task and process parameters, organizational design, or information technology [102]. With regard to characteristics and technology, information overload is often defined as "[too much] information presented at a rate too fast for a person to process" [152] and implies the danger of getting lost in data that may be irrelevant to the current task at hand and of data being processed and presented in an inappropriate way [200]. In past crisis informatics research, several prototypes and techniques were explored to mitigate information overload in large-scale emergencies, whereof some are outlined in Table 6.1. The first intuitive step to find relevant (or to filter out irrelevant) information is the use of search engines that facilitate simple keyword-

Table 6.1 Overview of different techniques deployed to mitigate information overload in crises, disasters, or emergencies

Technique	Description
Search engine	Formulation of simple keyword search or complex Boolean search query engine, including operators such as "AND", "OR", and "NOT", which are embedded into web interfaces or provided by search APIs [167].
Metadata filtering	Filtering of information by metadata, such as language, location, social media platform, or time, which is often combined with search functionality [197].
Interactive visualizations	Use of interactive visualizations, such as charts, maps, timelines, or word clouds, to reduce the displayed data to a specific subset by a specific gesture [314].
Message classification	Use of supervised machine learning models to classify information as relevant or irrelevant for a specific emergency [142] or to categorize them into humanitarian information types [10].
Message clustering	Categorization of text documents into similar groups using similarity metrics and unsupervised machine learning techniques, which do not require labeled data for training [107].
Information summarization	Automatic and real-time algorithms that use extraction or abstraction techniques to provide a general information summary of a disaster event [373].

based or complex Boolean search queries. While these are often embedded in social media platforms such as Facebook or Twitter directly, developers can use platform search APIs to integrate their results into supportive third-party applications [167]. In such applications, search engines are often combined with additional functionality allowing the filtering of information by metadata, such as language, location, social media platform, or time [197]. While this functionality is often provided by specific forms, interactive visualizations, such as charts, maps, timelines, or word clouds, can be used to reduce the displayed data by a specific gesture [314]. For instance, if a pie chart displays the numbers of positive, neutral, and negative sentiment messages, a click on the positive "wedge" could trigger attached list or map views that only show messages with positive sentiment.

While search engines and metadata filtering can be useful measures for reducing information overload, too restrictive search queries can lead to the problem that emergency services may miss out on relevant information. This is especially true for location-based filtering since only a small amount of social data posts contain geocoordinates. Here, machine learning algorithms can help to find relevant information after data is collected. For instance, message classification techniques often apply supervised machine learning to binarily classify social media posts as relevant or irrelevant for a specific emergency [142] or to categorize them into humanitarian information types, such as affected individuals, infrastructure and utilities, donations and volunteering, caution and advice, sympathy and support, other useful information, or not applicable [10]. However, these techniques are often tailored to specific emergencies, thus not being generally applicable, and rely on the time-intensive labeling of data and model training. This stands in contrast to unsupervised message clustering techniques that categorize text documents into similar groups using similarity metrics and unsupervised machine learning techniques; hence, labeled data for training is not required [107]. Considering the human capacity of information processing, Miller [284] suggests "organizing or grouping the input into familiar units or chunks" to overcome such limitations. In accordance, further research suggests that 'chunking' social media messages by specific tools positively influences emergency managers' intention to use social media during emergencies [349]. However, clusters might not be self-explanatory and require a useful summary of cluster content or at least describing labels [135]. On the one hand, information summarization approaches might be used for the automatic and real-time extraction or abstraction of a dataset or subset to provide an information summary [373]. On the other hand, the automatic labeling procedure outlined in Sect. 6.5 can be relevant for obtaining a concise overview of the cluster contents.

6.2.2 Clustering, Embeddings, and their Application in Crisis Informatics

Clustering is performed by unsupervised machine learning methods, which can be applied when no class is to be predicted [107]. Thus, the data should rather be grouped into natural clusters [487]. These groups are most frequently found with some kind of similarity measure for comparing the data. In contrast to supervised machine learning techniques, labeled data is not required, which makes it an interesting field of study. Authors like Imran et al. [167] already pointed out that the considerable amount of labeled data constitutes a challenge in disaster situations. There are many algorithms suited for clustering data, such as k-means [146] and mean-shift clustering [69]. Further existing work looked at clustering techniques in general [503], reviewed techniques in contrast to k-means [175], and with respect to Twitter data [13]. However, k-means is a widely used algorithm that is also applied in crisis informatics [10] because it is simple and computationally efficient [175]. Since clustering methods rely on numerical data as input, it is important to find a good numerical representation of the textual nature of a tweet. More precisely, it would be necessary to map the Twitter posts into a latent space that has an inherent structure based on contextual similarity. This means that similar posts get a similar number vector, and different posts are distant in this continuous space.

This challenge can be addressed with embedding models. The goal of representing text as numerical vectors with meaning can be abstracted as the goal to represent words as numerical vectors with meaning. This dates back to the 1960s and is based on the distributional hypothesis (Harris, 1954), which can be interpreted as "a word is characterized by the company it keeps" (Firth, 1957). With the Word2vec model from Mikolov et al. [283], word embeddings became one of the biggest trends in Natural Language Processing (NLP) research until the present. Mikolov et al. [283] proposed a shallow neural network for building the embeddings, reaching state-of-the-art performance in various NLP tasks. As a next step, the words of a sentence or document can be further processed to get a single sentence or document embedding, respectively. More or less based on the Word2vec model, various word and document embedding approaches have been proposed in recent years. Many embedding models are self-supervised, meaning that they create their own training labels without any human annotator [257]. To model the distributional hypothesis, they need a large dataset for training. If a model has never seen a certain word (also called out-of-vocabulary word), it is hard to map it to a meaningful vector. Training a new model for every new disaster situation and other fields of use is not feasible; therefore, people often rely on pre-trained embeddings. A model like GloVe [330] that has been trained on large common crawls, Wikipedia pages, and Twitter data has vocabulary sizes varying from 400,000 to 2.2 million unique words. It seems worthwhile to examine if it would be more beneficial to train the embedding models using general or domain-related data, such as other past crisis datasets.

One of the main purposes of clustering is to gain insights from the underlying structure by discovering natural groupings [175]. These groupings can reach from being very obvious to being latent, depending on the data and the observer. With regard to crisis informatics, a brief overview of the different categorization or grouping possibilities for social media posts is shown in Table 6.2. Most existing work categorizes posts as either relevant or not relevant for a specific emergency, while some try to map predefined humanitarian categories. The different groupings show that supervised classifiers cannot cover all different possibilities, so an unsupervised clustering method might be helpful. Emergency situations are highly dynamic, so it is obvious that the goal is to not predefine the groups. Clusters from a clustering algorithm can and should be highly different, given new disaster situations.

Table 6.2 Overview of different grouping possibilities proposed in various papers

Paper	Grouping possibilities
[10]	"damage", "personal opinion", "caution and advice", "not relevant"
[10]	"after effect", "personal opinion", "updates", "other useful information", "not relevant"
[168]	"personal only", "informative (direct)", "informative (indirect)", "informative (direct or indirect)", "other"
[168]	"caution and advice", "casualties and damage", "donations of money, goods or services", "people missing, found or seen", "information source", "other"
[457]	"off-topic", "on-topic and relevant to situational awareness", "on-topic and not relevant to situational awareness"
[457]	"off-topic", "on-topic and not relevant to situational awareness", "social environment", "built environment", "physical environment"
[457]	"off-topic", "on-topic and not relevant to situational awareness", + 32 information types
[2], [142], [198], [244], [304], [416], [455]	"relevant/informative", "not relevant/informative"

6.2.3 Research Objectives

The review identified a variety of measures to reduce information overload (Table 6.1), highlighting that chunking or clustering information positively influences emergency managers' intention to use social media during emergencies [102, 349]. Apart from the emphasis on crisis situations, the evaluation performed in this paper can be seen as an intrinsic evaluation method for embedding-based clustering. Intrinsic evaluations are performed within the word vectors themselves, especially where no classifier is trained on them [26, 385]. One of the most popular evaluation methods in this sector is the word similarity task, where the cosine similarity of word embeddings is compared to human judgments [111]. Extrinsic evaluations, on the other hand, apply the embeddings on a downstream NLP task, for example, sentiment analysis or POS tagging [303]. The contextually most related work to ours is conducted by Alam, Ofli, and Imran [10]. They propose a system for clustering social media data in crisis situations using an embedding approach. The authors train a new Word2vec model on a crisis dataset. These contextual embeddings are then averaged to get a document vector. Afterward, the authors perform PCA to reduce the computational cost, which they unfortunately do not specify in numbers. On the lower dimensional data, they also perform k-means with an adaptive k-search approach. The resulting clusters are analyzed and labeled manually by humans. In contrast to their work, we consider different approaches and take a deeper look into the assessment of them. Our goal is to evaluate and answer specific research questions concerning the embedding creation since this is the most important step for clustering the posts.

Clustering in disaster situations is often found in relation to the temporal or spatial dimension [266, 333, 376]. In connection with textual data, state-of-the-art literature is, aside from Alam, Ofli, and Imran [10], lacking in this field. Yin et al. [511] cluster event-specific topics in emergencies but still use TF-IDF vectors instead of embeddings, which are, in most cases, proven to be superior. Abstracting the work from the crisis context, Dai, Bikdash, and Meyer [84] propose a clustering method in the area of public health surveillance. Their goal is not to find different groups but to model a classifier that binarily decides if a post is related or unrelated to the health topic. The authors use the Word2vec model to create embeddings but do not consider other possible models. Our work can be compared to Li et al. [244], where the authors conduct a large evaluation of different word and document embedding approaches for classification tasks in crisis situations. In a similar work, Khatua, Khatua, and Cambria [204] compare word embeddings in the context of virus outbreaks. The difference between the work of both and ours lies in the task setting and evaluation method. Their goal is to classify posts in a supervised way that leads to an extrinsic evaluation method for assessing the embeddings, whereas we work in an unsupervised setting for clustering and propose an intrinsic evaluation method for embeddings. Similar to their work, we apply embeddings learned from previous disasters to examine if they lead to a better separation of the data (RQ1). Beyond that, we put a special emphasis on language-invariance and time-criticality based on RQ2 and RQ3.

6.3 Architecture and Embedding Models

Our process of clustering social media posts is shown in Fig. 6.1. The unstructured text is processed in an NLP pipeline, which tokenizes the tweets according to the tokenizer used for the embedding model. This is necessary as different embedding models use different tokenization techniques. In the next step, an embedding creation approach is applied. These approaches comprise two differing groups of models. The sentence (or document) embedding models are directly building vectors for the whole tweet, while the word embedding models just create word vectors for every word in the tweet. If a sentence embedding model is chosen, the vectors are immediately fed into the clustering algorithm. If we take a word embedding model, the word embeddings of a tweet are averaged (Avg), or the minima and maxima (MinMax) vectors are concatenated. We chose k-means [146] as the clustering algorithm. K-means is dependent on the hyperparameter k, which specifies how many clusters should be formed.

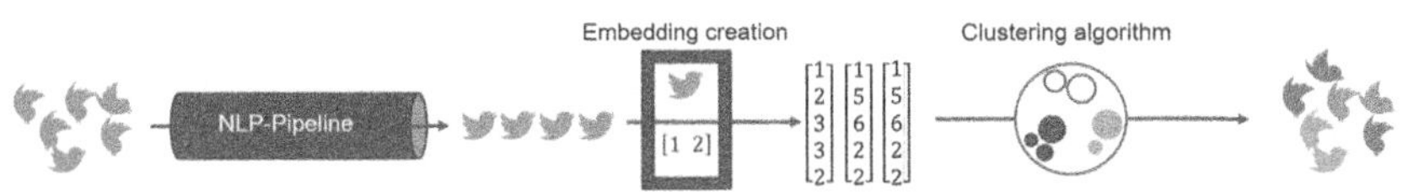

Fig. 6.1 The clustering process, beginning with an NLP-pipeline for tokenization, over the embedding creation, to the clustering algorithm

6.3.1 Software and Hardware Architectures

As a software basis for the clustering algorithm and evaluation metrics (see Sect. 6.3 and Sect. 6.5), we chose scikit-learn. The implementation of the embedding models depends on the individual instructions given by the authors or relevant libraries of the models. The same applies to the tokenization steps that are required for preprocessing the data according to the models. The system used for the implementation and evaluation has an Intel Xeon processor with a single core at 2.3 GHz. A Nvidia Tesla K80 with 12 gigabytes of RAM is used for computing matrix-based calculations of several representation models. Furthermore, 25 gigabytes of RAM are available for storing the representations efficiently.

6.3.2 Used Embedding Models for Comparison

Many embedding approaches have been developed since the proposal of Word2vec. For a suitable selection of approaches, we considered the work of Li et al. [244] and new emerging trends such as the state-of-the-art embedding model BERT by Devlin et al. [89]. The models used here are listed below with a brief description, while detailed descriptions can be found in the original papers.

- **Word2vec (W2V)**—word embeddings [283]. This model is one of the first that used shallow neural networks to create word embeddings. Due to its popularity, many different datasets were used by various authors to train the network. We are using three pre-trained models:
  - Twitter Word2vec model: Trained with 400 million tweets [128].
  - Crisis Twitter Word2vec model: Trained with 52 million crisis-related tweets [170].
  - Crisis Twitter Word2vec model: Trained with 364 million crisis-related tweets [8, 9].
- **GloVe**—word embeddings [330]. This model combines the idea of Word2vec with word occurrence statistics. The Twitter model pretrained on 2 billion tweets is used.
- **FastText**—word embeddings [38, 186]. The main difference of this Word2vec extension is that it splits the words into n-grams, which has the advantage of getting good vectors for rare and out-of-vocabulary words. Another advantage is that FastText is already pre-trained in 157 languages, which is especially important in disaster situations.
- **InferSent**—sentence embeddings [79]. This model is, in contrast to the previous models, trained on a supervised task. The authors are using a bidirectional LSTM, followed by a comparison layer and fully connected layers, to solve the SNLI task [42]. We are using the GloVe and FastText pretrained variants.

- **Universal Sentence Encoder (USE)**—sentence embeddings [484]. This model extends the approach of InferSent to two more tasks (question answering and translation) with different architectures. We use the base and large pretrained model.
- **Sent2Vec**—sentence embeddings [320]. The goal of Sent2Vec is to assign each word a word embedding so that the average of all of them in the document constitutes a good vector. We are using the Twitter-learned embeddings.
- **Sentence-BERT (SBERT)**—sentence embeddings [356, 513]. Bidirectional Encoder Representations from Transformers (BERT) [89] is considered to be the state of the art in the embedding creation task. The word embeddings are created in a contextualized manner, i.e., that not only the word itself is considered but also the context in which the word appears. Since finding the most similar sentence in 10,000 sentences requires about 65 hours [356], we take advantage of the BERT extension from Reimers and Gurevych [356], who state to solve this in 5 seconds, while maintaining the accuracy of BERT. We evaluate the base and large model.

6.4 Evaluation of Embedding Models

6.4.1 Datasets and Measurements

In order to answer our research questions, we decided to use a German and an English flooding dataset for evaluation. First, the German data is based on the 2013 European Floods, which had a severe impact on Germany [199] and contains about 4,000 posts related to the flooding. Second, the English set was crawled during the 2013 Colorado floods and contains about 1,000 posts [313]. Both sets were chosen because they refer to a similar scenario, are different in language, and also contain labels. Labeled data can be helpful for further work concerning an external evaluation. Furthermore, it is important that both sets are gathered during a flooding scenario in different languages, which is required to answer RQ2.

Evaluating clusters is a difficult task because the formation of the clusters is ambiguous. We decided against an external evaluation. The external evaluation assesses the clustering performance with an existing ground truth. By having such a ground truth, the benefits of clustering would be missed. Especially in the disaster setting, it can be difficult since several different groupings can be formed, as we have seen in Sect. 6.2. Instead, we chose an internal evaluation, where the clusters themselves are analyzed by the means of a measurement. This approach comes with the problem that the clustering algorithm can be optimized on this measurement. But since we are only using k-means, this concern is reduced. As measurement criteria, we use the Silhouette Coefficient [370], Calinski-Harabasz Index [52], and Davies-Bouldin Index [86]. The Silhouette Coefficient measures how fitting an object is to its cluster compared to the other clusters. The measure ranges from -1 to 1; a higher score is better. The Calinski-Harabasz Index measures the ratio of the sum of between-cluster distances and of inter-cluster distances. Again, a higher score relates to better clustering performance. The Davies-Bouldin Index, on the other hand, indicates better clustering performance with a lower score, which is achieved when the groups are farther apart and less dispersed.

6.4.2 Results for the English and German Datasets

The results of the clustering task for the English and the German datasets are shown in Table 6.3 and Table 6.4. The three best scores for each evaluation metric are highlighted. We choose k based on Sect. 6.2.2, where already several grouping possibilities are listed. To have some kind of adaptive evaluation, we test each model with a k ranging from 4 to 10. The majority of algorithms performed best when using a k of 5. The evaluation results shown in the tables are related to a k-means clustering with a k of 5, representing 5 clusters.

For the English task, both Word2vec models based on a crisis dataset with the average sentence embeddings performed the best when inspecting the Silhouette Score (Crisis Word2vec 1: 0.7) and the Davies Bouldin Score (Crisis Word2vec 2: 0.22). The Calinski Harabasz Score of both is also particularly good, taking second and third place behind the English FastText model (254.256). Regarding

Table 6.3 Evaluation of the English clustering task; the three best scores per metric (bold) and models with none (red), one (yellow), or multiple (green) high scores are marked

Embedding Model	Variant		Embedding Creation		Cluster Creation		Silhouette		Calinski Harabasz		David Bouldin	
W2V.Twitter	Avg	MinMax	99.06	38.75	0.47	0.96	0.07	0.05	102.93	55.50	1.71	2.85
W2V.Crisis1	Avg	MinMax	51.45	25.58	**0.28**	0.79	**0.70**	0.10	**172.99**	72.39	**0.22**	1.62
W2V.Crisis2	Avg	MinMax	127.86	131.08	**0.32**	0.68	**0.69**	0.09	**143.99**	76.61	**0.22**	**1.53**
Glove	Avg	MinMax	292.11	256.57	**0.40**	0.48	0.08	0.14	111.33	125.36	2.60	2.32
FastText	Avg	MinMax	166.44	163.73	0.54	0.76	0.06	**0.16**	64.51	**254.26**	2.67	1.61
USE	Base	Large	**1.36**	**0.08**	0.81	0.87	0.03	0.02	36.19	36.16	4.14	4.16
InferSent	Glove	FastText	112.99	91.49	4.15	5.33	0.05	0.05	48.08	45.87	3.37	3.56
Sent2Vec	Unigrams		260.64		0.99		0.03		43.77		3.47	
SBERT	Base	Large	25.68	70.76	0.87	1.24	0.06	0.07	65.07	66.48	2.83	2.99
SBERT.sts	Base	Large	**23.65**	70.26	1.02	1.20	0.04	0.05	49.42	47.32	3.74	3.89

Table 6.4 Evaluation of the German clustering task; the three best scores per metric (bold) and models with none (red), one (yellow), or multiple (green) high scores are marked

Embedding Model	Variant		Embedding Creation		Cluster Creation		Silhouette		Calinski Harabasz		David Bouldin	
W2V.Twitter	Avg	MinMax	101.67	71.82	1.31	2.73	0.23	0.23	549.62	639.45	**1.15**	2.06
W2V.Crisis1	Avg	MinMax	54.40	**50.17**	**1.15**	2.07	0.22	0.13	927.60	474.89	1.69	2.59
W2V.Crisis2	Avg	MinMax	128.56	129.21	**1.15**	1.88	**0.24**	0.15	912.67	518.65	**1.41**	2.43
Glove	Avg	MinMax	291.58	254.34	**0.91**	1.31	**0.28**	0.19	**1282.74**	542.23	1.87	2.24
FastText	Avg	MinMax	157.67	154.99	1.52	1.62	0.15	**0.27**	**1160.52**	**8873.64**	2.34	**1.18**
USE	Base	Large	**1.53**	**0.28**	2.03	2.04	0.11	0.11	357.96	358.17	2.95	2.96
InferSent	Glove	FastText	127.93	105.07	15.08	22.00	0.14	0.11	417.04	371.29	2.79	2.27
Sent2Vec	Unigrams		260.67		2.94		0.14		443.13		2.66	
SBERT	Base	Large	95.61	301.08	3.09	3.74	0.08	0.09	407.61	514.45	2.83	2.84
SBERT.sts	Base	Large	109.67	305.15	3.42	4.52	0.08	0.08	261.76	228.70	3.10	3.24

the overall clustering quality, it is also assumable that the FastText model performs good, even if the models that were learned on the crisis dataset seem to be better suited. Most of the other embedding techniques reach less promising results. Both variants of Universal Sentence Encoders (USE) are consistently the worst at creating embeddings that are good for forming clusters with the k-means method. In turn, they are the fastest algorithms with regard to the embedding creation time. Comparing the crisis Word2vec models in relation to this evaluation metric, the first is more than two times faster than the second, resulting from the fact that it was trained on a smaller dataset. The FastText model is about 110 and 40 seconds slower than the two better-performing models. In terms of cluster creation time, only InferSent is significantly deviating, but this can be considered negligible with about 5 seconds.

Inspecting the results of the German dataset, the FastText model again achieves the highest Calinski Harabasz Score. In contrast to the first task, both disaster-related embedding models are not as good with regard to the Silhouette Score and the Davies Bouldin Score. GloVe has the best Silhouette Score (0.276) and Word2vec, which is based on a general Twitter dataset, has the best Davies Bouldin Score (1.149). Inspecting the three best scores of each quality evaluation metric, it seems that the FastText model followed by the GloVe embeddings is best suited for the clustering task in German. As in the previous task, Universal Sentence Encoders are the fastest methods but come with a low clustering quality in comparison to most other embedding methods; only SBERT produces lower results. As observed in the previous task, the cluster creation time of all algorithms except InferSent is similar. Yet, the creation time of up to 22 seconds can no longer be considered as negligible.

It should be noted that the absolute results for the two different tasks cannot be compared with one another because the specific scores are not comparable for different datasets. For the English task, the two Word2vec models, which were learned with crisis data, performed best. The clustering of the German dataset, on the other hand, showed that the domain-dependent models were not able to maintain these results. This is most likely due to a loss in generalizability since the training was primarily performed with English crisis data. It can be observed that the model that was trained with more data (Crisis Word2vec 2) also seems to be somewhat better when inspecting the language shift. This is especially important because messages in crisis situations are often communicated in different languages apart from English. To this criterion, the FastText model might generalize best since it is available in 157 different languages. The model itself performed well on the first task, taking third place behind the domain-dependent models with regard to the clustering quality, and best on the second task. It is interesting to see that in the case of the Word2vec model, average embeddings seem to perform much better than the concatenation of the minimal and maximum word embedding, whereas, in the case of the FastText model, it is vice versa. This indicates that these specific embeddings learned by the FastText model are more meaningful and unique with regard to the whole tweet. Universal Sentence Encoders have by far the fastest embedding creation processes for both tasks. These speed advances are accompanied by poor cluster quality results. The well-performing Word2vec and FastText models require between 50 to 170 seconds, which might be fast enough for many applications. If a faster model is needed, it may be sensible to reduce the dataset on which, for example, the Word2vec model is trained. However, such a reduction in the data leads to a reduction in the generalizability for new data.

6.5 Towards Automatic Cluster-Labeling

Mitigating information overload in emergencies is a complex problem. While it is required to improve the performance (i.e. to allow a near real-time application) and quality of clustering, emergency managers must be able to make sense of the generated clusters to facilitate the practical value of clustering in emergencies [420]. Descriptions or explanations of computed clusters can be achieved in several ways, such as information summaries, labels, or word clouds (see Sect. 6.2.2). Therefore, we derived criteria that are necessary for effective techniques to describe and explain clusters in emergency situations in a human-friendly way, which are enlisted in Table 6.5. The criteria are based on the problem of information overload, the conditions in emergency situations, and the human view of explanations, that are put into context with parts of our literature review.

Table 6.5 Criteria for human-friendly cluster explanations in emergency situations

Criteria	Description
Concise	Information explanations of the membership of tweets to a cluster need to be concise in the same way humans tend to prefer selected causation explanations over a complete list [294].
Covering	A good representation of the cluster content should be comprehensive for the tweets in it. Otherwise, valuable information could be overlooked, which could cost lives in emergency situations [142].
Non-redundant/unique	The explanation of one cluster should be unique, as it can be extracted from the cluster description process by Siroker and Miller [407].
Non-laborious	Information explanations should at best be formed automatically. Manual actions require time, which is limited in disaster situations [198].
Invariant	The explanations should be invariant to domain shifts, since emergency situations are highly dynamic [244].
Truthful	The explanations of the clusters should be truthful. Truthfulness is important for human-friendly explanations but as stated by Molnar [294] not as important as having concise explanations.

A simple approach to extracting an explanation is to show the user the tweet that is closest to the centroid of each cluster. However, this may not cover the other tweets in the cluster. Another approach would be to show the user several samples of posts in the cluster, leading to a higher coverage but, in turn, to a lower conciseness. An approach with very concise explanations of the clusters is present in the work of Alam, Ofli, and Imran [10], where emergency personnel have to label the different groups after the clustering process. These labels are short titles for the content present in the clusters, comparable to the grouping names in Table 6.2. However, this approach contradicts the non-laborious criterion. Since disaster situations have to be handled in a real-time manner by emergency personnel, we want to propose a suggestion to automate this process. The idea is to utilize the latent space of the embedding vectors, which is already created when the posts are clustered. For each group, we try to construct posts that are as generic as possible but representative of the group as well (invariant and covering). These posts are mapped into the embedding space by using the same embedding method we used when performing the clustering. Then, the artificial posts are vectors with the same dimensions as the vectors of the real posts. In this way, we can assign them to the most fitting clusters. In turn, this leaves us with the opportunity to provide the fitting cluster with the label we already know from the corresponding self-created tweet (truthful criterion). For the demonstration of the approach, we created constructs of posts for the groups identified by Imran et al. [168]. To ensure that the posts fulfill the criteria to be relevant to the disaster and plausible to the group, we inspected the dataset used by Imran et al. [168]. We extracted the posts for each group, removed the stop words, and calculated the word frequencies. From the list of words and their frequencies, we deleted all posts that were too specific to the crisis. Based on the remaining ones we identified the words that were common in the tweets and sensible for the group to examine the posts containing them. This way, we created the generic posts that are shown in Table 6.6.

Table 6.6 Generic posts for different humanitarian categories

Group names	Generic posts
Caution and advice	· [disaster_name] + " warning—Stay safe—take precautions"
Casualties and damage	· "Buildings are damaged and destroyed. #" + [disaster_name] · "Several people were injured #" + [disaster_name] · "Several people were killed #" + [disaster_name]
Donations	· "You can make a donation to the " + [disaster_name] + " relief" · "Please provide goods, support or other donations for victims of #" + [disaster_name]
People missing, found, or seen	· "Several people are missing or unaccounted #" + [disaster_name] · "Please help find this person—contact us for any pointers #" + [disaster_name] · "Several people have been located and are alive #" + [disaster_name]
Information Source	· "Photos of " + [disaster_name] + "http://t.co/random#report#documenting" · "News: A video of the " + [disaster_name] "

For example, a generic tweet for "Casualties and damage" can contain important terms like "buildings," "damaged," and "destroyed," which are very often stated in damage-related posts. This construct is ended by a hashtag concatenated with the disaster-related keyword that was chosen by the user or by the disaster name if an event detection system has been executed beforehand (invariant to other situations). This specific information is important because otherwise, this tweet could have never been part of the original dataset. These constructed posts are particularly beneficial since they contain all the necessary information, and no human effort is needed by the time they are used.

Eventually, it is possible that one cluster has more than one label (multi-labeling), giving an overview of several topics that could be contained in it. In this way, we can have much more labels and self-constructed posts than groups for many different cases. In the end, we contradict the uniqueness criterion by accepting different clusters to have the same label. If clusters have no label, the cluster and the posts in it are called "not identifiable" or "other".

6.6 Discussion and Conclusion

In this paper, we examined techniques for clustering social media posts in emergencies based on their textual content as a means to reduce information overload. More specifically, we evaluated 19 different embedding methods that are suitable for building clusters of similar postings. The answers to the research questions that were proposed in the beginning are based on the evaluation in Sect. 6.4.

To what degree are domain-dependent embeddings helpful for clustering the dynamic data in emergencies (RQ1)? To answer this question, we evaluated two Word2vec models that were trained with domain-dependent crisis data and compared them to models that were trained without any specific context. Regarding the English task, the domain-dependent models are, in fact, superior in two of three clustering scores. However, by inspecting the German task, the domain-independent but language-specific model FastText is achieving higher scores. The two crisis Word2vec models are still performing reasonably well on the German task, but it is noticeable that they are not able to generalise so satisfactorily with regard to the language shift since the training data primarily consisted of English data. This concerns the second research question, which asks for the best language-invariant models.

Which embeddings are more invariant with respect to the language of the data (RQ2)? It can be observed in both tables that FastText with MinMax sentence embeddings is a particularly good choice when dealing with the problems at hand, even reaching the best scores for the German task. Pre-trained FastText models are available in 157 different languages, which makes them, apart from their good score, a sensible option for this case. As stated before, the other models, like the pretrained Word2vec or SBERT variants, produce inferior results on the German dataset since they were trained on English data and have never seen most of the German words before. This means that they are not able to infer any meaningful vectors for non-English words, resulting in a worse clustering performance than a model that was trained with data from different languages.

Which embedding methods are suitable for the time-critical analysis of Twitter data in emergency situations (RQ3)? The third research question unites the other research questions and combines them with the time-criticality of disaster situations. To address this, we separately measured the embedding and cluster creation time in the evaluation. The embeddings created by InferSent set aside, cluster creation time seems to be almost invariant with respect to the other models. Inspecting the embedding creation times, USE was by far the fastest method for clustering the data, taking less than two seconds. However, since this method does not perform well, we advise taking a disaster-pretrained Word2vec model when dealing with English data and FastText when dealing with other languages. While the Word2vec-based models seem to be a bit faster, both can take up to three minutes in our evaluation.

6.6.1 Practical and Theoretical Implications

In this paper, we provide an overview and comparison of embedding models across two languages and propose an intrinsic embedding evaluation task (C1), give advice on the implementation of clustering and embedding approaches (C2), derive criteria for cluster explanations and propose a method for automatic post-labeling (C3), contribute with findings on the applicability of general and domain-dependent embedding models (C4), and propose a system for reducing information overload in social media streams (C5).

Comparison of embedding models across two languages and the proposal of an intrinsic embedding evaluation task (C1). The paper provides a comprehensive overview and comparison of the performance and quality of 19 embedding methods for clustering. By comparing datasets from two languages, it outlines language-specific conditions for the performance and quality of embedding models. While most existing approaches in crisis informatics examined clustering based on the temporal or spatial dimension [266, 333, 376], this work contributes to the implementation and comparison of text-based clustering approaches. With this evaluation, we propose a generally new intrinsic evaluation task that gives insights into the embeddings and is easily adaptable for other domains.

Advice on the implementation of clustering and embedding approaches (C2). The evaluation of the proposed clustering method shows that the results are highly dependent on the input data, i.e., the embedding representations. This can be seen as an advantage of the proposed system as other works in this field do not evaluate multiple models or neglect a comparison of domain dependence and generalization capabilities of their systems [78, 83, 84]. However, these methods consider other important factors that are further described in the next section. While existing work on embedding models focused on an extrinsic evaluation method on classification tasks [204, 244], our intrinsic evaluation on the clustering task revealed further advice for the implementation of embedding models. In our evaluation, disaster pretrained embeddings performed well but were not able to generalize to languages other than that of the pretrained data. When dealing with other languages, it is sensible to either use a model being trained for this case (FastText) or retrain the model with additional language data. In our tests, we evaluated the Sentence-BERT model by Reimers and Gurevych [356], which failed to produce satisfactory results.

Derivation of criteria for cluster explanations and the proposal of a method for automatic cluster labeling in emergency situations (C3). In 6.5, we derived criteria for human-friendly cluster explanations in emergency situations. Furthermore, in contrast to a manual labeling process [135], we proposed a method for automatically labeling the clusters. This method relies on self-constructed tweets that could reflect humanitarian categories [168], are as general as possible with respect to the event, and are as fitting as possible to the group at the same time. These posts are then mapped into the embedding space, where they are assigned to the best-matching clusters.

Applicability of general and domain-dependent embedding models (C4). Domain-dependent embeddings are helpful for clustering if the domain is fitting. If the domain is not fitting (for example, due to a language shift), a highly general model would be better suited. This stands in contrast to the findings of Khatua, Khatua, and Cambria [204], who, in their extrinsic evaluation, come to the conclusion that domain-dependent embeddings mostly outperform generic pre-trained embeddings. It is similar to the result of Li et al. [244], who observed that their crisis-specific embeddings are more suitable for specific crisis tasks. We can deduce that depending on the unsupervised task, it is important to consider different embedding models. Concluding from a different path of reasoning by testing language invariance, we can also say that the currently existing crisis-specific embeddings are not able to generalize as well to other situations. This fact leads to further research opportunities in which an embedding training dataset consisting of various disaster situations and different languages could be created.

A system for reducing information overload in social media streams (C5). The proposed system (Fig. 6.1), in combination with the insights from the evaluation and the automatic cluster-labeling procedure, is suitable for reducing the overload of user-generated content in social media by finding meaningful and representative clusters. Crisis and emergency personnel can utilize this implementation to gain an overview of the online discourse and pick the right set of posts for their specific response activities [197]. In addition to this area of application, the system is also beneficial for all purposes that require compressed information on social media, such as brand management in the field of business analytics. Similarly, the clustering system can be beneficially employed in the cybersecurity domain to foster the process of detecting security events. Furthermore, our findings can enrich the clustering scheme by Alves et al. [16] or similar works in this domain.

6.6.2 Limitations and Outlook

For future work, it would be interesting to see if newer embedding models based on a large corpus of different disaster situations, such as CrisisLexT26 [313], would be beneficial to the proposed problem. However, as these are only available in a limited number of languages, it might be sensible to create a novel dataset containing different disaster situations combined with general Twitter data that may contain more languages. In this way, the new advances of the state-of-the-art embedding models can be combined with domain dependence, which works well according to our evaluation. Furthermore, the inclusion of a wider range of Twitter data leads to a higher generalizability for other languages. It would also be interesting to create more multilanguage evaluations since we only evaluated German and English datasets. Especially evaluations for low-resource languages might lead to further insights that are important for dealing with information overload in crisis situations. While this evaluation focuses on Twitter messages, which are inherently limited to 280 characters, the evaluation of larger texts from social media such as Facebook might lead to different results.

The general clustering framework can be enriched with insights from different works that emphasise other factors of analysis. For example, a combination of the incremental clustering approach proposed by Comito, Forestiero, and Pizzuti [78] should be beneficial when considering the data as incoming streams. Moreover, testing different clustering algorithms, as done by Curiskis et al. [83], might even increase the performance. Concerning the prospect of automatically post-labeling the clusters, it would be helpful to build as many generic posts as possible. In our work, this method can only be seen as a prospect, requiring further evaluations and modifications to prove the concept. A possible way would be to measure the labeling quality externally with ground truth data. However, as emphazised earlier, this can lead to wrong conclusions. It could be combined with qualitative user research with emergency managers, for instance, using interviews or scenario-based walk-throughs, to evaluate the efficiency—in terms of mitigating information overload—and usability of the clustering prototype. Finally, this paper focused on textual content and did not analyse information overload based on multimedia files. Existing studies highlight the application of image filtering techniques for deduplication and relevance assessment in crises [10] or unsupervised image segmentation for damage assessment [227], which could be used to complement text-based methods against information overload.

Acknowledgements This work has been co-funded by the German Federal Ministry of Education and Research in the project CYWARN (13N15407) and by the LOEWE initiative (Hesse, Germany) within the emergenCITY center.

ActiveLLM: Large Language Model-based Active Learning for Textual Few-Shot Scenarios

7

Abstract

Active learning is designed to minimize annotation efforts by prioritizing instances that most enhance learning. However, many active learning strategies struggle with a 'cold-start' problem, needing substantial initial data to be effective. This limitation reduces their utility in the increasingly relevant few-shot scenarios, where the instance selection has a substantial impact. To address this, we introduce ActiveLLM, a novel active learning approach that leverages Large Language Models such as GPT-4, o1, Llama 3, or Mistral Large for selecting instances. We demonstrate that ActiveLLM significantly enhances the classification performance of BERT classifiers in few-shot scenarios, outperforming traditional active learning methods as well as improving the few-shot learning methods ADAPET, PERFECT, and SetFit. Additionally, ActiveLLM can be extended to non-few-shot scenarios, allowing for iterative selections. In this way, ActiveLLM can even help other active learning strategies to overcome their cold-start problem. Our results suggest that ActiveLLM offers a promising solution for improving model performance across various learning setups.

Original Publication *Markus Bayer, Justin Lutz, and Christian Reuter. "ActiveLLM: Large Language Model-based Active Learning for Textual Few-Shot Scenarios"*. In: Transactions of the Association for Computational Linguistics *(2025)*.

Supplementary Information The online version contains supplementary material available at https://doi.org/10.1007/978-3-658-48778-2_7.

M. Bayer, *Deep Learning in Textual Low-Data Regimes for Cybersecurity*, Technology, Peace and Security I Technologie, Frieden und Sicherheit, https://doi.org/10.1007/978-3-658-48778-2_7

DOI: 10.48550/ARXIV.2405.10808. *arXiv: 2405.10808*. URL: https://doi.org/10.48550/arXiv.2405.10808[1]

Note Supplementary material containing the full prompt design, dataset and model descriptions, an example prompt and response, and additional experiments can be found in Appendix A.1 of the Electronic Supplementary Material.

Material
Experimental Code: https://github.com/PEASEC/ActiveLLM/

7.1 Introduction

The selection of training examples significantly impacts the performance of models. Even Large Language Models (LLMs), such as GPT-3, show high variances based on training example selection [527]. This dependency is even more pronounced in smaller models such as BERT [527], which are sometimes preferred for their cost and resource efficiency as well as data protection benefits over APIs. In order to combine the advantages of both the lightness of BERT-like models and the capabilities of LLMs, we explore the classical concept of Active Learning (AL) for BERT-like models, extended by LLMs such as GPT-4 [315].

AL encompasses strategies integrated into the labeling process to select instances with a high learning impact [529]. Typically, a model is trained iteratively during the annotation process to query instances for labeling by the annotators (oracle). While AL can yield significant learning improvements with mid-sized datasets, it often encounters a 'cold-start' problem [68], rendering it unsuitable for low-data regimes. The cold-start problem occurs in many AL methods because they lack sufficient data at the start of data labeling to accurately measure informativeness and select informative samples. For example, the uncertainty strategy selects the most uncertain instances based on an iteratively learned classifier during annotation. However, the classifier cannot make accurate uncertainty guesses without enough data.

Furthermore, as most AL strategies require a model to be trained iteratively, the use of pre-trained models such as BERT can result in very high delay times during annotation, making the process unusable in real-world annotation scenarios. In so-called 'model-mismatch scenarios', where the instance selection model (query

[1] Minor typographical errors in the original publication have been corrected for this dissertation.

model) differs from the model used for the final application (successor model), AL yields limited gains [529].

To address these challenges of few-shot and model-mismatch scenarios, we propose ActiveLLM, an AL method leveraging LLMs (see Fig. 7.1). ActiveLLM can select instances with high learning impact even without initial supervised data and requires no training during the annotation process. This method serves both as a standalone AL approach and as a solution to the cold-start problem in other AL strategies. Therefore, our contributions are as follows:

(C1) Novel AL Method: We introduce a novel AL strategy using LLMs such as GPT-4, tailored for few-shot learning to overcome the cold-start problem.
(C2) Efficient and Scalable Methodology: Our approach decouples the query process from the successor model dependency, enhancing scalability and establishing a high degree of practicality compared to conventional AL strategies.
(C3) Rigorous Evaluation: We conduct comprehensive evaluations to demonstrate that ActiveLLM is not constrained to a specific LLM. It exhibits superior

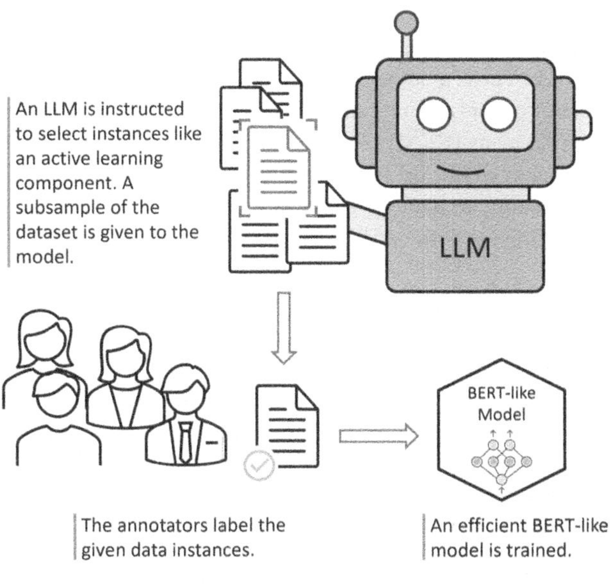

Fig. 7.1 Depiction of ActiveLLM

performance compared to other AL and few-shot learning methods and can be utilized alongside these methods to achieve beneficial outcomes.

The code for this study is freely available[2].

7.2 Related Work

7.2.1 Active Learning

The rationale behind AL is that a system selectively chooses instances to be labeled, thereby reducing the labeling effort [393]. In this context, annotators serve as oracles, responding to queries from the AL system regarding specific instances. This system utilizes a query strategy that often hinges on measures of uncertainty or diversity. Traditionally, the querying is performed by the machine learning model itself, which undergoes iterative training during the labeling process. For instance, the model might request labels for instances where its uncertainty is highest, using approaches such as Least Confidence (LC) [76], Margin of Confidence [443], or Prediction Entropy (PE) [394]. It is then retrained with the newly annotated data.

In addition to uncertainty-based methods, diversity-based approaches also play a critical role in AL. These methods aim to select instances that are diverse to ensure the labeled dataset covers a wide range of feature space. For example, techniques like Embedding KMeans (EKM) [386, 517] leverage the model's ability to embed instances into a latent space. By applying clustering algorithms such as KMeans in this space, the system can identify and select the most diverse examples, thereby enriching the training set with varied and representative data points.

7.2.1.1 Active Learning with Transformer Models

While AL methods have demonstrated their utility across various traditional machine learning models, their application to transformer models remains challenging. The study by Ein-Dor et al. [100], among the first to explore AL with pre-trained transformer models such as BERT, shows that AL can enhance the performance of BERT classifiers, particularly in low-data scenarios and when the initial dataset is rich in relevant class instances. However, most of the other works indicate mixed results. For example, Jacobs et al. [174] combine uncertainty and diversity-based methods, utilizing SentenceBERT embeddings to refine the instance selection process. Their findings indicate that while AL reduces labeling effort, its effectiveness with BERT

[2] https://github.com/PEASEC/ActiveLLM/

models is mixed, and the improvements are modest compared to older NLP models. Studies by Grießhaber, Maucher, and Vu [134], Houlsby et al. [154], Schröder, Niekler, and Potthast [387], Seo et al. [392], and Yuan, Lin, and Boyd-Graber [517], contribute to a similar mixed picture of AL methods for transformer models. In Table 7.1, we present an overview of works utilizing AL for transformer models, categorizing them by query strategies, tested models, and model matching/mismatching scenarios.

Table 7.1 Overview of different works applying AL to transformer models. "Model" describes the successor model trained from the active learning process. "Matching" indicates if the successor model matches the query model

Reference	Query Strategies	Model	Matching	Comparison To Random
[100]	LC, Perceptron Ensemble, EGL, Core-Set, D-AL	BERT	✓	Improvements in imbalenced low-data scenarios
[174]	VR, PE, BALD, Diversity Heuristics	BERT	✓	Mixed results
[517]	ALPS, PE, BERT-KM BADGE	BERT	✓	Mixed results
[387]	PE, Breaking Ties, LC, C-AL	BERT distilRoBERTa	✓	Small improvements in later iterations
[392]	BATL, LC, ALPS, PE, BALD, D-AL, Core-Set, BADGE	DISTRE, BERT SCIBERT	✓	Mostly improvements
[134]	BALD	BERT base	✓	Mixed results
[397]	MNLP, VR, BALD	BERT base, DistilBERT base, ELECTRA base	✓	Significant improvements
[397]	MNLP, VR, BALD	BERT, DistilBERT, ELECTRA	✗	Significant improvements when mismatch is of the form: distilled transformer (query) and normal transformer (successor)

(Continued)

Table 7.1 (Continued)

Reference	Query Strategies	Model	Matching	Comparison To Random
[448]	LC, MNLP, Mahalanobis Distance	(Distil-)BERT, (Distil-)RoBERTa, (Distil-)ELECTRA, XL-Net	✗	Significant improvements when successor model (from distilled mismatch setting) is used for pseudo-labeling an independent model
[306]	MNLP, BERT-KM, BADGE, ALPS	XLM-RoBERTa, miniLM	✗	Significant improvements when successor model (from distilled mismatch setting) is included in query process by giving delayed feedback to the distilled model

Research Gap: The most significant gap in AL research is the use of LLMs, such as GPT-4, to improve AL with BERT-like transformer models, which, to our knowledge, has not been explored previously.

7.2.1.2 Model-Mismatch Scenarios

As transformer models often require substantial computational resources and time [447], the training runtimes for those can be prohibitively long during labeling, making them impractical for real-world applications of AL. Schröder, Niekler, and Potthast [387] noted that the incremental training of these models can be so time-consuming, that it can negate the cost savings from reduced labeling effort. Therefore, the scope of AL has been extended to include scenarios where the querying model may differ from the final (successor) model—a phenomenon known as model mismatch—or situations where no model is used for querying.

But again many studies [24, 387, 397] report that a model mismatch between the query and successor model often leads to unsatisfactory results [529]. While efforts by Shelmanov et al. [397] and Nguyen et al. [306] attempt to address this by using smaller, similar models for querying, there remains a dependency on the model type.

Similarly, Tsvigun et al. [448] offer a promising approach for decoupling the query and successor models, although a distilled version is still required in the process.

Research Gap: In contrast to the use of smaller, resource-efficient models for querying, we propose leveraging much larger language models. This approach may seem counterintuitive at first due to concerns about long runtimes, but as these models are capable of zero-shot learning, no training is required [400]. Furthermore, these models can be accessed in real-time through several free chat interfaces, requiring no machine learning expertise, making them ideal for practical applications.

7.2.1.3 Cold-Start Problem

While AL can significantly streamline the learning process by reducing labeling efforts, most strategies encounter a cold-start problem. In the absence of sufficient initial labeled data (label seed), the AL system may struggle to make informed predictions about uncertainty or diversity [517]. This presents a major challenge for low-data regimes, particularly in few-shot scenarios with very few available data instances. Yuan, Lin, and Boyd-Graber [517] propose ALPS, which is one of the only works that directly addresses the cold-start problem in AL with transformer models. ALPS leverages BERT's pre-existing masked language modeling objective along with clustering to select instances. The authors hypothesize that this strategy provides more reliable confidence scores in the initial stages of model training than a classifier head. However, the results are mixed which is also later confirmed by Nguyen et al. [306].

Research Gap: The cold-start problem is a major challenge in AL research, especially for few-shot scenarios. This work aims to overcome this challenge by proposing an AL strategy that needs no initial labeled data, enabling improvements even at the very beginning of the AL process. This allows the method to also be used to address the cold-start problem of other AL methods by providing them with initial data.

7.2.2 Few-Shot Learning

Few-shot learning refers to methods for scenarios where models need to learn from a limited number of training instances. These scenarios have become increasingly important due to their importance in real-world problems [516] and the increasing capabilities of deep learning models. While LLMs such as GPT-3 inherently have robust few-shot or zero-shot capabilities, our focus is on enhancing few-shot learning in smaller models such as BERT. With specific few-shot learning methods like ADAPET [431] these smaller language models can attain similar or even better few-shot performance than GPT-3.

One approach to this is to formulate tasks as cloze-style tests, where certain words in a text instance are missing and the language model is employed to fill in the blanks. This way, no classifier needs to be trained on top of the language model, leading to a more effective utilization of the language models [123]. Such an approach is exemplified by the ADAPET method [431], as well as in the works of Gao, Fisch, and Chen [123], Schick and Schütze, [383], and Zhang et al. [522]. PERFECT, introduced by Mahabadi et al. [272], eliminates the need to manually formulate such cloze tasks by incorporating task-specific adapters and specialized label embeddings. Similarly, in SetFit by Tunstall et al. [449], no cloze tasks are required, as a sentence transformer [356] is fine-tuned on the available training data. The instances are then encoded with the resulting model and a regular classification head is trained on them.

While not few-shot methods per se, transfer learning and data augmentation are often employed to address few-shot scenarios. Transfer learning allows models to leverage knowledge from related domains [372], providing a strong foundation in few-shot scenarios. Data augmentation, particularly using LLMs [92, 465, 502], has been used to generate high-quality synthetic data in low-data regimes [534]. Similarly, in this study we investigate whether AL can be useful for few-shot learning, especially by overcoming the cold-start problem.

Research Gap: In general, LLMs dominate the machine learning field, in particular in few-shot learning, but there are numerous reasons why researchers and practitioners still rely on smaller BERT-like models, especially for inference, where concerns such as memory limitations, API costs, and privacy often preclude the use of LLMs [40, 378, 469]. To the best of our knowledge, our approach is the first to leverage the capabilities of LLMs to support instance selection for training smaller BERT-like models. In doing so, we combine the flexibility and efficiency of smaller models with the strategic guidance of LLMs.

While Ein-Dor et al. [100] and Grießhaber, Maucher, and Vu [134] also assess the performance of AL with BERT in low-resource scenarios, their datasets are still considerably larger than those in few-shot scenarios. However, in scenarios with very few instances, example selection has the greatest impact. In this work, we aim to improve state-of-the-art few-shot learning methods with our AL approach.

7.3 Method

ActiveLLM is a pool-based sampling method that operates in batch mode, meaning that it selects a subset from a pool of unlabeled data for querying an oracle. It employs instruction-tuned LLMs as query models, while allowing the choice of a successor model to be independent of these models. A demonstration of the process is given in Fig. 7.1, where the LLM is instructed to select instances, which are then labeled and used to train a BERT-like model, allowing any such model to be trained.

The design of the prompts is critical for achieving optimal results with instruction-tuned LLMs. To this end, we have crafted detailed prompts in various configurations to determine their effectiveness in this context. We differentiate between two modes of ActiveLLM. The first addresses few-shot scenarios, where ActiveLLM is executed only once on the dataset. The second is suitable for general scenarios involving iterative querying, which also incorporates feedback from previous iterations.

7.3.1 Few-Shot Learning Mode

The prompts are constructed so that the language model receives a description of AL, including exact details of its tasks, while also keeping the context length small, as this is a limiting factor for current models [255]. Given a specific task, ActiveLLM creates a prompt consisting of AL role allocation, instructions on the selection process, a description of how to format the output, followed by a batch of unlabeled instances.

An excerpt of the prompt design is illustrated in Fig. 7.2. The parameters displayed on the right indicate certain text phrase constellations which can be included, excluded, or adjusted:

Guidelines: This includes a phrase indicating that the model is given not only a set of instances but also guidelines for the task.

Instances: This parameter determines how many instances the model should label. We test different values in Section 7.4.2.3.

Advice: Initial experiments showed that the LLMs often respond with explaining general strategies for choosing instances in an AL manner. Hence, we included the parameter ‘advice,’ which incorporates these strategies.

Consider yourself in the position of an active learning component to help a human annotator. You have to choose the instances that the annotator has to label. You are given {the guidelines for the task and} a set of instances of a dataset. You can only choose {32} instances. {You would ideally want to choose those instances that would provide the most informative and diverse data for the model. Here are some strategies to consider: *advices*}	Guidelines Instances Advice
{Label Guidelines for the human annotator: *guidelines*}	Guidelines
{The output format should be a list of the instances that you would label, separated by a comma.} {Please think step by step about what you would do to select the instances to label. After this provide the list of instances that you would label, separated by a comma.}	No CoT Normal CoT
{Please describe instance by instance why you would select or not select it for labelling. Do not stop before you successfully found {32} that you would suggest to label. After this provide the list of the instances that you would label, separated by a comma.} For example, if you would label the instances 1, 4, 5 then the output should be: 1, 4, 5. The following instances are given to you (separated with "\n ##### \n"):	Explanation Instances

Fig. 7.2 Prompt design of ActiveLLM in few-shot learning mode

No CoT/CoT/Explanation: One of the most common and beneficial prompt engineering strategies is the use of Chain of Thought (CoT) prompting [480]. We experiment with no CoT, 'normal' CoT ("think step by step"), and the task to explain the thoughts on each instance.

Finally, a batch of unlabeled instances is appended to the prompt. We hypothesize that, contrary to common AL strategies, large batches or even all instances do not need to be evaluated by the query model to find instances with high learning impact. We test different values of this parameter in Sect. 7.4.2.2.

7.3.2 Iterated Querying Mode

In the second variant, we design ActiveLLM similar to other AL strategies, allowing the LLM to query new instances repeatedly. For this, it is sensible to include data that was selected in previous iterations. As the prompt should not become too large, we consider three additional parameters (see Appendix A.11 and Figure A.2 of the Electronic Supplementary Material for the full prompt):

No Recap: Indicating the baseline, where no instances from previous queries are recalled.

Recap: The instances from previous queries are directly included in the prompt.

Index Recap: To reduce the context size, the indices of the instances from previous queries are included in the prompt.

For the latter two scenarios, we included instructions in the prompt to indicate that instances from previous iterations are incorporated. In both cases, we did not include the annotated labels, as our experiments showed that they were generally disregarded by the language models. An example of a specific prompt and LLM answer is given in Appendix Figure A.3 of the Electronic Supplementary Material.

7.4 Experiments

7.4.1 Description

Our experimentation perspective with ActiveLLM focuses on addressing the following questions:

1. How should the prompt be designed? (Sect. 7.4.2)
2. Can the chosen prompt be applied to other models and datasets? (Sect. 7.4.3)
3. How does ActiveLLM compare to other AL strategies? (Sect. 7.4.4)
4. Can ActiveLLM improve state-of-the-art few-shot learning methods? (Sect. 7.4.5)
5. How does the method perform in non-few-shot-scenarios? (Sect. 7.4.6.1)
6. Is ActiveLLM capable of overcoming the cold-start problem in other AL strategies? (Sect. 7.4.6.2)

As is common in research, our experiments simulate the AL cycle. To do this, we take a subset of the training set (labels removed) for the respective task and allow the LLM to select a certain number of instances based on the prompt given in Sect. 7.3. These instances are then assigned the true labels as if they had been annotated by a perfect annotator. For comparison, the baseline method uses random sampling, where the first instances from a shuffled dataset are selected.

7.4.1.1 Datasets

For the initial prompt engineering experiments, we are interested in a less common dataset, as a LLM might be less biased by data leakage. We chose the Specialized CTI dataset [27], where Twitter posts are classified according to their relevance to specific CTI events (further details in Appendix A.1.2 of the Electronic Supplementary Material). In the general applicability tests, we used the best-performing prompt from the prompt engineering phase on the commonly-used GLUE benchmark, sampling 32 instances for each task to create a few-shot scenario. For comparison with the few-shot method SetFit, we utilize the AGNews dataset [525], consisting of 32 instances, for which an adaptation is available. The SST-2 dataset [409], previously used in the general applicability section, is also employed in the final experiments. To measure the performance we use the test sets for CTI and AGNews, and on the validation sets for GLUE. Unless otherwise specified (see Sects. 7.4.2.3 and 7.4.6), we select 32 instances per task, either via AL or random sampling for the baseline, which represents a common few-shot learning setting [272, 384, 431].

7.4.1.2 Models and Hyperparameters

In our experiments, we primarily employed GPT-4 [315] as the query model for ActiveLLM, as it achieved top results at the time the main experiments were conducted. However, in Sect. 7.4.3 we also experiment with other LLMs: o1 [316], GPT-4o, GPT-3.5 [48], Llama 3 70B, Gemini-Ultra 1.0 [20], Mistral Large, and Mixtral 8x7B [180] (ActiveLLM is renamed according to the used LLM—e.g. GPT-4: ActiveGPT4). To underscore the practicality of our method, we utilize chat versions of the LLMs to simulate realistic interactions within chat environments (further details in Appendix A.1.3 of the Electronic Supplementary Material).

As a successor model for the selected instances of ActiveLLM, we chose BERT [89] and RoBERTa [262]. While foundational works such as from Devlin et al. [89] typically train BERT-like models for only 3–5 epochs, other studies [123, 296] suggest that higher epoch counts are necessary for stable and optimal performance in low-data regimes. Consequently, we have opted for 25 epochs for our experiments. Otherwise, we adhere to the default Huggingface transformer parameters, with a learning rate of 1e-3, a weight decay of 0.01, and a dropout probability of 0.1.

For comparison, we include in each experiment a random sampling baseline where the first instances from a shuffled dataset are selected. In addition, we compare ActiveLLM with the most commonly used AL strategies for transformers identified in our literature overview (see Table 7.1): LC, BALD, EKM, and PE. The strategies are implemented in a model-matching scenario (i.e., training the same query and successor model), can query the entire dataset of each task, and select 5 instances until the selection of 32 is reached. ActiveLLM, on the other hand, is implemented in a model-mismatch scenario, can only query a subset of the full training set, and selects the 32 instances in one query (unless otherwise specified). To implement the common AL strategies, we use the small-text library [386], which provides a unified framework for pool-based AL with support for transformer-based models.

7.4.2 Prompt Engineering Experiments

– *How should the prompt be designed?*

To answer the first question, we experiment with different configurations of ActiveLLM, such as whether to include AL strategy advices or guidelines, as shown in Fig. 7.2. We then evaluate the effects of varying batch sizes for the unlabeled data used in the prompts. Finally, we compare different selection sizes, i.e. the number of instances the LLM should select. While we want to find the most appropriate prompt, our goal is not to engineer every detail of it, which is left open for future work and practitioners.

7.4.2.1 Configuration Results

Table 7.2 presents the results of various prompt configurations. For these, we use a batch of 300 presented unlabeled instances. Prompt configurations A and B differ in whether the advice on how to select the instances is already included in the prompt. We observed that GPT-4 tends to reiterate the advice even when it is provided in the prompt. Furthermore, the results suggest that it is preferable not to include the advice in the prompt, likely due to the increased context size when the advice is repeated in both the prompt and the response.

Table 7.2 Prompt engineering results of the baseline and ActiveGPT4 on the CTI dataset sampling 32 instances. The results are averaged over 25 runs (5 random dataset and 5 random model initializations). A (advice), G (guidelines), C (no advice, but guidelines), 1 (no CoT), 2 (CoT: step-by-step), 3 (CoT+: explain each instance)

	A	CoT	G	F1 (SD)
Baseline	–	–	–	0.6781 (0.029)
Prompt A1	✓	✗	✗	0.6522 (0.100)
Prompt A2	✓	✓	✗	0.6951 (0.102)
Prompt A3	✓	✓+	✗	0.6876 (0.108)
Prompt B1	✗	✗	✗	0.7135 (0.060)
Prompt B2	✗	✓	✗	**0.7214** (0.053)
Prompt B3	✗	✓+	✗	0.6271 (0.139)
Prompt C2	✗	✓	✓	**0.7329** (0.053)

We also test the three CoT variants: A1 and B1, where there is no CoT; A2 and B2, which use the standard CoT approach ("think step by step"); and A3 and B3, which instruct to include explanations for each instance. The common "think step by step" instruction yielded the best results. This might be because this approach allows GPT more reasoning in structured steps, while not increasing the context size too much.

Finally, we experimented with including labeling guidelines in the prompt. As anticipated, this prompt configuration produced the best results, likely because it enabled the model to better differentiate between instances. Although labeling guidelines are often created prior to the labeling process, they are rarely included in the dataset for common tasks. While the guidelines are available for the CTI task, its inclusion in the prompt should be seen as an exploration of what can be achieved rather than the default case for ActiveLLM. Therefore, in future evaluations, we will utilize ActiveGPT4 with prompt B2, i.e., no guidelines or additional advice on AL, only employing the simple "think step by step" CoT prompting.

7.4.2.2 Presented Batch Size Results

In the subsequent experiment, we explore varying batch sizes of unlabeled instances presented to ActiveGPT4. Fig. 7.3 displays the results for 50, 100, 200, 300, and 400 examples. There is a noticeable trade-off between the context length and the number of presented examples. While a larger pool of examples generally allows

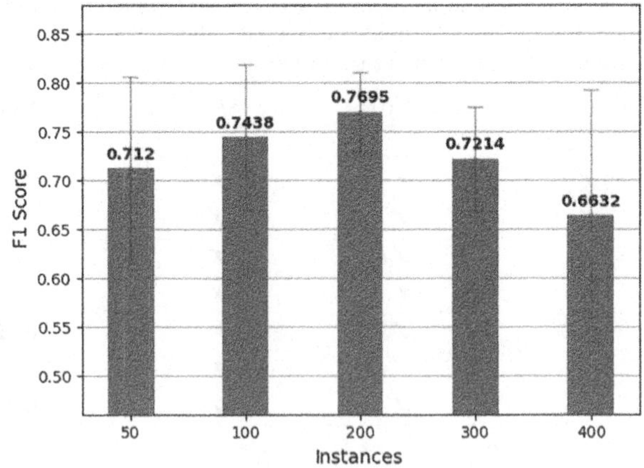

Fig. 7.3 Varying size of the batch of unlabeled instances from which ActiveGPT4 can select 32 instances. All results are averaged over 5 runs on the CTI dataset (F1)

for a more diverse set, an increase in context length tends to degrade the model's performance. It can be observed that the optimal prompt with 300 examples, as used in the previous section, could be significantly enhanced by reducing the size to 200. Consequently, for the primary experiments in Sect. 7.4.3, we limited the number of instances presented to the model to 200, contingent on the model's capacity to handle such a context size.

7.4.2.3 Selection Size Results

In our final experiment on prompt engineering, we investigate the impact of the selection size—how many instances the model should choose—on the outcomes. Many research works [272, 384, 431] typically use 32 examples for few-shot scenarios. We evaluate whether smaller or larger sizes might be more effective for ActiveGPT4. Figure 7.4 illustrates varying selection sizes for the CTI dataset. Selecting 32 examples per query appears to be a sensible choice. Notably, performance declines with up to 90 examples, suggesting that the model may struggle to process such a large number of examples effectively. However, performance increases beyond this point, as expected, because a larger training set is inherently beneficial.

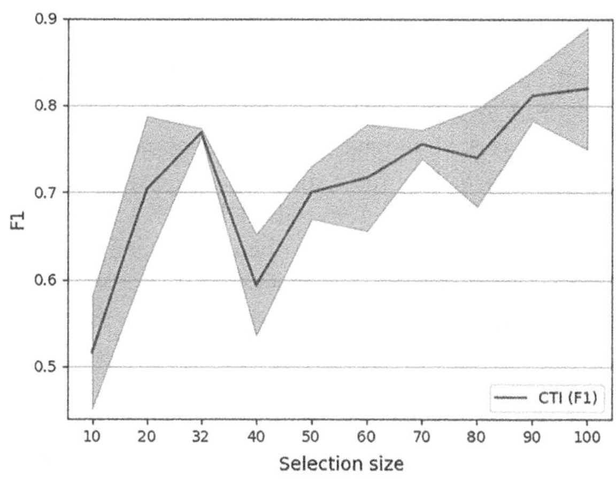

Fig. 7.4 Varying selection size of examples to be selected by ActiveGPT4. All results (F1) are averaged over 5 runs on the CTI dataset

As part of the following experiments, we also tested the selection size on the SST-2 dataset. The results are equivalent and can be found in Appendix Figure A.1 of the Electronic Supplementary Material.

7.4.3 General Applicability

– Can the chosen prompt be applied to other models and datasets?

In this section, we aim to assess how ActiveLLM equipped with various LLMs and the prompt configuration B2 from Sect. 7.4.2.1 performs on the more common datasets of the GLUE benchmark.

The results of this experiment are given in Table 7.3. ActiveGPT4, ActiveGPT4o, Activeo1 and ActiveMistralLarge are given 200 unlabeled instances as identified in the batch size experiments (Sect. 7.4.2.2). Llama 3, GPT-3.5, Gemini-Ultra and Mixtral can not process such extensive prompts and are therefore given 100 instances. All LLMs show that they are very good query models for ActiveLLM, outperforming the baseline in most cases. However, we recommend using ActiveLLM with GPT-4 or GPT-3.5, as they consistently improve results across all tasks. An interesting finding is that Mistral Large with 200 examples performed worse than its 100 example counterpart in five out of eight tasks. The tasks where Mistral Large performs better with 200 instances are those with generally shorter text length, suggesting that choosing a batch size based on token length may be better than a static batch size. This would also explain the variation in improvements across tasks and models, as some tasks involve many more tokens per instance, and some models perform better with larger prompts. In addition, the differences between the models may be due to the fact that the prompt design phase was carried out using GPT-4. Other models may work better and give more consistent results with a prompt design specifically for them. The most significant improvements can be observed with ActiveMistralLarge (200) and ActiveGPT4 on WNLI and SST-2, respectively, achieving improvements of 17.19 and 17.11 percentage points over the baseline.

Overall, ActiveLLM, utilizing GPT-4 and GPT-3.5, proves to be a robust approach for enhancing performance in few-shot scenarios across all tasks. While it may even produce better results with other LLMs, the improvements may be inconsistent and should be compared to a baseline.

To verify that our results are not specific to BERT, we repeated the experiments from Table 7.3 using RoBERTa. The results, shown in Appendix Table A.3 of the Electronic Supplementary Material, confirm that ActiveLLM also improves performance on a different BERT-based architecture. As with BERT, GPT-4 and GPT-3.5 give more consistent results than LLaMA, Mistral/Mixtral, and Gemini-Ultra.

Table 7.3 BERT-base few-shot results (32 instances) of ActiveLLM with different LLMs and of other AL strategies (LC, BALD, EKM, and PE) on GLUE tasks. 'Active 200/100' describes the size of the batch of unlabeled instances. All results are averaged over 5 runs (accuracy). Standard deviations are given in Appendix Table A.2. of the Electronic Supplementary Material.
**No result available after 48 hours of execution.

		QNLI	QQP	RTE	SST2	WNLI	MNLI-(m/mm)	MRPC	COLA
	Baseline	55.11	63.67	50.83	56.03	35.77	34.92/35.01	67.55	63.89
200	GPT-4	58.94 ↑	63.99 ↑	53.57 ↑	**73.14** ↑	40.28 ↑	35.45/35.14 ↑	68.33 ↑	64.70 ↑
	GPT-4o	**62.76** ↑	63.65 ↓	48.88 ↓	66.31 ↑	41.13 ↑	32.23/32.80 ↓	69.75 ↑	65.41 ↑
	o1	62.10 ↑	64.00 ↑	54.01 ↑	59.08 ↑	39.43 ↑	33.52/33.94 ↓	63.87 ↓	**69.91** ↑
	Mistral Large	60.23 ↑	65.26 ↑	**55.60** ↑	60.64 ↑	**52.96** ↑	36.59/35.92 ↑	67.06 ↓	67.44 ↑
100	Llama 3 70B	56.57 ↑	65.82 ↑	52.27 ↑	65.71 ↑	40.28 ↑	37.19/36.68 ↑	66.86 ↓	68.15 ↑
	Mistral Large	60.68 ↑	64.82 ↑	49.48 ↓	66.86 ↑	46.48 ↑	37.41/36.58 ↑	69.07 ↑	67.56 ↑
	GPT-3.5	55.11	**65.88** ↑	55.09 ↑	61.15 ↑	51.55 ↑	**37.44/37.62** ↑	67.70 ↑	68.63 ↑
	Gemini-Ultra	57.60 ↑	61.18 ↓	54.58 ↑	49.08 ↓	43.10 ↑	35.74/35.77 ↑	**69.95** ↑	69.22 ↑
	Mixtral 8x7B	61.13 ↑	65.27 ↑	50.97 ↑	68.00 ↑	41.41 ↑	33.62/33.75 ↓	68.14 ↑	66.96 ↑
AL	LC	56.76 ↑	58.65 ↓	54.44 ↑	56.86 ↑	42.82 ↑	36.48/36.18 ↑	66.96 ↓	69.26 ↑
	BALD	59.91 ↑	n.a.*	52.13 ↑	69.17 ↑	44.79 ↑	36.14/36.34 ↑	68.53 ↑	64.24 ↑
	EKM	52.44 ↓	59.62 ↓	47.65 ↓	69.77 ↑	46.49 ↑	32.87/32.75 ↓	61.08 ↓	66.60 ↑
	PE	53.75 ↓	58.65 ↓	54.44 ↑	56.86 ↑	42.82 ↑	37.11/37.20 ↑	66.96 ↓	69.26 ↑

7.4.4 Comparison

– How does ActiveLLM compare to other AL strategies?

We are interested to see how ActiveLLM performs against other AL methods. In Table 7.3 we also included the results for the most used AL strategies with transformers identified in our literature overview (see Table 7.1): LC, BALD, EKM, and PE. While these methods are implemented with the best preconditions with a model-matching scenario and access to the whole dataset of each task, they are often not able to improve over the baseline and in every case worse than an LLM-variant of ActiveLLM. This outcome is expected, as these active learning strategies face a cold-start problem. Therefore, in Sect. 7.4.6.2 we test whether ActiveLLM can be used to overcome the cold-start problem of these techniques. Furthermore, the practical use of these methods is limited, as the selection of examples to be labeled required at least several minutes and, in the case of larger datasets, several hours (see Table 7.4). For instance, with BALD on the QQP dataset, the approach did not yield any result even after 48 hours of execution. In contrast, ActiveLLM eliminates the need for model training during the active learning process and only considers a small subset of the unlabeled data, enabling query times of just a few seconds while still achieving higher improvements.

Table 7.4 Runtime of the AL strategies (LC, BALD, EKM, and PE) on the GLUE tasks from Table 7.3. The runtime (in minutes) refers to the total time required to select 32 instances over 6 AL iterations.
**No result available after 48 hours of execution

		QNLI	QQP	RTE	SST2	WNLI	MNLI	MRPC	COLA
	Average	265.26	279.11	11.01	170.34	5.17	627.12	12.31	25.29
AL	LC	81.58	271.40	6.39	53.11	3.55	297.00	5.50	11.07
	BALD	812.26	n.a*	22.34	520.11	8.37	1610.23	30.54	68.26
	EKM	85.16	289.39	7.23	53.53	4.40	306.43	6.08	10.02
	PE	82.02	276.54	6.47	53.01	4.37	293.24	5.58	10.22

Standard deviations for the results of Table 7.3 are given in the Appendix Table A.2 of the Electronic Supplementary Material. While the values vary across tasks and models, the overall trend indicates that ActiveLLM does not amplify performance instability, and in many cases, can reduce it substantially.

7.4.5 ActiveLLM and Few-Shot Learning

– *Can ActiveLLM improve state-of-the-art few-shot learning methods?*

While showing significant improvements in few-shot scenarios compared to random sampling and other active learning strategies, one might ask why not directly use few-shot learning methods. Therefore, we evaluate the performance of ActiveGPT4 both individually and in combination with the state-of-the-art few-shot learning methods ADAPET, PERFECT, and SetFit. For each method, we have to use a different dataset for which an implementation is available: ADAPET with CTI, PERFECT with SST-2[3], and SetFit with AgNews.

The results of this experiment are shown in Fig. 7.5. ActiveGPT4 significantly outperforms the few-shot learning methods ADAPET and PERFECT (+ 17.64% + 24.29%). While not outperforming SetFit on its own, ActiveLLM, when combined with each few-shot learning method, significantly improves all of these methods.

[3] The original implementation only supports RoBERTa-Large.

This shows that few-shot learning methods can be improved through the selection of instances, and in some cases, this selection is even more important than the use of few-shot learning methods.

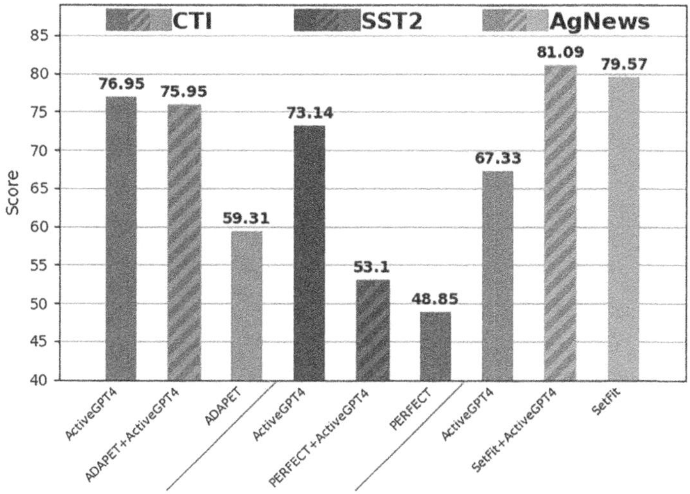

Fig. 7.5 Comparison of ActiveGPT4, both independently and in combination with few-shot methods, across different tasks (32 instances). The methods include ADAPET (evaluated using BERT-base on CTI with F1), PERFECT (using RoBERTa-Large on SST2 with accuracy), and SetFit (using BERT-base on AGNews with accuracy)

7.4.6 Non-Few-Shot-Scenarios

In our last experiments, we investigate how ActiveLLM performs in non-few-shot scenarios and if it can be used to overcome the cold-start problem of other AL strategies. For this purpose, we select a sample size of 25 examples per AL iteration and evaluate the process up to 300 examples.

7.4.6.1 Iterated Querying

– *How does the method perform in non-few-shot-scenarios?*

Figure 7.6 indicates that ActiveGPT4 is particularly beneficial in few-shot scenarios up to about 100 instances. Beyond this point, it performs comparably to random sampling. Furthermore, the figure highlights the importance of recalling instances that have already been selected, as the no-recap mode performs the worst. Providing the model with the indices of these instances appears to be the most efficient method, as it does not overly complicate or enlarge the prompt.

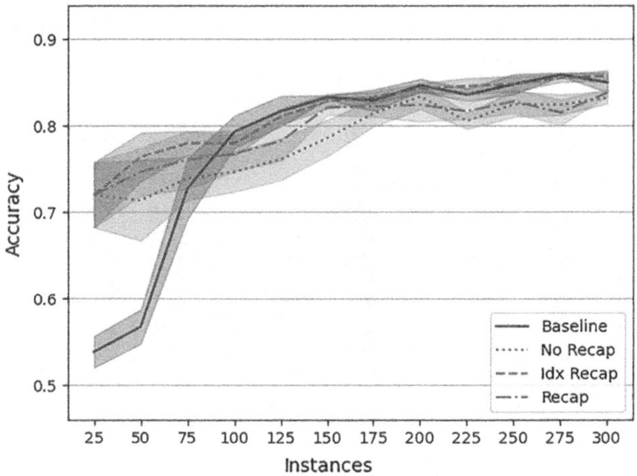

Fig. 7.6 Iterated querying on the SST-2 dataset with different strategies of recalling already labeled instances. All results are averaged over 5 runs (accuracy)

These results motivate the subsequent experiment to investigate whether ActiveLLM could be a viable solution to overcome the cold-start problem of learning strategies.

7.4.6.2 Overcoming the Cold-Start Problem

– Is ActiveLLM capable of overcoming the cold-start problem in other AL strategies?

As outlined in Sect. 7.1, a pervasive issue in AL research is the initial poor performance, which leads to random or worse selections. However, as deep learning models increasingly excel in low-data regimes, the quality of initial data becomes progressively crucial. Observing that ActiveLLM masters few-shot scenarios, we used the first 50 examples selected by ActiveGPT4 (index recap mode) as the starting point for PE, LC, BALD, and EKM.

Once again, a selection size of 25 was used, and the experiments continued until reaching a total of 300 examples. The results in Fig. 7.7 show that ActiveLLM alone outperforms all other methods within the first 50 examples. Using these examples as a starting point for LC and PE outperforms cold-start strategies, with performance plateauing at around 125 instances. For EKM and BALD, this effect cannot be reproduced, as the cold-start problem is not as severe with 50 examples as it is for LC and PE. Nonetheless, the experiment demonstrates that, in cases of a cold-start problem, ActiveLLM is an effective method for addressing it.

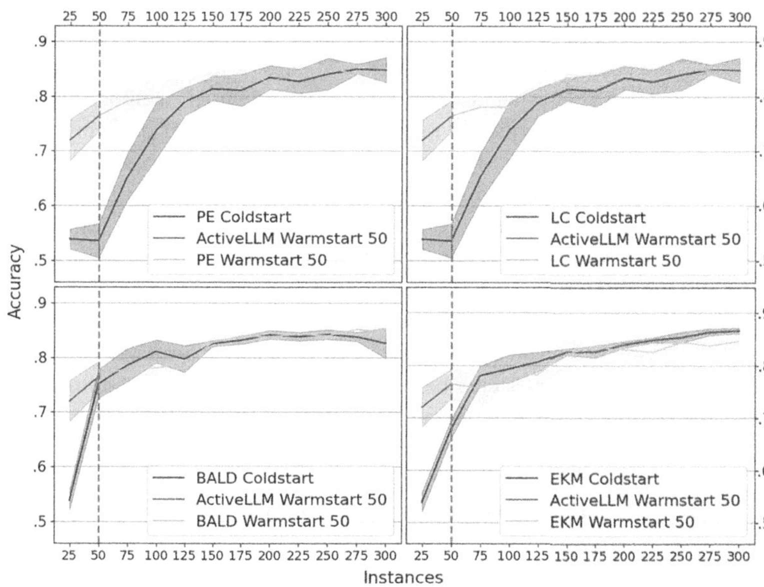

Fig. 7.7 Comparison of BALD, EKM, LC, and PE with and without selected instances by ActiveGPT4 as seed label set on SST-2. All results are averaged over 5 runs (accuracy)

7.5 Discussion & Conclusion

Utilizing instruction-tuned LLMs presents several advantages for AL with BERT-like transformers. Since these LLMs exhibit substantial zero-shot capabilities, they overcome the cold-start problem. This is crucial for low-data regimes and few-shot scenarios, which are becoming increasingly important [516]. Realized in the

ActiveLLM method, we show that this procedure indeed excels in few-shot scenarios, outperforming the baseline and common AL strategies for transformers. Furthermore, in our experiments, ActiveLLM significantly outperforms state-of-the-art few-shot learning methods, ADAPET and PERFECT, while also enhancing the performance of SetFit, ADAPET, and PERFECT when used in combination. We also demonstrate a variant of ActiveLLM with iterative querying, which is applicable to non-few-shot scenarios and incorporates feedback from previous iterations. In a combined setting, we show that ActiveLLM overcomes the cold-start problem inherent in conventional AL strategies.

Moreover, ActiveLLM is particularly suitable for practical scenarios as it effectively eliminates the dependency between the query and the successor model. In our experiments, the AL methods LC, BALD, EKM, and PE required several hours to complete (see Table 7.4), whereas even the more efficient AL methods commonly discussed in the literature typically require several minutes per querying iteration [517]. While some of the LLMs used in ActiveLLM require significantly more resources, our approach eliminates the need for model training during the AL selection process, as required by many other AL methods. Additionally, only a small portion of the unlabeled data is considered, still resulting in significant improvements while requiring just a few seconds for querying. Furthermore, the current cost-free availability of these LLMs through chat interfaces eliminates the need for extensive resources or financial investment. If cost-free availability persists, it democratizes AL, making it accessible to a wider audience rather than limiting it to hypothetical scenarios that assume unlimited time, resources, or funding.

7.5.1 LLM-based Active Learning Strategies

The question remains as to how LLMs choose instances. While they clearly explain their approach, including avoiding redundancy, selecting diverse instances, and considering ambiguities (see Figure A.3), it is nearly impossible to assess whether the models truly perform as claimed. However, LLMs are known for their ability to identify topics and patterns in texts, which suggests that diversity-based sampling, a common approach in other procedures, is a valid method. Due to this property, they can also identify differences between instances, enabling them to recognize unique, representative, frequent, or highly information-dense examples. Similarly, we expect these models to perform well on the tasks themselves, which is why they could be relatively proficient at selecting a balanced set of all classes. However, these models cannot inspect their own internals, which means they are not performing uncertainty sampling based on these internals. Related to uncertainty, LLMs might be very well suited for identifying difficult or ambiguous examples. Occasionally, the models mention that they try to filter out any instances that are anomalies that would teach the successor model incorrect patterns. In iterated querying mode, the language models often directly mention strategies regarding the feedback, explaining that for selecting the instances, they incorporate past instances and try to avoid redundancy, but also track very ambiguous instances from which more data might be needed in the successor training.

7.5.2 Limitations and Future Work

While LLMs such as GPT-4 offer greater capabilities, they often require substantial computational resources, incur high API costs, and raise concerns regarding data privacy and control [40, 378, 447, 469]. In contrast, BERT-like models remain an important choice for many practical applications due to their efficiency, open availability, and competitive performance when fine-tuned on domain-specific tasks. Although ActiveLLM leverages LLMs, their use is limited to the labeling phase and is not required after training. As such, the resulting BERT-like classifier can be deployed without relying on external APIs or continued access to proprietary LLMs. Nonetheless, it is important to acknowledge that the integration of LLMs during training may still pose privacy challenges with regard to the data exposed to them. Our approach therefore offers a trade-off: it combines the strengths of LLMs for data-efficient training while preserving the lightweight and self-contained nature of traditional transformer classifiers at inference time. It might also be of interest to use the LLM directly as an oracle in our framework, i.e., to have it label the selected instances directly. However, as this study focuses on the selection of training examples, we chose not to introduce additional potential points of failure. Instead, we refer readers to the work of Walshe et al. [460], and Wang et al. [475], who explore the use of LLMs in the annotation process. We also look forward to applying ActiveLLM to LLMs themselves. As reported by Zhang, Feng, and Tan [527], the selection of training examples for in-context learning can have a significant impact on LLM performance, making ActiveLLM a promising method in this context as well.

While the experiments confirm the strong performance of ActiveLLM's general design, we look forward to further exploring fine-grained LLM pipeline and prompt optimization techniques, including modular and programmable approaches [203]. In particular, the context size is critical for ActiveLLM. If the prompts get too large, the LLMs tend to forget the task or are not able to reason over the various instances. Although some models were reported to be able to handle long context inputs, we found that they were not able to reason over long contexts. This is in line with the findings of Hsieh et al. [157] who argue that simple retrieval-based tests, like the common needle-in-a-haystack test, are not suitable for testing long context understanding. In our results, we noticed differences for tasks with longer instances on average. We also hypothesize that the length may be the reason why LLMs sometimes omit certain instructions or information, such as the provided labels in our tests of the iterated querying mode. We are interested in considering not only the instances as a length restriction but also the token count in future work. Furthermore, in our experiments we wanted to show that the general prompt optimized in Sect. 7.4.2 is applicable to other models and datasets. Another approach would be to consider the prompt as adaptive and tune it for each model on the respective development set. While this introduces a new hyperparameter, we expect it to improve performance in most cases. Furthermore, optimizing the prompt for GPT-4 may also explain why certain types of LLMs perform worse.

A common problem in the field of LLMs is that many established and open benchmarks have leaked into the training data. This could bias the results of our work, as a model may already be familiar with the dataset. While we cannot completely avoid this phenomenon, we have tried to use a less known dataset for prompt engineering, as well as smaller and older models, which are less susceptible to data leakage.

When we describe ActiveLLM as much more efficient and less resource-intensive for practitioners than other AL methods, we assume the availability of cost-free chat interfaces or low-cost APIs. Certainly, the LLMs that ActiveLLM relies on consume considerable resources, even if they are only used for a very short time during sample selection for labeling. In terms of practicality, our experiments also focus on the use of the chat versions of the LLMs. However, directly using the models, e.g., with the APIs, might yield different results. While this might only play an insignificant difference for Llama or Mixtral, some chat interfaces, like those from OpenAI or Google, hide parameters, preceding prompts, and exact version declarations. We see our results as a baseline that might be improved with different parameters or without a preceding prompt in the future.

Acknowledgements This research work has been funded by the German Federal Ministry of Education and Research and the Hessian Ministry of Higher Education, Research, Science and the Arts within their joint support of the National Research Center for Applied Cybersecurity ATHENE and by the German Federal Ministry of Education and Research in the projects CYWARN (13N15407) and CYLENCE (13N16636). The calculations for this research were conducted on the Lichtenberg high performance computer of the TU Darmstadt.

A Survey on Data Augmentation for Text Classification

8

Abstract

Data augmentation, the artificial creation of training data for machine learning by transformations, is a widely studied research field across machine learning disciplines. While it is useful for increasing a model's generalization capabilities, it can also address many other challenges and problems, from overcoming a limited amount of training data, to regularizing the objective, to limiting the amount data used to protect privacy. Based on a precise description of the goals and applications of data augmentation and a taxonomy for existing works, this survey is concerned with data augmentation methods for textual classification and aims to provide a concise and comprehensive overview for researchers and practitioners. Derived from the taxonomy, we divide more than 100 methods into 12 different groupings and give state-of-the-art references expounding which methods are highly promising by relating them to each other. Finally, research perspectives that may constitute a building block for future work are provided.

Original Publication Markus Bayer, *Marc-André Kaufhold, and Christian Reuter. "A Survey on Data Augmentation for Text Classification". In:* ACM Comput. Surv. *55.7 (2023), 146:1–146:39.* DOI: https://doi.org/10.1145/3544558. URL: https://doi.org/10.1145/3544558[1]

[1] Minor typographical errors in the original publication have been corrected for this dissertation.

M. Bayer, *Deep Learning in Textual Low-Data Regimes for Cybersecurity*, Technology, Peace and Security I Technologie, Frieden und Sicherheit, https://doi.org/10.1007/978-3-658-48778-2_8

8.1 Introduction

An increase in training data does not necessarily result in a solution for the learning problem. Nevertheless, the quantity of data remains decisive for the quality of a supervised classifier. Originating from the field of computer vision, many different methods exist to artificially create such data, which are referred to as data augmentation. For images, transformations such as rotations or changes of the RGB channel are useful, as the resulting model should be invariant for these. Similar to computer vision, speech recognition uses procedures that change sound or speed. In contrast, research on data augmentation in Natural Language Processing (NLP) faces the difficult task of establishing such universal rules for textual data transformations which, when executed automatically, maintain labeling quality [209, 481]. Research in this area was, therefore, much more limited before 2019, despite existing extensive areas of application [188]. Nowadays, this challenge remains but is being addressed by many scientists from different research fields as more possibilities and complex mechanisms open up. Within these fields, researchers strive to meet various goals, e.g., generating more data for low-data regimes, balancing imbalanced dataset classes, or securing against adversarial examples. Thus, textual data augmentation comes in many contrasting forms that will be grouped and explained in this survey. We will provide an in-depth analysis and also relate the methods to the state-of-the-art, as they now face another challenge due to the advent of transfer learning. For example, Longpre, Wang, and DuBois [265] demonstrate that many data augmentation methods cannot achieve gains when using large pre-trained language models, as they are already invariant to various transformations. They hypothesize that data augmentation methods can only be beneficial if they create new linguistic patterns that have not been seen before. Keeping this in mind, the survey is closed with a meta-perspective on the methods. This survey is therefore intended to contribute to data augmentation and general text classification by highlighting the following aspects:

- Goals and applications (C1). We highlight the goals and applications of data augmentation that we derive from the comprehensive review. These have only been presented to a limited and incomplete extent in previous research papers.
- Comprehensive survey on data augmentation in text classification (C2). Our survey provides a holistic overview of the data augmentation field in text classification. While methods for other NLP disciplines are mentioned, the listing is not complete, nor are the methods set in relation to each other as the text classification data augmentation methods are.

- Data-structure-driven taxonomy and method-oriented categorization (C3). The text classification data augmentation methods are clustered according to a data-structure-based, high-level taxonomy and then subdivided into more fine-grained method groups. This is also present in the surveys from Shorten and Khoshgoftaar [401] and Wen et al. [482] and is adapted for the text classification domain.
- Method-driven overview and in-depth details (C4). The textual data augmentation methods are explained clearly and concisely while including necessary details for delimitation and comparison. Contrasting to other works, our extensive study contains 12 groups with more than 100 different approaches.
- State-of-the-art review (C5): Within the literature survey, we examine the latest state-of-the-art considerations, for example, the limited benefit of textual data augmentation methods with large pre-trained models that are often neglected in current works.
- Relating methods (C6). Throughout this survey, the methods are set in relation to conception and performance comparisons while taking the underlying models and application contexts into account.
- Future research perspectives (C7). We identify future research opportunities that are either necessary for a state-of-the-art comparison or sensible to look into because of current challenges and promising directions for textual data augmentation.

The survey paper is structured as follows: The paper introduces the foundations of data augmentation in Sect. 8.2. This section is then broadened by considering the goals and applications. Sect. 8.3 is subdivided into the various data augmentation groups and contains the explanations, as well as tabular overviews of the methods. In Sect. 8.4, an analysis of the data augmentation methods from a more global perspective is given and various future research directions are discussed. Sect. 8.5 outlines the limitations of data augmentation and provides a conclusion for this survey.

8.2 Background: Foundations, Goals, and Applications of Data Augmentation

In many machine learning scenarios, not enough data is available to train a high-quality classifier. To address this problem, data augmentation can be used. It artificially enlarges the amount of available training data by means of transformations [434]. In the well-known LeNet by LeCun et al. [237], early versions of data augmentation have already been observed. The notion of data augmentation comprises various research in different sub-areas of machine learning. Many scientific works merely relate data augmentation to deep learning, yet it is frequently applied in the entire context of machine learning. Therefore, this paper adopts the notion of data augmentation as a broad concept, encompassing any method that enables the transformation of training data. However, following common understanding in research, semi-supervised learning is not regarded as a form of data augmentation and is only thematized if sensible in this survey.

An important term relating to data augmentation is label preservation, which describes transformations of training data that preserve class information [81]. For example, in sentiment analysis, an entity replacement within a sentence is often sufficient for label preservation, but randomly adding words may alter the sentiment (e.g., an additional "not" could invert the meaning of a sentence). In many research works, label preservation is adapted to also cover transformations changing the class information, if the label is adjusted correctly. Additionally, many transformations do not maintain the correct class in every case but with a high probability. Shorten and Khoshgoftaar [401] define this probability as the safety of a data augmentation method. When this uncertainty is known, it could be directly integrated in the label. Otherwise, methods like label smoothing [300] can model a general uncertainty.

The goals of data augmentation are manifold and encompass different aspects. As mentioned above, training data is essential for the quality of a supervised machine learning process. Banko and Brill [25] show that only the creation of additional data can improve the quality of a solution in the confusion set disambiguation problem, while the choice of the classifier does not lead to a significant change. The model selection and development will remain a crucial aspect of machine learning. Yet, scholars suggest that in some situations, the choice for higher investments in algorithm-choice and -development instead of corpus-development should be carefully considered [25]. Closely connected to this is the big data wall problem, which Coulombe [81] mentions in his work on data augmentation. It describes that big companies benefit from the special advantage of having access to a large amount of training data. Consequently, the already large GAFAM-Companies (Google/Alphabet, Amazon, Facebook/Meta, Apple, and Microsoft) expand their predominance over smaller businesses due to their data superiority. An ideal data augmentation method could approach these points and decrease the dependency of training data even though full elimination is not likely.

Additionally, creating training data for various classification problems is accompanied by high labeling costs. In many instances, assessment and labeling by experts are necessary to prevent incorrect training examples. These aspects can, for example, be especially stressed concerning the field of crisis informatics [30, 321]. Creating relevant classifiers for emergency services and responders is only possible during crises and requires resources and time from personnel needed elsewhere to, e.g., act as first responders, therefore in the worst-case costing lives [198]. Similarly, training data for medical image processing is very valuable. Due to the rareness of certain diseases, the privacy of patients, and the requirement of medical experts, it is particularly challenging to provide medical datasets [401]. In a related sense, many domains, such as cybersecurity, have a time-critical factor that requires training data to be collected as quickly as possible so that, in terms of the cybersecurity domain example, threats can be responded to quickly [193, 195]. With regard to such classification problems, data augmentation could help minimize the required amount of data needed to be labeled and to solve interlinked problems.

Data augmentation is particularly significant for the field of deep learning. Work such as that by Minaee et al. [288] has already extensively investigated the quality of deep learning algorithms in text classification, but there are many application scenarios where there is not enough data to produce high quality classifiers. For example, Srivastava et al. [417] have also demonstrated that deep neural networks in general are particularly powerful but encompass a tendency to overfit; faced with unseen instances, they might generalize badly. This observation can be illustrated with help of the bias variance dilemma. On the one hand, deep learning algorithms are, due to their deep and non-linear layers, very strong models with a lower bias-error. On the other hand, they show a high variance for different subsets of training data [127]. This problem can be solved by arranging the algorithm to prefer simple solutions or by providing a bigger amount of training data. The first option is aimed at methods of regularization, such as dropout or the addition of a L2 norm via the model's parameters in the loss-function. The second option is frequently realized by means of data augmentation and could, in this context, also be considered as a type of regularization. According to Hernández-García and König [151], data augmentation is a preferred regularization method, as it achieves generalization without degrading the models' representational capacity and without re-tuning other hyperparameters. While other methods reduce the bias error, data augmentation's objective is to keep it constant and is used to solve the problem at the root [401]. Nonetheless, data augmentation still depends on the underlying classification problem and can therefore not be effectively applied in all circumstances.

In the context of deep learning models, so called adversarial examples/attacks are generated more and more frequently. These small changes in the input data, which are almost unrecognizable to humans, mislead the algorithms to make wrong predictions [99]. Table 8.1 shows two different genuine examples in which the smallest changes in the texts alter the classification prediction. Alzantot et al. [17] further present an algorithm that generates semantically and syntactically similar instances of training data, successfully outwitting sentiment analysis and entailment models. With the help of adversarial training, these automatic adversarial example generators can be used as data augmentation methods, as done, for example, in [488], [434], [406], or [442], in order for models to be less susceptible to such easy alterations.

Table 8.1 Examples for Adversarial Attacks adapted from Ebrahimi et al. [99]

Original text	Altered text
South Africa's historic Soweto township marks its 100th birthday on Tuesday in a mood of optimism.	South Africa's historic Soweto township marks its 100th birthday on Tuesday in a moo**P** of optimism.
57% **World**	95% **Sci/Tech**
Chancellor Gordon Brown has sought to quell speculation over who should run the Labour Party and turned the attack on the opposition Conservatives.	Chancellor Gordon Brown has sought to quell speculation over who should run the Labour Party and turned the attack on the o**B**position Conservatives.
75% **World**	94% **Business**

If the amount of data is taken into consideration, it stands out that certain classification problems are often heavily unbalanced, for instance, only a small amount is relevant (positive) while the irrelevant (negative) data is prevalent [172]. For example, in an entire corpus for topic classification or crisis identification, only few data actually relate to the topics or the crisis in question. Zhong et al. [535] term a dataset as unbalanced, if the distribution of classes within it is not approximately equally balanced. Data augmentation may help to enhance the amount of data for a certain class in order for balanced class distributions to be present and thus for a classifier to be able to be modelled more robustly [120, 539].

Data augmentation can also be helpful in sensitive domains. Dealing with confidential or privacy-related data, one can decrease the usage of real-world data by crafting artificial data. It is even possible to only train the algorithm on the newly created data, in order to prevent drawing any conclusions on non-artificial training data from a deployed model. For example, Carlini et al. [55] have demonstrated a method for extracting training data from large language models that could contain private information. For training such a privacy ensuring model, special data augmentation techniques that are able to anonymize the data have to be used.

Data augmentation exists in different types and areas of application. A taxonomy of the types in the textual domain can be seen in Fig. 8.1. The augmentation methods can be divided into the transformation of raw data (data space) and processed representations of data (feature space) [401]. These representations are transformed types of data, for example, activation vectors of a neuronal network, the encoding of an Encoder Decoder Network, or LSTM hidden states, respectively embeddings of data. Abstracting from the textual realm, in many cases, data augmentation depends on the underlying problem (text classification, image recognition, etc.); and is therefore applied in different ways in different areas. Procedures generic enough to be used across different areas are for the most part limited to the feature space.

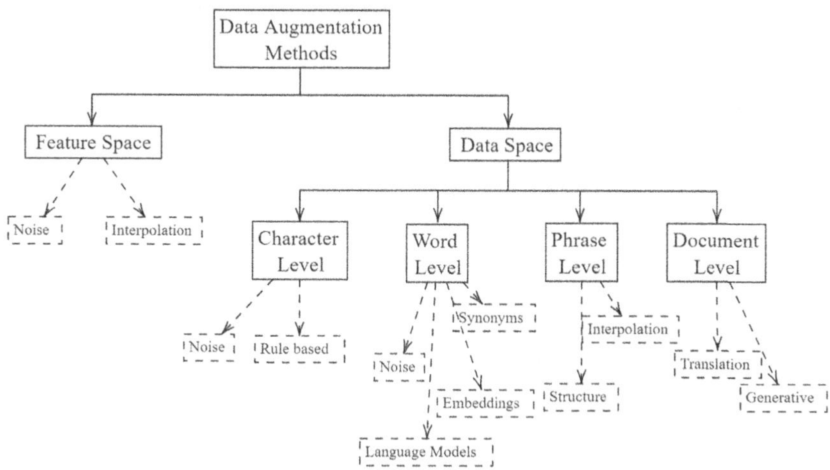

Fig. 8.1 Taxonomy and grouping for different data augmentation methods

The most substantial research on data augmentation exists in the field of computer vision. This is due to the intuitive construction of simple label preserving transformations. Data augmentation methods in computer vision are, among other, geometric transformations [406, 434], neural style transfers [82, 177, 343], interpolation of images [144], random partial deletions [499], and generative adversarial network (GAN) data generation [99]. Sophisticated techniques can additively improve the accuracy baseline for different problems by around 10 to 15 percent [499]. Another area of application for data augmentation is speech processing. Researchers have successfully used acoustic transformations of the input data. Ko et al. [208] have achieved up to 4.3 points better accuracy values by modifying speed. Furthermore, interfering with vocal tract length [499] or adding noise [144] may also enhance the quality of classifiers. The application of data augmentation in the textual realm is considered a difficult task, since textual transformations preserving the label are difficult to define [209, 481]. Nevertheless, several simple and sophisticated methods have been developed in this and adjacent research areas.

8.3 Textual Data Augmentation Methods

In the following, different data augmentation methods for textual data are summarized, explained, and subdivided in different groupings. Mainly methods focusing on the application of text classification are included, although augmentation methods for other tasks in the textual realm are also mentioned if they fit into the group. In this survey, text classification is considered a problem of the field of NLP, where units such as sentences, paragraphs, or documents are categorized into class labels [288]. For example, generative or sequence-tagging tasks, where either text has to be generated or the words of the units have to be tagged individually, are not regarded as tasks in this sense. This means that augmentation methods for tasks such as topic classification, sentiment analysis, or spam identification are focused, described, and analyzed in detail. Other tasks, on the other hand, like generative question answering, part of speech tagging, or machine translation are only mentioned in a non-comprehensive way. Therefore, in the context of text classification, our paper provides a comprehensive overview containing the necessary details for researchers and practitioners. For a more general perspective on NLP augmentations (including sequence-tagging, parsing, text generation, etc.), we recommend the reader to have a look at the work of Feng et al. [113], which is not as detailed in text classification as our work but presents a broader task view. In contrast to this task-driven view by Feng et al. [113], we are taking a method-oriented perspective while conducting a data-structure-driven, high-level categorization (see Fig. 8.1). Contrary to other surveys in the field of data augmentation, we focus on setting the augmentation methods into context by comparing the conception and performance, with regard to the underlying models and application context. In this way, the listed augmentation groups contain an explanation with details on the differences within the group and a comprehensive overview of how the methods differ and which results they produce. This allows the reader to gain insights into which data augmentation technique might be most promising for the own use-case and what specifics need to be considered, while it is also possible to follow the data-structure based taxonomy. In the end, we discuss important future research directions by setting all methods into context, which can help accelerate developments in this field.

In the next section, data augmentation methods relevant in textual contexts are summarized and grouped. Generally, the methods are described in a sensible order for the specific group. In groups with many similar approaches, we summarize the most important information in tabular form. We also extract information regarding improvements. The improvement indications are intended to give a quick overview of how well a method may perform but are not in-depth informative or comparable on their own. For a more detailed perspective, the models and datasets are also displayed. This should provide a more holistic perspective, although in-depth information has to be extracted from the respective papers themselves.

8.3.1 Data Space

8.3.1.1 Character Level

Noise Induction

The addition of noise to the input data is one of the data augmentation methods with the smallest alterations, especially when applied on a character level. As explained in more detail further on, the induction of noise can also be used at the word level as well as in the feature space.

In this context, the basic idea of the method of Belinkov and Bisk [33] is to add artificial and natural noise to the training data so that, in their case, neural machine translation (NMT) models are less susceptible to adversarial examples. As artificial noise, Belinkov and Bisk [33] describe operations like the random switching of single letters ("cheese" → "cehese"), the randomization of the mid part of a word ("cheese" → "ceehse"), the complete randomization of a word ("cheese" → "eseehc"), and the replacement of one letter with a neighboring letter on the keyboard ("cheeae"). Similarly, Feng et al. [113] randomly delete, swap, and insert characters of texts (the prompt portion) that are used for fine-tuning text generators. For this, they moreover ignore the first and last character of a word. To measure the suitability for text generators, they intrinsically measure the diversity, fluency, semantic context preservation, and sentiment consistency. The applied method is better than the baseline in every respect. These augmentations are also usable in the text classification domain. Ebrahimi et al. [99] used an existing model, trained with the initial dataset, to generate adversarial examples. They use the direct input data to flip a letter if the change increases the loss of the existing model. If a new model is trained with the additional data once again, the error rate is improved and the success of adversarial attacks is significantly mitigated. Furthermore, they compare their approach with the adversarial method from the previously mentioned work of Belinkov and Bisk [33] and the feature space method from Miyato, Dai, and Goodfellow [292] (see Sect. 8.3.2.1). Based on a CharCNN-LSTM on the AG News dataset, they achieve the best improvement in accuracy by obtaining an additional 0.62%. While the method of Miyato, Dai, and Goodfellow [292] improved the score by only 0.24 points, it is interesting to see that the method of Belinkov and Bisk [33] even decreased the accuracy by 0.33 points. Coulombe [81] describes the induction of weak textual manipulations through the aforementioned change, deletion, and addition of letters in words and, in addition, the alteration of upper and lower case and the modification of punctuation. The highest absolute accuracy improvement by 2.5% can be seen in comparison to the best functioning baseline. However, the evaluation was performed with basic architectures and no embeddings, wherefore further studies are needed to validate the usefulness in a current setting.

Natural noise, as defined by Belinkov and Bisk [33], covers spelling mistakes that are common in the respective language, using spelling mistake databases. Each word associated with a common mistake is replaced with the misspelled word, and if there is more than one, the mistake is randomly sampled. Belinkov and Bisk [33] receive varying BLEU scores with their artificial and natural noise methods; most noise operations made the model more robust against attacks with similar operations. Most importantly, natural noise almost consistently worsens a translation model regarding the baseline. Analogous to the natural noise defined by Belinkov and Bisk [33], Coulombe [81] also adds common spelling mistakes in the textual data and achieves good improvements when added to classifiers. The best baseline (XGBoost) was improved by an additional 1.5%. With such transformations, learners are better able to deal with spelling mistakes in prospective texts, even if mistakes are not present in the original training dataset. This variant of data augmentation can for example be of interest when dealing with texts originating from social networks.

Rule-based Transformations

Coulombe [81] implements rule-based transformations through the use of regular expressions. According to him, such rules are not easy to establish, since many surface-level transformations require deeper changes to preserve the grammar, and other transformations depend on the language. Valid transformations are, amongst others, the insertion of spelling mistakes, data alterations, entity names, and abbreviations. Coulombe concretely implements the transformation of verbal short forms to their long forms and vice versa ("I am" ↔ "I'm"). In the English language, this is semantically invariant if ambiguities are preserved [81]. With this form of data augmentation Coulombe achieves very good results [81]. The best baseline model (XGBoost) additionally gained 0.5% in terms of accuracy.

8.3.1.2 Word Level

Noise Induction

Noise induction can also be applied on the word level. For example, the method of Xie et al. [499] encompasses two noise patterns. With "unigram noising", words in the input data are replaced by another word with a certain probability. By the method of "blank noising", words get replaced with "_". By the adoption of both patterns, the authors achieved improved results in their experiments.

Li, Cohn, and Baldwin [250] are using syntactic and semantic methods as well as word dropout for the generation of noise. Syntactic noise is realized via the shortening of sentences and methods such as the alteration of adjectives or the relativization of modifiers, while semantic noise is generated by the lexical substitution of word synonyms (8.3.1.2). In contrast to these two methods, word dropout is more clearly comparable to noise. Random input neurons or rather words get masked out during the training of the network. According to the authors, their proposed methods achieve an improvement. Especially a combination of all methods promises an improvement of up to 1.7 points in terms of accuracy.

Two of the four sub-methods of the Easy Data Augmentation (EDA) method by Wei and Zou [481], i.e., random swap and deletion, should also be mentioned as methods of noise induction. In experiments, a combination of both sub-methods led to improved performance of the used classifier. EDA is very popular in the research field and is used as a method for comparisons in the works of Qiu et al. [336], Huong and Hoang [162], Anaby-Tavor et al. [18], Kumar, Choudhary, and Cho [223], Bayer et al. [28], Feng et al. [112], Luu, Nguyen, and Nguyen [268], and Kashefi and Hwa [190]. Wei and Zou [481] report that for a small dataset these two sub-methods gain higher improvements than the other two sub-methods that are based on synonym replacement and insertion (see Sect. 8.3.1.2). Nevertheless, Qiu et al. [336], Anaby-Tavor et al. [18], Bayer et al. [28], and Luu, Nguyen, and Nguyen [268] also report some cases in which EDA as a whole data augmentation method decreases the classification score. This result can be expected, as the methods random swap and deletion are not label preserving, for example, for sentiment classification: "I did not like the movie, but the popcorn was good" → random_swap→ "I did like the movie, but the popcorn was not good". While Wu et al. [493] also use random swap and random deletion, they propose random span deletion, where consecutive words are deleted. This technique would lead to a worse label preservation, but it is only used for language modelling with contrastive learning (see Sect. 8.3.4).

The training instances of one batch must have the same length when being fed into a neural network. For this purpose, the sequences are often zero-padded on one side. Rizos, Hemker, and Schuller [365] propose a specific noise induction method to augment the training data by shifting the instances within the confines of their padding so that the padding is not solely on one side. Evaluated by means of a hate speech detection dataset, the authors show that this method achieves additive performance gains of more than 5.8% (Macro-F1). Sun and He [428] also translate the instances by adding meaningless words either at the beginning or at the end. Unfortunately, they do not evaluate the impact of this method in isolation.

Xie et al. [498] propose a TF-IDF based replacement method in which they are replacing uninformative words of an instance with other uninformative words. As the authors are combining this technique with round trip translation (see Sect. 8.3.1.4) and unsupervised data augmentation, it is not clear to which degree it benefits the task. Similarly, Choi et al. [71] replace casual features/words that are a determining factor for the label. In the contrastive learning scheme, they mask these words to generate counterfactual examples as well as other non-casual words to generate normal augmentations. More details and results of this procedure can be found in Sect. 8.3.4.

More noise data augmentation methods related to other tasks can be found in the works of Cheng, Jiang, and Macherey [70], Li, Cohn, and Baldwin [250], Wang et al. [476], Andreas [19], Guo, Kim, and Rush [137], Kashefi and Hwa [190], Sun and He [428], and Kurata, Xiang, and Zhou [226].

Synonym Replacement

This very popular form of data augmentation describes the paraphrasing transformation of text instances by replacing certain words with synonyms. One of the first applications of this substitution in the field of data augmentation was introduced by Kolomiyets, Bethard, and Moens [214]. They substituted temporal expressions with potential synonyms from WordNet [285]. As the authors argue, the replacement of one original token in a sentence will mostly preserve the semantics. Based on the time expression recognition task, the authors propose replacing the headword, since temporal trigger words are usually found there. While this application, however, showed no substantial improvements, the authors also proposed a language model replacement method that was more suited for the task at hand (see Sect. 8.3.1.2).

In later years, many researchers experimented with word replacements based on thesauri. The works of Li, Cohn, and Baldwin [250], Mosolova, Fomin, and Bondarenko [297], Wang et al.[474], and many more partially or primarily execute synonym substitution in this way. Differences between the studies concern the specific words that are substituted, the synonyms that come into question, and the utilization of different databases. For example, Zhang, Zhao, and LeCun [526] and Marivate and Sefara [275] choose the synonyms for substitution on basis of the geometric distribution by which the insertion of a rather distant synonym becomes less probable. Furthermore, several approaches exclude stop words or words with certain POS-tags from the set of words considered for replacement. Interesting is also the second sub-method of EDA by Wei and Zou [481], where synonyms are not replacing specific words, but are randomly inserted into the instance. The replacement method, synonym selection, database, and improvements of the various approaches are listed in Table 8.2.

Table 8.2 Overview of different synonym replacement approaches

	Synonym Database	Replacement Method	Synonum Selection	Model Base	Dataset	Improvements
[214]	WordNet	Headword replacement	Not stated	Logistic Regression	TempEval	−1 (F1)
					Reuters (12)	−0.6
					Wikipedia (1)	−0.1
[526], [524]	mytheas (LibreOffice) WordNet-based	Randomly chosen number of words based on geometric distribution.	Randomly based on geometric distribution.	Character CNN		[526]/[524] (Acc.)
					AG News	−0.38 / −0.57
					DBP.	+0.05 / +0.13
					Yelp P.	−0.03
					Yelp F.	+0.36
					Yahoo A.	+0.22 / +0.65
					Amazon F.	+0.1 / +0.1
					Amazon P.	−0.17 / −0.17
[250]	WordNet	Substitutable words are nouns, verbs, adjectives, or adverbs that are not part of a named entity. Each word is replaced with a certain probability.	The remaining probability of substitution is shared among the synonyms based on a language model score.	CNN	MR	+0.8 (Acc.)
					CR	+1.2
					Subj	+0.5
					SST	+0.1
					MR/CR	+0.9
					CR/MR	+0.3

(Continued)

Table 8.2 (Continued)

	Synonym Database	Replacement Method	Synonum Selection	Model Base	Dataset	Improvements
[81]	WordNet	Only adverbs and adjectives, sometimes nouns, more rarely verbs.	Most similar companion information of the synonym with the context of the chosen word.	XGBoost MLP (2 hidden layer)	IMDB	+0.5 (Acc.)
						+4.92
[297]	WordNet	No pronouns, conjunctions, prepositions, and articles for replacement. Choosing uniform randomly.	Uniform random	CNN with word embeddings	Toxic Comment Classification	−0.09/-0.21 (AUC)
[474]	HIT IR-Lab Tongyici Cilin (Extended) (Chinese)	No time words, prepositions, and mimetic words. Chi-square statistics method.	Chi-square statistics method	Character CNN-SVM	Hotel R.	~ +1 (Acc.)
					Laptop R.	~ +1
					Book R.	~ +0.25

(Continued)

Table 8.2 (Continued)

	Synonym Database	Replacement Method	Synonum Selection	Model Base	Dataset	Improvements
[275]	WordNet	Verbs, nouns, and their combination. Geometric distribution.	Geometric distribution	DNN	AG News	~ +0.4 (Acc.)
					Sentiment	~ +0
					Hate Speech	~ −0.8
[187]	WordNet & Thesaurus.com	For Minibatch: Augmentation with probability, POS-tag replacement, replacement of one word per sentence that maximizes loss.	Synonym that maximizes the loss.	Kim CNN	TREC	+1.2 (Acc.)
[481]	WordNet	No stop words. Choosing n random words to be replaced (SR) or from which the synonyms are inserted at a random position (RI)	Uniform random	CNN		SR / RI (Acc.)
					Classification tasks (500)	~ +1.9 / ~ +2.0
					(2000)	+1.2 / ~ +0.9
					(5000)	~ +0.7 / ~ +0.6
					(full)	~ +1.0 / ~ +0.9

(Continued)

Table 8.2 (Continued)

	Synonym Database	Replacement Method	Synonum Selection	Model Base	Dataset	Improvements
[209]	WordNet	Replacement of a word based on a certain probability.	Temperature hyperparameter learned while training.	CNN	SST-5	−0.6 (Acc.)
					SST-2	+0.5
					Subj	0
					MPQA	+0.2
					RT	+0.1
					TREC	−0.4
[336]	WordNet	Replacement of a word based on a certain probability.	Temperature hyperparameter learned while training.	TextRCNN	ICS	−0.26 (Macro F1)
					NEWS	+1.63
[428]	Not stated	Filtering words according to their POS-tag. Selecting a fixed or variable number of words.	Specific or variable number of synonyms.	LSTM-CNN	Tan NLPCC	Results only in combination with other methods
[67]	WordNet	Not stated	Not stated	BERT	IMDB (40)	−0.87
					SST-5 (40)	−0.87 (Acc.)
					TREC (40)	+0.01

(Continued)

Table 8.2 (Continued)

	Synonym Database	Replacement Method	Synonum Selection	Model Base	Dataset	Improvements
[160]	WordNet	No stop words. 10% of documents randomly selected.	Not stated	M-BERT	CodiEsp-P CodiEsp-D	−0.7 (F1) +0.6 (F1)
[311]	WordNet	Keywords replaced are ordered by their RAKE score (e.g., the probability of being a keyword).	Randomly selected. Replacement only with same POS-tag.	No model (intrinsic evaluation with different metrics)	Yelp-LR (small subset of Yelp Reviews)	+0.001 (SStd) +0.015 (SBLEU) −0.018 (UTR) −0.02 (TTR) −0.016 (RWords) 0 (SLOR) −0.007 (BPRO) 0 (SDiff)
[112]	WordNet	No stop words. Uniform random replacement until 20% of the words in a sentence are changed.	Uniform random	CNN	Yelp P.	Only against other data augmentation methods

Also to be emphasized is the more sophisticated integration into the learning process, as described by Jungiewicz and Smywinski-Pohl [187]. The authors replace words with synonyms only if the replacement with the chosen synonym maximizes the loss of the current state of the classifier model. Apart from this, there are approaches that adapt the general idea of thesauri-based replacements to perform augmentation on specific tasks, for example, in Kashefi and Hwa [190] and Feng et al. [112].

Embedding Replacement

Comparable to synonym substitution, embedding replacement methods search for words that fit as good as possible into the textual context and additionally do not alter the basic substance of the text. To achieve this, the words of the instances are translated into a latent representation space, where words of similar contexts are closer together. Accordingly, these latent spaces are based on the distributional hypothesis of distributional semantics [115, 145], which is currently mostly implemented in the form of embedding models. The selection of words that correspond to this hypothesis and are, thus, near in the representation space, implies that the newly created instances maintain a grammatical coherence, as displayed in Fig. 8.2. Besides this advantage, Rizos, Hemker, and Schuller [365] argue that the "method encourages the downstream task to place lower emphasis on associating single words with a label and instead place higher emphasis on capturing similar sequential patterns." Benefits of this data augmentation technique in comparison to the synonym substitution method are that techniques based on the distributional hypothesis are more comprehensive and the context of texts is considered. This means that substitutions are not limited by a database, like WordNet, and that grammatically more correct sentences can be generated [23]. Furthermore, the general form of this approach can be beneficial for languages which have no access to a large thesauri but a lot of general text resources, on the basis of which the self-supervised embedding models can be easily trained [81].

Wang and Yang [473] use this kind of augmentation to better classify annoying tweets. They utilize k-nearest-neighbor to identify the most suitable embeddings as a substitution of the training data words. Compared to the baseline, they achieve an additive improvement of up to 2.4 points in the F1-Score with logistic regression. Marivate and Sefara [275], Rizos, Hemker, and Schuller [365], Huong and Hoang [162], and others utilize the embedding replacement in very similar ways. The greatest differences in terms of the method exist in the selection of words to be replaced (e.g., POS-tag based) and the selection of the replacing words based on the embeddings. An overview of the differences can be found in Table 8.3.

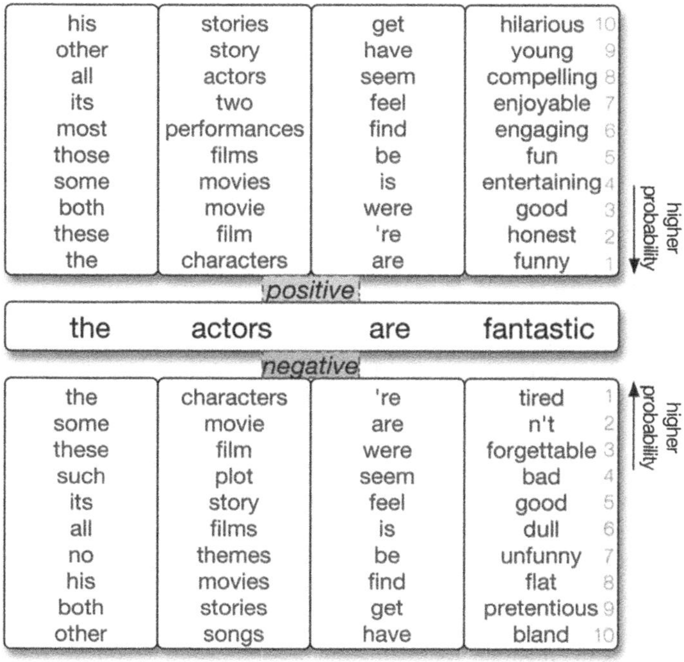

Fig. 8.2 Example of a sentence with predicted words that can be used to replace a word in the sentence [209]

A major factor for poor results is that the embedding replacement does not necessarily guarantee the preservation of the contextual meaning of the instances. This, in turn, could lead to distortions of the label; e.g., "the movie was fantastic" and "the movie was horrible" are valid transformations but the sentiment is the opposite. A way to address this issue is the use of the counter-fitting method of Mrksic et al. [298] for synonym embedding substitution, as carried out by Li, Cohn, and Baldwin [250]. Counter-fitting is an approach that depicts word embeddings on the basis of a target function in a way that similarities between synonyms are rewarded and similarities between antonyms are sanctioned [298]. Li, Cohn, and Baldwin [250] extend this approach by selecting the most fitting words with a higher possibility for the replacement. This is done by leveraging a language model that can give an indication on how well a given word fits into a sequence. However, the authors achieve rather mixed results with this method. The counter-fitting method offers considerably less replacement possibilities, since embeddings have to be trained on the downstream task, leading to a smaller coverage of their corpora words. Alzantot et al. [17] use this method in combination with a language model filtering in their adversarial example generator. They extend the approach by only incorporating the words that are maximizing the target label prediction probability (label preservation) of an already trained classifier. The authors report no improvements in terms of the task testing set, but they show that the model is safer regarding adversarial attacks. Embedding replacement methods are moreover used in specific task-dependent ways, such as by Kashefi and Hwa [190].

Replacement by Language Models
Language models represent language by predicting subsequent or missing words on the basis of the previous or surrounding context (classical and respectively masked language modelling). In this way, the models can, for example, be used to filter unfitting words, as already discussed in Sect. 8.3.1.2 in relation to the work of Alzantot et al. [17]. The authors generate similar words with GloVe embeddings and the counter-fitting method and utilize a language model to choose only words with a high probability of fit. In contrast to embedding replacements by word embeddings that take into account a global context, language models enable a more localized replacement [275]. Utilizing the subsequent word prediction, language models can also be used as the main augmentation method. Kobayashi [209] is, for example, using an LSTM language model to identify substitution words. However, language models do not only substitute words with similar meaning, but also with words that fit the context in principle [209]. This trait is encompassed with a greater risk of label distortion. To prevent the attachment of wrong labels to the new training data due to changed semantics, Kobayashi [209] modifies the language model so that it allows the integration of the label in the model for the word prediction (label-conditional language model). Inspired by this approach, Wu et al. [491] alter the architecture of the language model BERT [89] in a way that it is label conditional (c-BERT). In an evaluation with different tasks the authors showed that in comparison to Kobayashi [209] and other approaches they were able to considerably increase the performance of a classifier (see Table 8.4). However, the c-BERT approach also has the disadvantage that the language model is fixed when applied, and in the case of low-data regimes, the augmentation might no longer be label preserving [160]. For this reason, Hu et al. [160] include the c-BERT method in a reinforcement learning scheme, which learns the task in a normal supervised fashion but is also able to simultaneously fine-tune the c-BERT LM. With this adaption, the authors

Table 8.3 Overview of different embedding replacement approaches

	Replacement Selection	Embedding Selection	Model Base	Dataset	Embedding Model	Improvements
[473]	Not stated	K-nearest-neighbor and cosine similarity	Logistic regression	Petpeeve dataset	Twitter W2V	+1.7
					UrbanDictionary W2V	+0.3 (F1)
					GoogleNews W2V	+2.4
[275]	Random	Random with probability proportional to cosine similarity.	DNN	AG News	Wikipedia W2V	~ +0.5
				Sentiment	Wikipedia W2V	~0 (Acc.)
				Hate Speech	GloVe Twitter	~ −0.3
[365]	Every word	Cosine similarity threshold + POS-tag matching	CNN+LSTM/GRU	HON	Word2Vec Hate Speech	−22.7 (Macro F1)
					FastText Wikipedia	+1.0
					GoogleNews W2V	−3.3
					GloVe Common Crawl	+0.3
				RSN-1	GloVe Common Crawl	−0.2
				RSN-2	GloVe Common Crawl	0

(Continued)

Table 8.3 (Continued)

	Replacement Selection	Embedding Selection	Model Base	Dataset	Embedding Model	Improvements
[428]	1. Method: Filtering words according to their POS-tag. Selecting a fixed or variable number of words. 2. Method: Replacing adverbial phrases (Chinese related).	Own similarity measure and specific or variable number of replacements	LSTM-CNN	Tan	W2V self-pretrained	Results only in combination with other word level augmentation methods
[250]	Substitutable words are nouns, verbs, adjectives, or adverbs that are not part of a named entity. Each word is replaced with a certain probability.	Embeddings are found with counter-fitting. Candidates are replaced with a probability. The remaining probability is shared among the embeddings based on a language model score.	CNN	MR CR Subj SST MR/CR CR/MR	GoogleNews W2V	−0.6/−4.2 (Acc) +0.1/ −3.7 +0.2/ −1.4 −0.4/−4.2 +1.9/+0.4 +0.1/ −3.0

(Continued)

Table 8.3 (Continued)

	Replacement Selection	Embedding Selection	Model Base	Dataset	Embedding Model	Improvements
[162]	Not stated	Cosine similarity	Random Forest, Naïve Bayes, SVM	Vietnamese comments	W2V Vietnamese	Results only in combination
[17]	Random sampling with probabilities proportional to the neighbors each word has within the counter-fitted embedding space + exclusion of common articles and prepositions.	1. K-nearest-neighbors with Euclidean distance + counter-fitting method. 2. Google LM to filter out words. 3. Selection of the word that will maximize the target label prediction probability.	LSTM	IMDB	GloVe	Adversarial training: No improvements but safer model
[183]	Only for multi-piece words. Random probability for replacement.	Random embedding of the k-nearest- neighbor	Small transformer model	Various GLUE tasks	GloVe	No augmentation baseline comparisons
[331]	No stop words or symbolic and numerical data	Cosine similarity threshold of 0.97	Manhattan LSTM model	Thai text similarity task	Thai2fit (Thai language)	+1.71

Table 8.4 Evaluation results of the state-of-the-art language substitution method c-BERT

Pub.	Method	Dataset	Improvements(Accuracy)
[491]	c-BERT	+ 0.8(CNN)/+1.3(RNN)	SST-5
		+0.2(CNN)/+0.5(RNN)	SST-2
		+0.5(CNN)/+0.4(RNN)	Subj
		+0.5(CNN)/+0.7(RNN)	MPQA
		+0.8(CNN)/+0.6(RNN)	RT
		+0.8(CNN)/+0.2(RNN)	TREC
[337]	c-BERT with consistency training	MLNI-m	+0.4(RoBERTa-Base)
[18]	c-BERT	ATIS	−1.9(BERT)/−0.8(SVM)/−5.8(LSTM)
		TREC	+1.1(BERT)/+1.1(SVM)/+6.5(LSTM)
		WVA	+0.2(BERT)/+0.5(SVM)/+2.4(LSTM)
[18]	c-BERT	ATIS	−1.9(BERT)/−0.8(SVM)/−5.8(LSTM)
		TREC	+1.1(BERT)/+1.1(SVM)/+6.5(LSTM)
		WVA	+0.2(BERT)/+0.5(SVM)/+2.4(LSTM)
[160]	c-BERT integrated in reinforcement learning scheme	SST−5(42)	+1.17(BERT)/+2.19(normal c-BERT)
		IMDB(45)	+1.97(BERT)/+1.97(normal c-BERT)
		TREC(45)	+0.73(BERT)/+0.87(normal c-BERT)
[183]	c-BERT and embedding substitution for multiple-pieces words	MNLI-m	+2.3(TinyBERT)
		MLNI-mm	+1.9(TinyBERT)
		MRPC	+3.4(TinyBERT)
		CoLA	+21.0(TinyBERT)

significantly outperform the original c-BERT approach in a low-data regime setting. The results can be found in Table 8.4 together with the results of Anaby-Tavor et al. [18], who evaluated c-BERT as comparison, and Qu et al. [337], who employed the c-BERT model with supervised consistency training (see 8.3.4) on the MLNI-m task.

Jiao et al. [183] apply the already improved method by Wu et al. [491] and further adjust it in their work on TinyBERT. In doing so, the scholars reflect on the fact that the quality of the data generated with BERT is poor if many multiple-pieces words are included. To mitigate this problem, they propose to perform a embedding substitution on the base of GloVe embeddings [330] for such words. Further language model augmentations for different tasks are proposed by Gao et al. [122], Ratner et al. [352], Fadaee, Bisazza, and Monz [106], and Kashefi and Hwa [190].

8.3.1.3 Phrase and Sentence Level

Structured-base Transformation

Structure-based approaches of data augmentation may utilize certain features or components of a structure to generate modified texts. Such structures can be based on grammatical formalities, for example, dependency and constituent grammars or POS-tags. Such approaches are therefore more limited to certain languages or tasks. Sahin and Steedman [375] are, for example, concerned with the augmentation of datasets from low resource languages for POS-tagging. By the method of "cropping", sentences are shortened by putting the focus on subjects and objects. With the "rotation" technique, flexible fragments are moved. The authors state that this method is dependent on certain grammatical sentence structures in different languages and probably only generates noise in the English language. Both methods are well suited for a multitude of low resource languages. They are also tested by Vania et al. [451] for the augmentation of training data for dependency parsers for low-resource data. Feng, Li, and Hoey [114] propose a method for changing the semantics of a text while trying to preserve the fluency and sentiment. Given a set of phrases (replacement entities) to every instance, the so-called Semantic Text Exchange method first identifies phrases in the original text that can be replaced by a replacement entity based on the constituents. Then phrases similar to the identified phrases are replaced by a masked token. Subsequently, this is filled by an attention-based language model so that the similar words better suit the replacement entity. Feng et al. [112] adapt this approach by automatically selecting the 150 of the 200 most frequent nouns from the Semantic Text Exchange training set as replacement entity candidates and splitting their Yelp Review dataset into windows, as the method is only suitable for short texts. In an analysis with this dataset Feng et al. [112] reported that the Semantic Text Exchange method decreases fluency, diversity, and semantic content preservation.

An important work was proposed by Min et al. [287] who show that inversion (swapping the subject and object part) and passivation result in a higher generalization capability in natural language inference. In fact, considering their work in comparison with preliminary work [129, 278, 437] suggests that BERT is able to extract the relevant syntactic information from the instances but is unable to use this information in the task, as there are too few examples in the MNLI dataset demonstrating the necessity of syntax. Here, even a limited utilization of Min et al. [287] data augmentation methods already helps to mitigate this problem.

Interpolation

In numerical analysis, interpolation is a procedure to construct new data points from existing points [418]. While the formal interpolation versions are found in the feature space section, a sensible definition of interpolation in the data space of text is difficult to construct. However, the substructure substitution (SUB^2) method by Shi, Livescu, and Gimpel [399] is considered as such in this context due to its resemblance to the feature space methods. SUB^2 substitutes substructures (dependents, constituents, or POS-tag sequences) of the training examples if they have the same tagged label (for example, "a [DT] cake [NN]" in an instance can be replaced with "a [DT] dog [NN]" of another instance). The variant adapted for classification views all text spans of an instance as structures and is constrained by replacement rules that can be combined or completely left out. The replacement rules are only replacing (1) same lengths spans, (2) phrases with phrases, (3) phrases of the same constituency label, and (4) spans that come from instances with the same class label. The authors show that their methods outperform the baseline when applied to low resource tasks. Their classification variant nearly doubles the accuracy on a subsample of the SST-2 and AG News datasets. Furthermore, they achieve better results than the language model augmentation c-BERT (Sect. 8.3.1.2). Similarly, Kim et al. [205] propose a data augmentation method based on lexicalized probabilistic context-free grammars that extracts grammar trees from an input sentence and combines/substitutes them internally with trees from other instances of the same class. Words are replaced with other words having the same POS-tag from the other sentences of the same class and WordNet synonyms. In this way, they can achieve a considerable performance improvement when applied in a few-shot, semi-supervised learning environment.

8.3.1.4 Document Level

Round-trip Translation

Round-trip translation[2] is an approach to produce paraphrases with the help of translation models. A word, phrase, sentence, or document is translated into another language (forward translation) and afterwards translated back into the source language (back translation) [7]. The rationale behind this is that translations of texts are often variable due to the complexity of natural language [81], which leads to various possibilities in the choice of terms or sentence structure. The process is depicted in Fig. 8.3.

The approach is promising because of its good inherent label preserving and highly valuable paraphrasing capabilities. By the translation of text, the content is preserved and only stylistic features based on the traits of the author are excluded or altered [338]. Some translation systems can propose several translation options; this is hinted in Fig. 8.3 ("k^2 paraphrases"). Yu et al. [514], Aroyehun and Gelbukh

[2] Even though Coulombe [81], Yu et al. [514], Xie et al. [498], Ollagnier and Williams [311], and others use the term backtranslation for their data augmentation works as well, these approaches are assigned to the round-trip translation approaches because they execute forward and back translation.

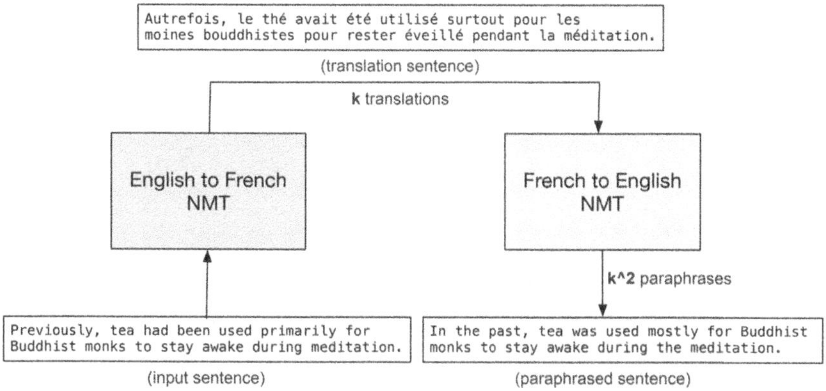

Fig. 8.3 Round-trip translation process [514]

[23], Coulombe [81], Kruspe et al. [218], and others use this technique to generate artificial training data. Their works differ with regard to the used language and the subsequently applied filtering methods. These filtering methods are important, as the process of the twofold translation may be faulty [23]. Furthermore, Xie et al. [498] as well as Chen, Yang, and Yang [67] change the normal beam search generation strategy to random sampling with a temperature parameter to ensure a greater diversity. Details on the different round-trip translation applications are presented in Table 8.5.

Generative Methods

Generative methods are becoming increasingly interesting in recent data augmentation research. As the capabilities of language generation increased significantly, the current models are able to create very diverse texts and can thus incorporate new information. Here, Qiu et al. [336] introduce a variational autoencoder (VAE) based on a method that is used for text generation in their system. VAEs are probabilistic neural network structures that consist of an encoder network, which transforms input data into a latent representation, and of a decoder-network, which transforms the latent representations back. The authors differentiate between unconditional and conditional VAEs. With unconditional VAEs, separate text generation models are trained for all classes, whereas with conditional VAEs (CVAEs), label information is fed into the model as an additional input. Furthermore, they distinguish between sampling from the prior distribution, which leads to greatly diverse instances, and the posterior distribution, which produces text that is semantically closer to the training data. With the unconditional VAE and sampling from the prior distribution, they

Table 8.5 Overview of round-trip translation approaches

	Model	Languages	Filtering	Model	Dataset	Improvements
[514]	Google's NMT [492]	en → fr → en	No filtering	Convolution and self-attention	SQuAD	+1.5 (EM) / +1.1 (F1)
[81]	Google Translate API	Not stated	Excluding identical instances. Similarity threshold based on lengths.	XGBoost	IMDB	+0 (Acc.)
				MLP (2 hidden layer)		+5.8
[23]	Google Translate API	en → fr, es, de, hi → en	No filtering	NBSVM	Aggression Detection	+0.19 (Macro F1)
				CNN		+5.31
				LSTM		+7.39
				BiLSTM		+5.6
				CNN-LSTM		+5.94
				LSTM-CNN		+19.45
				CNN-BiLSTM		+14.33
				BiLSTM-CNN		+6.87
[218]	Google Translate	Randomly selected	No filtering	Fusion CNN	TREC Incident Streams track	~ −1.2 (F1)
[275]	Google Translate API & Amazon translate	en → fr,	"We ensured that the [...] texts carry the same meaning as the source text"	DNN	AG News	~+0.33 (Acc.)
		de → en			Hate Speech	−2.3
[498]	WMT'14 English-French translation model	en → fr → en	No filtering	Randomly initialized transformer	Yelp-5	+1.65 (Acc.)

(Continued)

Table 8.5 (Continued)

	Model	Languages	Filtering	Model	Dataset	Improvements
[337]	WMT19 and released in FairSeq	en → de → en	No filtering	RoBERTa	MLNI-m	+0.9 (Acc.)
[109]**	Translation models from Britz et al. [100]	en → de, zh → en	No filtering	BERT	MNLI	+0 (Acc.)
					QNLI	+0.2 (Acc.)
					QQP	+0.4 (Acc.)
					RTE	+3.6 (Acc.)
					SST-2	+0.7 (Acc.)
					MRPC	+0 (F1)
					CoLA	+2.3 (Mcc)
					STS-B	+0.6 (Corr.)
[398]*	Not stated	Not stated	No filtering	Transformer base with consistency training	CoLA	+1.4 (Mcc)
					MNLI	+0.9 (Acc.)
					QNLI	+0.6 (Acc.)
					QQP	−0.2 (Acc.)
					RTE	+5.1 (Acc.)
					SST-2	+0.7 (Acc.)
					MRPC	+2.6 (F1)
					STS-B	+0.4 (Corr.)

(Continued)

Table 8.5 (Continued)

	Model	Languages	Filtering	Model	Dataset	Improvements
[211]	MarianMT	en → fr, de, es → en Chained: en → es → fr → en	Word sense disambiguation: Retaining of those in which the target word occurs exactly once (in both original and augmented instance).	MT-DNN	SemEval-2013 + SemEval-2015 + Senseval-2 + Senseval-3	No baseline comparisons

* Trained with consistency training
** Trained with contrastive learning

achieve the highest improvements of up to 2 F1-points (see Table 8.6). Malandrakis et al. [274] make similar efforts by evaluating VAEs for augmentation. While their objective is more narrowed, as they are interested in natural language understanding with limited resources, they analyze a broader variety of VAE augmentation variants. They also propose augmentation by CVEAs and unconditional VAEs with sampling from the posterior or prior distribution. Furthermore, they test two different learning objectives, where in the first the VAEs are used to reconstruct the input and in the second the VAEs take an instance of a particular class and try to construct another instance from that class. They also experiment with the addition of a discriminator network to the VAE that predicts the respective class from which an output appears to be. In intrinsic and extrinsic evaluations, the CVAEs with the reconstruction task are best performing. The discriminator variant achieves poor results, which stem from the little amount of available training data for the many different classes. Contrary to the improvements of Qiu et al. [336], the CVAEs outperform the VAE generation. An excerpt of the extrinsic evaluation is given in Table 8.6. However, it must be considered that the task at hand is very specific. VEAs are also a main component of the NeuralEditor proposed by Guu et al. [141] that generates new texts based on edition vectors. For the training of the generative model, they take pairs of instances x' and x in the training data that are lexically similar, encode the differences of them and noise into an edition vector z, and try to generate x based on x' and z. It should be noted that the lexical similarity is just a rough approximation of semantic similarity. This represents a potential source of error, as, e.g., instances could be negated which in turn weakens the label preservation capabilities. However, this suffices the purposes of the authors, as they only use the method for language modeling. Specifically, in this domain, they report improvements in terms of generation quality and perplexity. Raille, Djambazovska, and Musat [345] propose Edit-transformer, which is an adaptation of the NeuralEditor with the additional ability to function cross-domain, so that the learned edits of a large dataset can be transferred to a smaller dataset. Besides the improvements in speed and language modeling, they also apply their method on three different classification tasks. The results are shown in Table 8.6. Rizos, Hemker, and Schuller [365] create an RNN that, depending on a specific class, learns language modelling to generate training data thereafter. The class specific RNN for augmentation is primed with a random start word from the class specific training data. However, the authors state that this method produces the poorest results compared to embedding substitution and noise generation. In a similar sense, Ollagnier and Williams [311] perform text generation using a language model (LSTM-CNN). In contrast, they split each document in a minibatch into sentences, then generate new sentences for 30% of them and utilize 30% of the beginning of a given sentence as prompt.

Table 8.6 Overview of text generation approaches

Pub.	Method	Model	Dataset	Improv.
[336]	VAE	Ensemble of BiLSTM, TextCNN, TextRCNN, and FastText with XGBoost as top-level classifier	ICS (Zh)	+0.04 (F1)
			News Category Dataset (EN)	+2.02
	CVAE + prior sampling		ICS (Zh)	−0.13
			News Category Dataset (EN)	+1.55
	CVAE + posterior sampling		ICS (Zh)	−0.06
			News Category Dataset (EN)	+1.88
[274]	VAE	BiLSTM	Movie	+4.0 (Macro F1)
			Movie + Live Entertainment	−0.5
	CVAE + prior sampling		Movie	+5.9
			Movie + Live Entertainment	+1.7
	CVAE + posterior sampling		Movie	+5.6
			Movie + Live Entertainment	+0.6
[224]	CVAE	BERT	SNIPS (few shot)	+8.00
			SNIPS	+0.06 (Acc.)
			FBDialog (few shot)	+7.42
			FBDialog	+0.0

(Continued)

Table 8.6 (Continued)

Pub.	Method	Model	Dataset	Improv.
[345]	Transformer-based sentence editor	CNN	Subj (20%)	+1.71 (Acc.)
		CNN	Subj (100%)	+1.62
		CNN	SST-2 (20%)	+0.87
		CNN	SST-2 (100%)	−0.84
		LSTM	Amazon Reviews (1%)	+1.12
		LSTM	Amazon Reviews (4%)	+0.41
[365]	RNN LM with random start word priming	CNN+LSTM + GloVe++	HON	-1.8 (Micro-F1)
			RSN-1	+8.2
			RSN-2	−7.4
[311]	CNN-LSTM LM priming with 30% of a sentence	CNN-LSTM	CodiEsp-P	+3.1 (F1)
[428]	seqGAN	LSTM + pretrained embeddings	Tan's task	+1.06 (F1)
		CNN + pretrained embeddings		+0.9
		LSCNN + pretrained embeddings		+0.8
[249]	CS-GAN (GAN, RNN and reinforcement learning)	CNN	Amazon-5000	+1.6 (Acc.)
			Amazon-30000	−0.21
			Emotion-15000	+0.77
			NEWS-15000	+2.25

(Continued)

Table 8.6 (Continued)

Pub.	Method	Model	Dataset	Improv.
[467]	GPT-2 for rarer instances without filtering	Logistic regression/biLSTM/Bi-attentive classification + ELMo + GloVe	Alerting Information Feed Prioritization	No comparative results
[18]	CVAE	BERT	ATIS (5)	+7.3 (Acc.)
			TREC (5)	+0.8
			WVA (5)	−1.8
[18]	GPT-2 generation and classifier filtering	BERT	ATIS (5)	+22.4 (Acc.)
			ATIS (20)	~0
			ATIS (50)	~ +2.0
			ATIS (100)	~ +0.5
			TREC (5)	+4.0
			WVA (5)	+1.4
[4]	DistilGPT2 generation and classifier filtering	BERT	AG-NEWS	+0.61 (Acc.)
		CNN	CyberTrolls	+0.45
		BERT	SST-2	+1.63

(Continued)

Table 8.6 (Continued)

Pub.	Method	Model	Dataset	Improv.
[28]	GPT-2 with conditional fine-tuning, special prompting, and embedding filtering	ULMFit	SST-2 (100)	+15.53 (Acc.)
			SST-2 (700)	−0.19 (Acc.)
			Layoff	+4.84 (F1)
			Management Change	+3.42 (F1)
			Mergers & Acquisitions	+1.42 (F1)
			Flood	+0.25 (F1)
			Wildfire	+0.44 (F1)
			Boston Bombings	+2.44 (F1)
			Bohol Earthquake	+2.05 (F1)
			West Texas Explosions	+3.81 (F1)
			Dublin	–2.54 (F1)
			New York	+0.44 (F1)
[75]	GPT-2 with conditional fine-tuning, special prompting, and classifier filtering	RoBERTa	MediaEval	+0.55 (micro-F1)
		FlauBERT	CLS-FR	+0.57
[256]	GPT-2 with a reinforcement learning component for class conditional generation.	XLNet	Offense Detection (20%)	+1.3 (F1)
			Offense Detection (40%)	+4.3
			Sentiment Analysis (20%)	+1.2
			Sentiment Analysis (40%)	+1.4
			Irony Classification (20%)	+1.0
			Irony Classification (40%)	+2.3

(Continued)

Table 8.6 (Continued)

Pub.	Method	Model	Dataset	Improv.
[512]	GPT-3 with prompt-based generation and pseudo-labeling	BERT (base)	COLA (0.1%, 0.3%, 1.0%)	+7.9, +3.2, −2.4
			TREC6 (0.1%, 0.3%, 1.0%)	+15.6, +17.1, −6.5
			CR (0.1%, 0.3%, 1.0%)	+11.0, 17.3, 8.9
			SUBJ (0.1%, 0.3%, 1.0%)	+1.3, −1.8, −1.2
			MPQA (0.1%, 0.3%, 1.0%)	+12.9, 13.4, 3.8
			RT20 (0.1%, 0.3%, 1.0%)	+6.2, 13.6, 17.5
			SST-2 (0.1%, 0.3%, 1%, full)	+20.9, 19.3, 5.7, 2.9
		BERT (large)	SST-2 (0.1%, 0.3%, 1.0%)	+23.7, 14.6, 3.0

Sun and He [428] use the seqGAN architecture [515] to generate artificial data on basis of a GAN. Comparable to computer vision, seqGAN consists of a generator network creating texts and a discriminator network examining the authenticity of the generated texts next to the real instances. As the discriminator network can only prove the authenticity after a sequence of words and thus gives delayed feedback to the generator, the generator network is trained as a reinforcement learning agent. Utilizing the method as a data augmentation technique, the authors only receive minor improvements of classification quality. Partially inspired by SeqGAN, Li et al. [249] propose CS-GAN, which consists of a GAN, RNN, and reinforcement learning component for sentence generation. The model receives the information about the label as a prior for the generator, which is implemented by the RNN and RL components, which is then required by the discriminator to generate meaningful sentences. Subsequently, a classifier forces the output of sentences to fit the label. The results are listed in Table 8.6. Wang and Lillis [467], Anaby-Tavor et al. [18], Abonizio and Junior [4], Bayer et al. [28], Claveau, Chaffin, and Kijak [75], and Liu et al. [256] use the GPT-2 model by Radford et al. [340], which achieves very good results in text generation, to create new complete instances. Concerning the adoption of the method, Wang and Lillis [467] only describe that they use rare instances as dependent examples for the generation. Anaby-Tavor et al. [18], on the other hand, develop a method that increases the safety with regard to label preservation. In a first step, they further train the GPT-2 model with training data of a certain task (fine-tuning). In the process, they concatenate the respective label to every instance in order to facilitate the generation of new data for the respective class. Finally, a classifier determines which generated instances can actually be assigned to the class stated. The authors manage to achieve significant improvements in the classification of sentences. They show that their method outperforms conditional VAEs (unfortunately no sampling technique is described) and even EDA (Sect. 8.3.1.1) and c-BERT (Sect. 8.3.1.2) when applied to a severe low-data regime. The results of their LAMBADA approach and CVEA implementation are given in Table 8.6. Abonizio and Junior [4] try to improve this approach by concatenating three random samples as a prompt for the generation. Furthermore, they are using DistilGPT2 by Sanh et al. [379], which is substantially faster and smaller. As can be seen in Table 8.6, the method consistently outperforms the baseline. While LAMBADA and PREDATOR are only applicable to short texts as instances, Bayer et al. [28] design a GPT-2 based approach to augment short as well as long text tasks. In this way, very high label preservation and diversity is to be achieved by fine-tuning the language model on the class specific data, generating data prompted with specialized training data tokens, and a filtering method based on document embeddings. They can achieve high improvements for constructed and real-world low-data regimes. However, they also discuss limitations of their method and useful applications in terms of specific datasets and tasks. The results can also be seen in Table 8.6. Similarly, Claveau, Chaffin, and Kijak [75] fine-tune the GPT-2 model using the class-specific data and input a random word from the original texts for generation. Afterwards a classifier is applied to filter the generated data instances. They evaluate their approach using English and French datasets (see Table 8.6). Liu et al. [256] use a reinforcement learning component after the softmax prediction of the GPT-2 model to predict the tokens depending on the class for which the instance is to be generated. The authors tested their method with various model architectures. It consistently improved all of them in all tasks, especially the larger pre-trained models, like BERT and XLNet. The results for XLNet are shown in Table 8.6. Yoo et al. [512] are among the first authors using the much larger GPT-3 model by Brown et al. [47] for data augmentation, which has considerably better generation capabilities. Such large language models are expensive and hard to be fine-tuned on the training data, which is why their augmentation method GPT3Mix selects some examples from the dataset and incorporates them with the label into sensible prompts for the model to be conditioned on. The newly created instances are then extracted from the generated text and a pseudo-label probability is calculated with the GPT model. As shown in Table 8.6, the method achieves outstanding performance increases on scarce and one full dataset. The authors further demonstrate that their method is superior to other augmentation methods, such as EDA [481], round-trip-translation and Tmix [67], and that the performance increases if larger models for the classification are used. Nevertheless, given the size of the GPT-3 network and the correspondingly large training dataset, it might even be able to replicate some of the training (or even test) instances that were left out in the creation of a scarce dataset.

In the generative method, proposed by Lee et al. [239], a first step is to subdivide the data into slices (informed by or based on the labels). Then, a generative model is trained on these slices to predict an instance in the slice based on a subsample of instances in that slice. This model is subsequently used to generate new data for underrepresented slices by priming it with instances from it. This way, the authors gain several improvements in text classification, intent classification, and relation extraction tasks with state-of-the-art results for the latter two. Furthermore, Ding et al. [91] and Chang et al. [60] propose methods using generative models for tasks other than text classification.

8.3.2 Feature Space

Data augmentation in the feature space is concerned with the transformation of the feature representations of the input.

8.3.2.1 Noise Induction

As in the data space, noise can also be introduced in several variants in the feature space. For example, Kumar et al. [224] employ four such techniques for the ultimate goal of intent classification. One of those methods applies random multiplicative and additive noise to the feature representations, as shown in [226]. However, in contrast, they are not transforming the created representations back into the data space. Another method called Linear Delta calculates the difference between two instances and adds it to a third (all from the same class). The third method, which interpolates instances, is further elaborated in Sect. 8.3.2.4 (see Table 8.8). For their fourth method, the authors are adapting the Delta-Encoder by Schwartz et al. [389] for textual data. There, an autoencoder model learns the deltas between instance pairs of the same class, which is then utilized to generate instances of a new unseen class. In a normal testing setting, the methods only slightly improve the classification results, while in a few-shot setting all methods are highly beneficial. Several feature space data augmentation methods stem from the adversarial training research field. As explained in the background section, the models are trained with adversarial examples, i.e., little perturbed training data instances that would change the prediction or maximize the loss. This can be formally written as follows [538]:

$$\min_{\theta} \mathbb{E}_{(z,y)\sim D} \left[\max_{||\delta|| \leq \epsilon} \mathrm{L} \left(f_{\theta}(\mathbf{x} + \delta), y \right) \right] \tag{8.1}$$

where θ are the model parameters and δ describes the perturbation noise added to the original instances (within a norm ball). Further, D is the data distribution, y the label, and L a loss function. The training of the network (outer minimization) can still be solved by stochastic gradient descent (SGD), while the search for the right perturbations (i.e., inner maximization) is non-concave [538]. As described by Zhu et al. [538], projected gradient descent (PGD) [77, 131] can be used to solve this. Unfortunately, several convergence steps (K) to get a good result make it computationally expensive [538]. Shafahi et al. [395] and Zhang et al. [520] propose two methods that calculate the gradient with respect to the input (for PGD) on the same backward pass as the gradient calculations regarding the network parameters during a training step. This mitigates additional calculation overhead of PGD. In detail, Free adversarial training (FreeAT) by Shafahi et al. [395] trains the same batch of training examples K times so that several adversarial updates can be performed. You Only Propagate Once (YOPO) by Zhang et al. [520] accumulates the gradients with respect to the parameters from the K steps and updates the parameters accordingly. Zhu et al. [538] also propose a method called Free Large-Batch (FreeLB), which is similar to YOPO, as it also accumulates the parameter gradients. On several tasks, this method consistently exceeds the results of the baseline and the other two methods. The results of the GLUE dataset are given in Table 8.7. Miyato, Dai, and Goodfellow [292] and Miyato et al. [293] change the normal adversarial training rule so that no label information is needed and call it virtual adversarial training. Without going into exact details, virtual adversarial training regularizes the standard training loss with a KL divergence loss of the distribution of the predictions with and without perturbations, where the perturbations are chosen to maximize the KL divergence. While the virtual adversarial training method is suitable for semi-supervised learning, we are particularly interested in the supervised setting. Their method improves the supervised DBpedia topic classification task baseline classifier by 0.11 points of accuracy, leading to a +0.03 increase in accuracy in comparison to the conventional adversarial training method. Jiang et al. [181] propose the adversarial method SMART, which relies on the virtual adversarial training regularization. They introduce the utilization of the Bregman proximal point optimization with momentum to solve the virtual adversarial training loss, which prevents the model from aggressive updates [292]. The authors show in their experiments that the method significantly improves the baseline and is also able to achieve better results than the other methods discussed in this paragraph (for an overview, see Table 8.7). Furthermore, they demonstrate robustness enhancement and domain adaption capabilities in several evaluation applications.

Wang, Gong, and Liu [468] and Liu et al. [258] developed methods for enhancing the pre-training of language models with adversarial training. Wang, Gong, and Liu [468] simply generate adversarial examples on the output embeddings in the softmax function of the language models. Thereby they manage to reduce the perplexity of the AWD-LSTM and QRNN models on different datasets, which leads, for example, to a reduction of 2.29 points with respect to the Penn Treebank dataset with the AWD-LSTM model. However, it is not clear how the training of bigger pre-trained language models like BERT and RoBERTa would have been influenced by this method. This is addressed in the work of Liu et al. [258] with their method called Adversarial Training for Large Neural Language Models (ALUM), which introduces noise to the input embeddings. The authors build their system based on the virtual adversarial training by Miyato, Dai, and Goodfellow [292], as they noticed that it is superior to conventional adversarial training for self-supervision. Furthermore, they found out that they can omit the Bregman proximate point method and the adversarial training proposed by Jiang et al. [181] and Shafahi et al. [395] when they are using curriculum learning, where the model is first trained with the standard objective and then with virtual adversarial training. They report promising generalization and robustness improvements with the largest transformer models. For example, RoBERTa models can be improved with the ALUM continual pretraining by +0.7 on the MNLI task, while standard continual pretraining does not introduce further gains. The results on the GLUE dataset are given in Table 8.7. The authors also tested the robustness of the models with three different adversarial datasets, where ALUM achieves significant improvements in all tasks. In another evaluation setting they combine adversarial pretraining with adversarial fine-tuning. ALUM improves all evaluation scores of the standard pretrained models. This model reaches the best performances and significantly outperforms the other models in all tested tasks, e.g., with an increased accuracy of + 0.4 more than without tuning the SNLI dataset. The improvement on the MNLI task is given in Table 8.7.

With regard to the generative adversarial training methods of the feature space, it is also of interest to investigate how the newly created examples can be transformed into the data space to enable their inspection. This is done in the works of Liu et al. [254] and Wan, Wan, and Wang [462]. Wan, Wan, and Wang [462] attempt to improve the classification behavior of a grammatical error correction system by training with adversarial examples. Such an example, extracted from the application of loss-increasing noise in the hidden representation of a transformer encoder, is mapped to the data space by a transformer encoder that was trained autoregressively. Then they use a similarity discriminator based on the model to filter instances that are not similar to their initial counterparts. Liu et al. [254] also use a transformer autoencoder architecture to generate data space instances. In contrast to the work of Wan, Wan, and Wang [462], they generate the noisy instances from the input embeddings, subsequently filter instances based on unigram word overlap, and try to improve machine question generation and question answering tasks. Both methods significantly improve the baselines and other methods.

Given the constraint that adversarial training can be computationally expensive, Shen et al. [398] propose three simple and efficient data augmentation methods of the feature space (see Fig. 8.4). Token cutoff sets the entire embedding of a single word to 0, while the feature cutoff sets one embedding dimension of each word in the input to 0. The third method, span cutoff, employs token cutoff across a coherent span of words. With each method, several different, slightly modified instances can be created, which the authors see as different perspectives/views that can be integrated in a multi-view learning fashion through consistency training. This means that the model should predict similar outputs across different views (details can be found in Sect. 8.3.4). The authors evaluate their model on the GLUE task and compare it with adversarial training algorithms as well as round-trip translation. In three out of eight tasks, an improvement over all other methods is achieved (see Table 8.7). They extend the cutoff strategies to work with language generation, and thereby significantly outperform the baseline as well as the adversarial training method of Wang, Gong, and Liu [468].

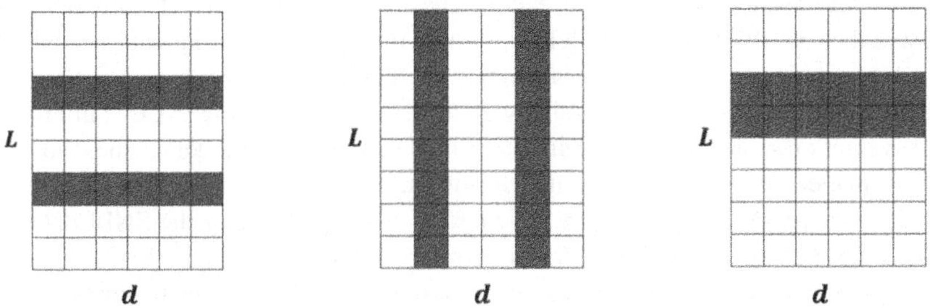

(a) ***Token Cutoff*** (b) ***Feature Cutoff*** (c) ***Span Cutoff***

Fig. 8.4 Visualization of the different cutoff methods [398]

Table 8.7 Comparison of different noise inducing methods on the GLUE task

	Model	SST-2 Acc	STS-B P/S Corr	MLNI-m/mm-Acc	QQP Acc	RTE Acc	QNLI Acc	MRPC F1	CoLA Mcc
Baseline	RoBERTa-L	96.4	92.4	90.2	92.2	86.6	94.7	90.9	68.0
Adversarial Training	PGD	96.4	92.4	90.5	92.5	87.4	94.9	90.9	69.7
	FreeAT	96.1	92.4	90.0	92.5	86.7	94.7	90.7	68.8
	FreeLB	96.7	92.7	90.6	**92.6**	88.1	95.0	91.4	71.1
	ALUM*	96.6	92.1	90.9	92.2	87.3	95.1	91.1	68.2
	ALUM	–	–	**91.4**	–	–	–	–	–
	SMART	96.9	**92.8**	91.1	92.4	**92.0**	**95.6**	92.1	70.6
Cutoff**	Token	96.9	92.5	91.0	92.3	90.6	95.3	93.2	70.0
	Feature	**97.1**	92.4	90.9	92.4	90.9	95.2	93.4	71.1
	Span	96.9	**92.8**	91.1	92.4	91.0	95.3	**93.8**	**71.5**

* only adversarial pre-training
** supervised consistency training

8.3.2.2 Interpolation Methods

For textual data, interpolation methods are mostly limited to the feature space since there is no intuitive way for combining two different text instances. Nevertheless, the application in the feature space seems reasonable, as the interpolation of hidden states of two sentences creates a new one containing the meaning of both original sentences [43, 67]. Besides this, from a learning-based perspective, interpolation methods have a high value for machine learning models. Possible explanations for the success of interpolation methods may stem from the balancing of classes, the smoothening of the decision border (regularization) [63], and the improvement of the representations [456]. For example, the Synthetic Minority Over-sampling Technique (SMOTE) approach in its original context was developed for the purpose of oversampling the minority class, which, as described in the background section, inherently leads to better classification performances. In fact, a balancing of a class can easily be achieved by simply copying the minority class. However, Chawla et al. [63] show that simple oversampling leads to more specific decision boundaries than applying SMOTE in the balancing of classes. Interpolation methods can smoothen the boundary, as shown in Fig. 8.5. Smoothened and more general decision borders signify that an algorithm can generalize better and, in relation to training data, is accompanied by less overfitting. In this context, when applying interpolation methods to representations of the input data, Verma et al. [456] empirically and theoretically prove that representations are flattened with regard to the number of directions with significant variance. This is desirable since data representations capture less space, meaning that a classifier is more uncertain for randomly sampled representations and a form of compression is achieved which leads to generalization [390, 441, 456].

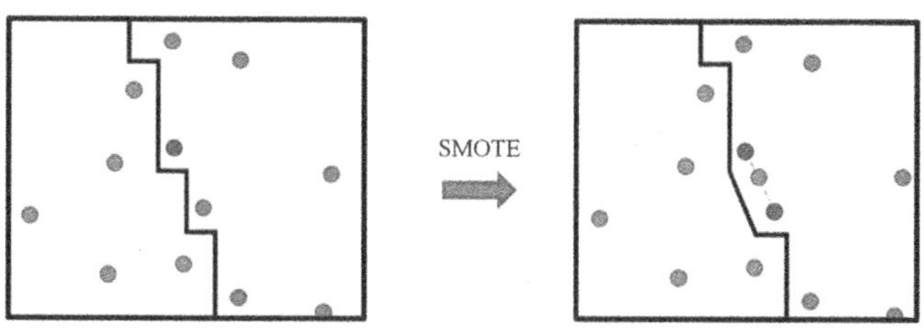

Fig. 8.5 Illustration of the interpolation method SMOTE

8.3.2.3 SMOTE Interpolation

SMOTE is an interpolation method of feature space representations of input data. With SMOTE, various neighbors close to a specific instance are searched within the feature space in order to be interpolated with the following formula:

$$\tilde{x} = x_i + \text{dist}(x_i, x_j)$$

where (x_i, y_i) is the source instance and (x_j, y_i) is a close neighbor with the same class label. dist(a,b) is a distance measure and $\lambda \in [0, 1]$. Unlike mixup, only instances of the same class get interpolated. The rationale behind the calculation of neighbors with the same class labels is that the interpolations tend to be class preserving, leading to a higher safety of the technique. However, this leads to a limited novelty and diversity of the created instances.

SMOTE is rudimentary illustrated in Fig. 8.5. In the illustration, a binary classification problem is shown, in which a learning algorithm has learned the specific decision border. To achieve a balanced class distribution, a new instance is added to the blue class by utilizing SMOTE. This addition achieves, apart from a balancing of the dataset, an adjustment of the decision boundary. The new boundary is less specific and thus contributes to more general decisions. SMOTE in combination with textual data augmentation is applied, for instance, in the work of Wang and Lillis [467]. Unfortunately, the authors do not describe how and at which point of the network the method is applied.

8.3.2.4 Mixup Interpolation

Mixup by Zhang et al. [521] is an interpolation method similar to SMOTE. In the simplest adoption, the convex interpolation is implemented with the following formulas:

$$\tilde{x} = \lambda x_i + (1 - \lambda)x_j, \text{ whereas } x_i, x_j \text{ are input vectors} \tag{8.2}$$

$$\tilde{y} = \lambda y_i + (1 - \lambda)y_j, \text{ whereas } y_i, y_j \text{ are one-hot-coded labels} \tag{8.3}$$

(x_i, y_i) and (x_j, y_j) are sampled from the training data and λ is either fixed in $[0, 1]$ or $\lambda \sim \text{Beta}(\alpha, \alpha)$, for $\alpha \in (0, \infty)$.

Mixup is a general technique that can be applied to all kinds of equal dimensional data. However, text cannot trivially be represented in equal dimensions [427]. As a very general method, Verma et al. [456] propose the idea of applying mixup within a randomly selected hidden layer of a neural network. Despite the fact that the authors only perform the tests on image datasets, this approach paves the way for the application of mixup in many textual related tasks. The results are very promising, and for textual evaluations we advise the reader to look at the methods described below (Table 8.8), which oftentimes can be seen as specializations of the approach by Verma et al. [456] for textual data. [275] state that they use mixup on representations of bag of word models, TF-IDF models, word embeddings, and language models. Unfortunately, the authors do not explicitly describe how the interpolation is performed. This raises questions about how to interpolate instances of different sizes, when, for example, word embedding vectors are used. Marivate and Sefara [275] report about 0.2, 0.4, and 0 points gain for the AG News, Sentiment 140, and Hate Speech detection task. In contrast, Qu et al. [337] describe the internal implementation of their interpolation. For the interpolation, they draw two instances from a mini-batch and linearly combine their input embedding matrices in the way described above. They improve the baseline on the MNLI-m task by an additional 0.6% in terms of accuracy. Guo, Mao, and Zhang [138] propose two variants, wordMixup and senMixup, where the interpolation is applied in the word embedding space and on the final hidden layer of the neural network before it is passed to a softmax layer. For wordMixup the sequences have to be zero padded so that the dimensions are the same. For senMixup this is not necessary, as the hidden embeddings generated are of the same length each. The improvement results of both methods with regard to the CNN model with pretrained GloVe embeddings (trainable), which is the best baseline, is presented in Table 8.8. Guo, Mao, and Zhang [138] further advances the wordMixup approach by using a nonlinear interpolation policy. The policy is constructed to mix each dimension of the individual word embeddings in a given sentence separately. Furthermore, the labels are also interpolated nonlinearly, while they are learned adaptively based on the mixed embeddings. This way, a much larger variety of generated examples can be created. While this procedure outperforms the other two variants in most tasks, it can also have a negative effect on the classification quality, as shown in Table 8.8. Similar to the senMixup method, Sun et al. [427] apply mixup to the output of transformer models. Furthermore, they only activate mixup in the last half of the training epochs to learn good representations first. The improvements on the GLUE benchmark are listed in Table 8.8. In a very similar way, Chen, Yang, and Yang [67] propose TMix, which is also able to interpolate the hidden representations of an encoder. Indeed, TMix is able to interpolate at every layer of the encoder, similar to Verma et al. [456]. Based on the work of Jawahar, Sagot, and Seddah [178], who analyzed the types of information learned in different layers of BERT, the authors narrowed down their approach and opted for 7, 9, and 12 as interpolation layers as they contain the syntactic and semantic information. The improvements of TMix are also shown in Table 8.8.

Table 8.8 Overview of different approaches of the replacement method 'mixup interpolation'

Method	Technique for textual application	Model	Datasets	Improv.
mixup in [275]	Not stated	DNN	AG News	+0.2 (Acc.)
			Sentiment 140	+0.4
			Hate Speech	0
[224]	Interpolation of the BERT CLS output	BERT-base-english-uncased	SNIPS (few shot)	+8.36 (Acc.)
			SNIPS	+0.0
			FBDialog (few shot)	+7.92
			FBDialog	+0.08
[337]	Interpolation of the embedding matrices	RoBERTa-base	MNLI-m	+0.6 (Acc.)
wordMixup in [138]	Interpolation of zero-padded word embeddings	CNN	Trec	+1.6 (Acc.)
			SST-1	+1.9
			SST-2	+0.2
			Subj	+0.3
			MR	+1.5
senMixup in [138]	Interpolation on the final hidden layer	CNN	Trec	+1.2 (Acc.)
			SST-1	+2.3
			SST-2	+0.3
			Subj	+0.5
			MR	+0.8

(Continued)

Table 8.8 (Continued)

Method	Technique for textual application	Model	Datasets	Improv.
Nonlinear Mixup in [138]	Nonlinear interpolation of padded word embeddings	CNN	Trec	+2.6 (Acc.)
			SST-1	+3.0
			SST-2	+2.3
			Subj	−0.5
			MR	+3.6
Mixup-Transformer in [427]	Interpolation after last layer of the transformer	BERT-large	CoLA	+2.68 (Corr.)
			SST-2	+0.81 (Acc.)
			MRPC	+1.72 (Acc.)
			STS-B	+0.89 (Corr.)
			QQP	+0.42 (Acc.)
			MNLI-mm	−0.01 (Acc.)
			QNLI	+0.13 (Acc.)
			RTE	+0.37(Acc.)

(Continued)

Table 8.8 (Continued)

Method	Technique for textual application	Model	Datasets	Improv.
TMix in [67]	Interpolation of the m-th BERT layer (7, 9, and 12 randomly chosen per batch)	BERT-base- uncased + average pooling + two-layer MLPs	AG News (10)	+4.6 (Acc.)
			AG News (2500)	+0.2
			DBPedia (10)	+1.6
			DBPedia (2500)	+0.0
			Yahoo! (10)	+2.4
			Yahoo! (2500)	+0.3
			IMDB (10)	+1.8
			IMDB (2500)	+0.5
TMix evaluated in [512]	Interpolation of the m-th BERT layer (7, 9, and 12 randomly chosen per batch)	BERT-base	SST-2 (0.1, 0.3, 1.0%)	−0.2, −1.5, −2.1
			COLA (0.1, 0.3, 1.0%)	+0.8, +2.4, −0.7
			TREC6 (0.1, 0.3, 1.0%)	−0.2, −1.4, +2.4
			R (0.1, 0.3, 1.0%)	−0.1, −0.5, −3.3
			SUBJ (0.1, 0.3, 1.0%)	−0.5, +0.4, −0.1
			MPQA (0.1, 0.3, 1.0%)	+0.2, 2.9, 0.0
			RT20 (0.1, 0.3, 1.0%)	+2.3, 0.6, −1.9

(Continued)

Table 8.8 (Continued)

Method	Technique for textual application	Model	Datasets	Improv.
Intra-LADA [66]	Interpolation of an instance with a randomly reordered version of itself	BERT-base- multilingual-cased + linear layer	CoNLL (5%)	+0.24 (F1)
			CoNLL (100%)	+0.03 (*)
			GermEval (5%)	+0.29
			GermEval (100%)	+0.04 (*)
Inter-LADA [66]	Interpolation of the nearest neighbors and sometimes randomly selected instances	BERT-base- multilingual-cased + linear layer	CoNLL (5%)	+1.32 (F1)
			CoNLL (100%)	+0.64
			GermEval (5%)	+0.49
			GermEval (100%)	+0.33
Intra-Inter-LADA [66]	Combination of Intra- and Inter-LADA	BERT-base- multilingual-cased + linear layer	CoNLL (5%)	+1.57 (F1)
			CoNLL (30%)	+0.59
			GermEval (5%)	+0.53
			GermEval (30%)	+0.78

* Included in the pre-training

Similarly, Chen et al. [66] propose an interpolation augmentation method in which the hidden layer representations of two samples are interpolated. However, they noticed that this method is not suitable for sequence tagging tasks. For this reason, they propose Intra- and Inter-LADA. Intra-LADA aims to reduce noise from interpolating unrelated sentences by only interpolating an instance with a randomly reordered version of itself. This way, they can increase the performance in every tested task (see Table 8.8). However, Chen et al. [66] also hypothesize that their Intra-LADA algorithm is limited in producing diverse examples. This limitation leads to Inter-LADA, which sets a trade-off between noise and regularization by interpolating instances that are close together. The closeness is estimated through kNN based on sentence-BERT [356] embeddings and extended by occasional sampling of two completely random instances. As it can be seen in Table 8.8, Inter-LADA oftentimes performs better than Intra-LADA. The combination of both can further improve the results.

8.3.3 Combination of Augmentation Methods

In augmentation research, a common technique is to combine several data augmentation methods to achieve more diversified instances [150]. Here, combination can mean either the application of multiple separate or stacked methods. For the first kind, Sun and He [428] propose word-level and phrase-level methods that they apply separately. While the results of the word-level and phrase-level methods differ insignificantly, the combination of both groups of methods produced very good results. Similarly, Li, Cohn, and Baldwin [250] combined their proposed methods, which led to better results for the in-domain evaluations. In the work of Bonthu et al. [39] round-trip translation, random swap, random deletion, and random synonym insertion are separately combined, which leads to the best improvement of a LSMT classifier. Furthermore, in contrastive learning, it makes sense to use more than one data augmentation strategy since the goal is to learn meaningful representations that can be fostered by many different views. For example, Yan et al. [505] and Wu et al. [493] use several simple methods of data augmentation for the contrastive learning objective. Details on contrastive learning and the results of the works can be found in the next section. The method of stacking data augmentation techniques, on the other hand, is not always feasible. It is, for example, in most cases not possible to first apply a feature space method and then a data space method. Qu et al. [337] test stacking with round-trip translation, cutoff, and adversarial examples. Round-trip translation and the training with adversarial examples produce the best results. Marivate and Sefara [275] stack round-trip translation, synonym and embedding replacement with mixup. In two out of three evaluation settings, this procedure reduces the minimal error.

For the combination of augmentation methods, the meta-learning augmentation approach by Ratner et al. [352] is also of interest. It describes the utilization of a neural network to learn data augmentation transformations [401]. Specifically, Ratner et al. [352] use a GAN to generate sensible sequences of transformations that were defined beforehand. This approach is usable for image as well as text datasets and the authors show that it can achieve a significant improvement when applied to a relation extraction task with augmentations based on language model replacements.

8.3.4 Training Strategies

While semi-supervision is not considered as data augmentation in this work, it can still be sensibly combined through consistency training. In its origin, consistency training is used to make predictions of classifiers invariant to noise [498]. This can be enforced by minimizing the divergences between the output distributions of real and noised instances. Additionally, as only output distributions are included in the process, this consistency can be trained with unlabeled data. Several authors analyze how consistency training behaves when data augmentation methods are used for noise. This process can be illustrated by taking an instance whose label is unknown, applying a label-preserving data augmentation method, and then learning the model to predict the same label for both instances. In this way, the model can learn the invariances and is able to generalize better. Xie et al. [498] show that they achieve very good results by employing consistency training with round-trip translation and a TF-IDF-based replacement method, with an absolute improvement of 22.79% in accuracy on an artificially created low-data regime based on the Amazon-2 dataset with BERT base. They are also able to outperform the state of the art in the IMDb dataset with only 20 supervised instances. Chen, Yang, and Yang [67] even extend this approach within their MixText (TMix) system. First, they generate new instances with round-trip translation. Then, they guess the label of the original and augmented instances by taking a weighted average of the predictions of all of them. In the training, they randomly sample two instances and mix them together with TMix. If one of the two instances is from the original data, they are using the normal supervised loss, but if both instances are from the unlabeled or augmented data, they use the consistency loss, like Xie et al. [498]. Consistency training can also be applied in a supervised fashion as an additional term in the training objective to enforce identical predictions. This is, for example, used in the cutoff method by

Shen et al. [398]. They show in their ablation studies that this consistency term improves the accuracy results additively by 0.15%.

Qu et al. [337] combine supervised consistency training with contrastive learning. The contrastive learning scheme should bring the original and augmented instances closer together and the other instances further apart in the representation space. Contrastive learning can be applied in the pretraining phase of a language model so that meaningful representations are learned directly. Wu et al. [493] show that training a language model from scratch with this objective can lead to increased performances for downstream tasks. As augmentation methods the authors use word deletion, span deletion, random reordering and synonym substitution, as well as combinations in sets of two. The evaluation of several tasks shows that there is no clear best augmentation method. Fang and Xie [109] and Yan et al. [505] show that contrastive learning can also result in better sentence representations when using an already pretrained model and further training the masked language modeling task with contrastive learning. While the work of Fang and Xie [109] uses round-trip translation, Yan et al. [505] experiment with adversarial training, token mixing, cutoff and dropout. Qu et al. [337] and Choi et al. [71] even include contrastive learning into the supervised setting. As augmentation strategies, Qu et al. [337] use adversarial training combined with round-trip translation and Choi et al. [71] use counterfactuals based on language model substitution. Combined with consistency training, Qu et al. [337] achieve even further improvements. A comparing overview can be found in Table 8.9.

Other training strategies in which the order of how the training examples are presented to the learning algorithm is altered are for example employed by Liu et al. [258], Yang et al. [508], and Claveau, Chaffin, and Kijak [75]. Liu et al. [258] adopt a curriculum learning approach, where the algorithm first learns less difficult instances. Transferred to the data augmentation topic, the model is first trained with the original data and then with the augmented data. Yang et al. [508] reverse this step and first train the model with the augmented data and then with the original data. This way, the model can correct unfavorable behavior that it learned through noisy augmented data. They also tried an importance-weight loss in which the weights of the synthetic instances are lower but find that the other training method performs better.

Table 8.9 Overview of the contrastive learning works using data augmentation

Pub.	Method	Model	Dataset	Improv.
CERT [109]	Continued contrastive self-supervised learning (MLM) with round-trip translation.	Continued BERT training compared to BERT	MNLI	+0 (Acc.)
			QNLI	+0.2 (Acc.)
			QQP	+0.4 (Acc.)
			RTE	+3.6 (Acc.)
			SST-2	+0.7 (Acc.)
			MRPC	+0 (F1)
			CoLA	+2.3 (Mcc)
			STS-B	+0.6 (Corr.)
CODA [337]	Supervised contrastive learning with adversarial training and round-trip translation.	RoBERTa-base fine-tuning compared to RoBERTa-base	MNLI-m	+0.5 (Acc.)
			QNLI	+0.8
			SST-2	+0.5
			RTE	+3.3
			MRPC	+1.5

(Continued)

Table 8.9 (Continued)

Pub.	Method	Model	Dataset	Improv.
CLEAR [493]	Contrastive self-supervised learning (MLM) from scratch with word deletion, span deletion, random reordering, synonym substitution (also in combination). Span deletion + random reordering is shown here.	Transformer architecture by [453] trained from scratch compared to RoBERTa-base	MNLI	-0.5 (Acc.)
			QNLI	−0.3 (Acc.)
			QQP	+1.8 (Acc)
			RTE	+6.5 (Acc)
			SST-2	+0.1 (Acc)
			MRPC	+2.8 (F1)
			CoLA	+8.2 (Mcc)
			STS-B	+0.4 (Corr)
ConSERT [505]	Continued contrastive self-supervised learning (MLM) with adversarial training, token shuffling, cutoff, dropout.	Continued BERT-large training compared to SBERT-large-NLI	STS12	+0.99 (S. Corr.)
			STS13	+3.9
			STS14	+2.83
			STS15	+2.85
			STS16	+2.5
			STSb	+2.31
			SICK-R	+4.89
C^2L [71]	Supervised contrastive learning with counterfactual augmentation.	BERT-base finetuning compared to BERT-base	CF-IMDb	+0.7 (Acc.)
			CF-IMDb Revised	+3.3
			CF-NLI	−1.2
			CF-NLI RP	+2.2
			CF-NLI RH	+1.3
			CF-NLI RP & RH	+1.8

8.3.5 Filtering Mechanisms

Mechanisms that filter the generated instances are especially important for methods that are not perfectly label-preserving. A simple mechanism is, for example, employed by Liu et al. [254], who remove generated instances based on the overlap of unigram words with their original equivalents. Similarly, other metrics could also be used, e.g., Levenshtein distance, Jaccard similarity coefficient, or Hamming distance. Wan, Wan, and Wang [462] use a similarity discriminator, initially proposed by Parikh et al. [327], which also measures the similarity of two sentences.

The generative methods by Anaby-Tavor et al. [18], Abonizio and Junior [4], and Claveau, Chaffin, and Kijak [75] filter instances based on a classifier that was trained on the class data. This significantly reduces the diversity of samples, and the classifier cannot really be improved as it is already familiar with these instances. Bayer et al. [28] improve this by using embeddings to measure the quality of the generated instances with regard to the class and more importantly by incorporating the human expert in the loop who needs to determine the correct threshold. However, Yang et al. [508] consider another filtering mechanism in their work which does not require human assistance and is very sophisticated due to the inclusion of two perspectives. Generally, Yang et al. [508] propose a generative method that is suitable for increasing the dataset size for question answering tasks. While they propose to utilize language models for fine-tuning and generation of questions and answers, their filtering methods can be adapted for other data augmentation methods as well. A first filtering mechanism determines whether a generated instance is detrimental by measuring whether the validation loss increases when including the artificial instance. As this would require retraining the model with each generated example, the authors propose to use influence functions [210, 366] to approximate the validation loss change. Furthermore, they first train on the augmented instances and then on the original training data so that the model can adjust itself when unfavorable noise is included in the augmented instances. The other filtering mechanism tries to favor diversity by selecting examples that maximize the number of unique unigrams.

8.4 Discussion: A Research Agenda For Textual Data Augmentation

In the previous section, different data augmentation methods were grouped, explained, compared in terms of performance and put into context with each other. One has to keep in mind that the results reported by the authors of the approaches linked in this survey paper are restricted in their expressiveness and only show one perspective. Many results are limited to special kinds of models and datasets. Based on our findings, we identified an agenda for future research on data augmentation as follows:

8.4.1 Researching the Merits of Data Augmentation in the Light of Large Pre-trained Language Models

Generally, it is not possible to determine which augmentation method works best for a given dataset, nor predict which research direction will be the most appealing in the future. Nevertheless, some patterns in current approaches hint to the directions research can follow in order to overcome current obstacles and challenges. One of the most significant challenges, as formulated by Long-pre, Wang, and DuBois [265], concerns the usage of large pre-trained language models, which makes the utilization of several data augmentation methods obsolete. Large pre-trained models are currently state of the art, nevertheless, we advise taking further advancements and findings in the research landscape into account, as for example deep belief networks [62], capsule networks [533], or task-specialized networks e.g. for sentiment analysis [176, 270]. Experiments with BERT or other bigger language models are therefore of particular interest. Similarly, several studies [18, 28, 160, 254, 256, 512] have shown that methods only slightly transforming instances with random behavior, such as with synonym replacement (Sect. 8.3.1.2), EDA (synonym replacement, random swap, deletion, and insertion in one) (Sect. 8.3.1.2), or by inserting spelling errors (Sect. 8.3.1.1), tend to be less beneficial in this setting than more elaborate ones. Particularly adversarial training (Sect. 8.3.2.1), cutoff (Sect. 8.3.2.1), interpolation (Sect. 8.3.1.3 and 8.3.2.2), and some generative methods (Sect. 8.3.1.4) have shown significant improvements with large pre-trained language models. While replacement methods based on embeddings (Sect. 8.3.1.2) and especially language models (c-BERT) (Sect. 8.3.1.2) can also gain improvements in combination with those pre-trained models, several studies [18, 256, 275, 337] have shown that the previously mentioned methods can, in most cases, achieve improved results.

The described performance differences become apparent when approaching the challenge highlighted by Longpre, Wang, and DuBois [265] from an intuitive perspective. Large language models map data to a latent space with representations nearly invariant to some transformations. For example, synonym replacement methods only replace words that are by definition very close in the representation space, leading to instances that are almost identical [297]. As Longpre, Wang, and DuBois [265] hypothesize, data augmentation methods can only be helpful, if they are able to introduce new linguistic patterns. In such instances, using the mentioned methods and generative methods, in particular, might be sensible, as they are based on other large language models that can introduce a high novelty. However, the challenge proposed by Longpre, Wang, and DuBois [265] does not have to be universally true. For example, the SUB^2 method by Shi, Livescu, and Gimpel [399] only interpolates phrases from the training data and thus does not include unseen linguistic patterns but achieves high gains with a pre-trained model. Another interesting aspect concerns the experiments conducted by Yoo et al. [512] with which they demonstrate that their GPT-3-based generative augmentation method actually improves as the size of the pre-trained classifier increases. The authors hypothesize that larger classifiers have more capacity to better incorporate the GPT3Mix samples.

8.4.2 Improving Existing Data Augmentation Approaches

In general, most promising data augmentation methods have limits and challenges that may be overcome with further research. **Generative models** or their output needs to be conditional on the specific class. Otherwise, the created instances might not preserve the label. This conditioning is oftentimes reached by training a model, which in turn requires enough data to be consistent. Bayer et al. [28] have shown that the conditional model can best replicate the data class, if the problem definition and task data is relatively narrow. Tasks with a broad variety of topics in the data seem less suitable. This problem might be mitigated by adopting new conditioning methods. Currently, most approaches are extended by filter mechanisms. Existing mechanisms, as detailed in Sect. 8.3.5, have some drawbacks which might be reduced in the future. Another obstacle concerns generative models themselves, which can require many resources and time to create new instances [28]. Therefore, lightweight alternatives need to be tested in this setting, thus potentially preventing a high dependency on resources, which is referred to by Bayer et al. [28] as the high resource wall problem. Similarly, methods like **round-trip translation** are limited by the underlying model. For example, Marivate and Sefara [275] hypothesize that round-trip translation might not be appropriate for social media data, where translation errors increase. This problem will be addressed in the future, as machine translation models improve their translation capabilities for such difficult instances.

For **adversarial examples**, Liu et al. [258], hypothesize that good generalizability performance stems from the perturbation of the embedding space, rather than the input space. However, data space adversarial training methods should not be disregarded too quickly, as Ebrahimi et al. [99] show that their data space method achieves better results than the virtual adversarial training by Miyato, Dai, and Goodfellow

[292]. A general challenge for adversarial training is that it can disturb the true label space in the training data. For example, adversarial example generators often rely on the belief that close input data points tend to have the same labels [538]. Concerning the data space methods, this is often not true for natural language tasks, where few words or even characters determine the class affiliation (e.g., sentiment classification: "I can't believe I like the movie" → small_transformation→ "I can' believe I like the movie"). Whether this applies to the adversarial example generators in the feature space needs to be evaluated. If so, research needs to find a way to exclude cases where small transformations disturb labels and at best include cases where stronger transformations still preserve the labels. For this purpose, inspecting feature space methods would be helpful. However, such an inspection is difficult to conduct due to their high-dimensional numerical representation. The same applies to the feature space's **interpolation methods**, where a back transformation to the data space is not trivial. Though, certain approaches, such as those from Liu et al. [254] and Wan, Wan, and Wang [462], use techniques such as encoder-decoder architectures capable of transforming the newly created instances to the data space. An inspection of interpolated instances could lead to interesting insights. This opens another research direction where the **interpolation of instances in the data space** could be further investigated. A method that initially implements this behavior is SUB^2 (Sect. 8.3.1.3), which interpolates instances of the data space through sub-phrase substitutions. This, however, does not result in a high diversity, which is particularly interesting. In this regard, further analysis of the GPT-3 language model by [47] could be valuable, as it shows very interesting interpolation capabilities in the data space.

However, even **avoidably inferior methods** can achieve better results if they are integrated sensibly. The work of Jungiewicz and Smywinski-Pohl [187] can serve as an example. They perform synonym substitution only if it increases the loss of the model. This demonstrates that some data augmentation techniques proposed in the different groups are advanced, sometimes adopting existing methods and refining them.

We highlight some advanced works of the different groups in Table 8.10 to show which research directions can be considered in the future. It must be emphasized that these methods are not necessarily the best in their groups. The selection is made by the author team on the basis of the information gathered while writing this survey.

Table 8.10 Collection of some of the most advanced data augmentation techniques for text classification

	Group	Work	Method description	Improvement
Data Space	Character Level Noise	[99]	Flip a letter if it maximizes the loss	+0.62 Acc. (LSTM)
	Synonym Replacement	[187]	Only replace words with a synonym if it maximizes the loss	+1.2 Acc. (Kim CNN)
	Embedding Replacement	[17]	Choosing embeddings based on the counter-fitting method	−0.6 – +1.9 Acc. (CNN)
		[250]	Counter-fitting, language model selection, and maximizing the prediction probability	Safer model (LSTM)
	Language Model Replacement	[160]	c-BERT integrated in reinforcement learning scheme	+0.73 – +1.97 Acc. (BERT)
		[183]	c-BERT and embedding substitution forcompound words	+1.9 – +21.0 Acc. (TinyBERT)
	Phrase Level Interpolation	[399]	Substitutes substructures	+20.6 – +46.2 Acc. (XLM-R)*
	Round-trip Translation	[498]	Random sampling with a temperature parameter	+1.65 Acc.
	Generative Methods	[28]	Conditional GPT-2 with human assisted filtering	−2.54 F1 – +15.53 Acc. (ULMFit)*
		[256]	GPT-2 with a reinforcement learning component	+1.0 – +4.3 F1 (XLNet)*

(Continued)

Table 8.10 (Continued)

	Group	Work	Method description	Improvement
Feature Space	Noise	[181]	Virtual adversarial training with special optimization	+0.5 – +5.4 Acc. (RoBERTa-l)
		[258]	Virtual adversarial training with curriculum learning	-0.3 Corr. – +1.2 Acc. (RoBERTa-l)
		[398]	Embedding noising	+0.0 Corr. – +4.4 Acc. (RoBERTa-l)
	Interpolation	[427]	Interpolation after lastlayer of the transformer	-0.01 Acc. – +2.68 Corr. (BERT-l)
		[67]	Interpolation of a random BERT layer	+0.0 – +4.6 Acc. (BERT-b)*
		[66]	Interpolating neighborsand reordered versions	+0.53 – +1.57 F1 (BERT-b)*

* Results contain tests on low-data regime datasets

8.4.3 Establishing more Comprehensive Evaluation Criteria and Standards for Method Comparison

labelmethodspscomparisonspsstandardsspscriteria A general problem in data augmentation research concerns that mostly only improvements with regard to the prediction performance on specific datasets are presented. While this metric is likely the most important one, other metrics, such as time and resource usage, language variety, or configurability, are also important for practitioners as well as for researchers. For example, the generative approaches based on GPT-2 seem very promising when considering prediction performance gain. Nevertheless, language variety is narrowed down, as the model is primarily trained on English data. Furthermore, only few authors discuss the time required for the application of their data augmentation methods. The GPT-2 based method of Bayer et al. [28] takes up to 30 seconds for generating one example, leading to several computing days for a 10-times augmentation of a small dataset. For instance, in the context of crisis informatics this might take too long, as classifiers have to be created quickly for immediate incident management [192]. We therefore urge scientists developing data augmentation techniques to consistently describe the limitations of their approaches. For further data augmentation research, flexible standards should be established in order for methods to be compared more reliably, similar to other machine learning research fields, e.g. few-shot learning [44] or natural language generation [125]. It seems unrealistic that one or few general datasets can capture all peculiarities of data augmentation methods, especially not of those that one tailored to a specific problem. Nevertheless, a small benchmark that can be included in evaluations of upcoming data augmentation methods would be desirable. In the best-case, such a benchmark would address different data augmentation goals, consisting of two or more datasets, from which one replicates a few-shot learning setting and the other a normal learning setting. With the growing usage of generative models, it might also be sensible to consider using datasets that are not part of current training datasets for language models, as an incorporation of testing data would lead to wrong conclusions. The benchmark should not be too large, in order to ensure specific evaluations can still be carried out. Researchers that try to develop such a benchmark, could also consider to specify how much data augmentation should be performed and what models should be used. When determining which model should be used, it might be useful to create an updatable benchmark, as proposed by Gehrmann et al. [125], which can be modified according to more recent state-of-the-art models.

8.4.4 Enhancing the Understanding of Text Data Augmentation

Shorten and Khoshgoftaar [401] highlight that while for some image data augmentation techniques it is easy to understand how they might improve the dataset and derived classifiers, however, for other techniques this improvement has not been explainable yet. This also applies to the text regime, where for example, data augmentation methods that paraphrase text without changing the meaning are naturally sensible, while methods applied in the feature space are much more complex to capture. Already the visualization of the data of feature space augmentations created by, for example, adversarial examples or interpolation methods, is much more complicated than in the image domain. As previously elaborated, existing approaches try to convert representations back into the data space by using encoder-decoder architectures [254, 462]. Resulting data space representations could then be investigated and used to better understand underlying data augmentation methods. Furthermore, a more in-depth understanding of why and when data augmentation works needs to be established. With the rise of large language models the question emerges whether data augmentation methods paraphrasing input instances without incorporating new patterns may be obsolete [265]. Certain works have challenged this perspective, by demonstrating that even existing patterns can be beneficial for performance [399]. In this context, it is interesting to note that the augmentation method of Yoo et al. [512] provides better results when the size of the pre-trained language model increases.

8.4.5 Fostering the Usability of Data Augmentation Application

Most data augmentation methods are still research-based in their incremental development progress and therefore not suitable for every practitioner. A simple way to improve the usability is to publish code and, in the best-case, develop libraries that can be used out of the box for augmenting a text dataset. Dhole et al. [90] propose a first large framework to include many text data augmentation methods and filtering mechanisms. The library by Papakipos and Bitton [324] is not as big for textual data augmentation methods, but can be used for multiple modalities (audio, image, text & video). While these are very useful libraries, the amalgamation of many procedures comes with abstraction problems. For example, only individual data instances can be transformed and the augmentation procedure does not have access to the entire dataset, so that, for example, no interpolation procedures are implemented. In addition to creating libraries, it might be useful to explore augmentations with a good learning process integration. This can be considered as a criterion to simplify embedding the procedure in the general learning process. Resource utilization, speed, and general continuity in the learning process are crucial for this process. The first two criteria are becoming increasingly relevant as they are related to the current trend of data augmentation, i.e. the use of large underlying models that create a high resource and time execution overhead. As described above, this might be countered with utilizing more lightweight models. A low continuity in the learning process refers to the circumstance that a text data augmentation method is detached from the actual training process; or in the worst case, the learning procedure needs to be split into two halves. The former, also described as offline data augmentation by Feng et al. [113], means that the original data is augmented independently from the model training. A data augmentation technique is called online, if it is embedded into the learning process so that the artificial instances are stochastically included by the learning algorithm, which is, e.g., implemented in the work of Bonthu et al. [39]. The second form occurs, for example, when a feature space method needs to separate the normal network structure, in order to detach the encoder or embedding layer from the rest of the network. This results in a continuity problem of learning, so that, e.g., the encoder or embedding level cannot be trained further.

8.5 Conclusion

This survey provides an overview over data augmentation approaches suited for the textual domain. Data augmentation is helpful to reach many goals, including regularization, minimizing label effort, lowering the usage of real-world data particularly in privacy-sensitive domains, balancing unbalanced datasets, and increasing robustness against adversarial attacks (see Sect. 8.2). On a high level, data augmentation methods are differentiated into methods applied in the feature and in the data space. These methods are then subdivided into more fine-grained groups, from noise induction to the generation of completely new instances. In addition, we propose several promising research directions that are relevant for future work. Especially in this regard, a holistic view on the current state of the art is necessary. For example, the increasing usage of transfer learning methods makes some data augmentation methods obsolete, as they follow similar goals. Hence, there is a need for more sophisticated approaches that are capable of introducing new linguistic patterns not seen during pre-training, as suggested by Longpre, Wang, and DuBois [265].

While data augmentation is increasingly being researched and seems very promising, it also has several limitations. For instance, many data augmentation methods can only create high quality augmented data, if the original amount of data is large enough. Furthermore, as Shorten and Khoshgoftaar [401] describe, data augmentation is not capable of covering all transformation possibilities and eliminating all kinds of biases in the original data. Adopting the example of Shorten and Khoshgoftaar [401], in a news classification task, in which there are no articles containing sports, the standard data augmentation methods will most certainly also not create sport articles, even though this would be necessary. In contrast, data augmentation might induce new undesirable biases. For instance, language models like GPT can contain biases that are then propagated into the dataset [412]. The wide variety of techniques and some very sophisticated methods also bring another layer of complexity that needs to be understood. Moreover, data augmentation can be time consuming, meaning that not all methods are feasible for time critical machine learning development domains, e.g., in some areas of crisis informatics. An increased demand for resources, especially concerning training generative models, is inherent to data augmentation.

In order to mitigate some of the limitations and amplify the strengths of data augmentation, however, we proposed our research agenda, which comprises (1) researching the merits of data augmentation in the light of large pre-trained language models, (2) improving existing data augmentation approaches, (3) establishing more comprehensive evaluation criteria and standards for method comparison,

(4) enhancing the understanding of text data augmentation, as well as (5) fostering the usability of data augmentation application.

Acknowledgements This work has been co-funded by the German Federal Ministry of Education and Research in the project CYWARN (13N15407), by the LOEWE initiative (Hesse, Germany) within the emergenCITY center, as well as the German Federal Ministry of Education and Research and the Hessian Ministry of Higher Education, Research, Science and the Arts within their joint support of the National Research Center for Applied Cybersecurity ATHENE. The calculations for this research were conducted on the Lichtenberg high performance computer of the TU Darmstadt.

Data Augmentation in Natural Language Processing: A Novel Text Generation Approach for Long and Short Text Classifiers

9

Abstract

In many cases of machine learning, research suggests that the development of training data might have a higher relevance than the choice and modelling of classifiers themselves. Thus, data augmentation methods have been developed to improve classifiers by artificially created training data. In NLP, there is the challenge of establishing universal rules for text transformations which provide new linguistic patterns. In this paper, we present and evaluate a text generation method suitable to increase the performance of classifiers for long and short texts. We achieved promising improvements when evaluating short as well as long text tasks with the enhancement by our text generation method. Especially with regard to small-data analytics, additive accuracy gains of up to 15.53% and 3.56% are achieved within a constructed low-data regime, compared to the no augmentation baseline and another data augmentation technique. As the current track of these constructed regimes is not universally applicable, we also show major improvements in several real world low-data tasks (up to +4.84 F1-score). Since we are evaluating the method from many perspectives (in total 11 datasets), we also observe situations where the method might not be suitable. We discuss implications and patterns for the successful application of our approach on different types of datasets.

Supplementary Information The online version contains supplementary material available at https://doi.org/10.1007/978-3-658-48778-2_9.

M. Bayer, *Deep Learning in Textual Low-Data Regimes for Cybersecurity*, Technology, Peace and Security I Technologie, Frieden und Sicherheit, https://doi.org/10.1007/978-3-658-48778-2_9

Original Publication *Markus Bayer, Marc-André Kaufhold, Björn Buch-hold, Marcel Keller, Jörg Dallmeyer, and Christian Reuter. "Data augmentation in natural language processing: a novel text generation approach for long and short text classifiers". In:* Int. J. Mach. Learn. Cybern. *14.1 (2023), pp. 135–150.* DOI: https://doi.org/10.1007/S13042-022-01553-3. URL: https://doi.org/10.1007/s13042-022-01553-3[1]

Note Supplementary material containing data, architecture, and infrastructure descriptions, pseudocode of the algorithms, and insights into the generated augmentation data can be found in Appendix A.2 of the Electronic Supplementary Material.

Material
Application Code: https://github.com/PEASEC/A-novel-data-augmentation-generation-approach-for-long-and-short-text-classifiers

9.1 Introduction

Deep learning has attracted considerable attention due to increased computing power in combination with a higher availability of training data for a wide range of problems [425]. In some learning tasks, especially small-data regimes, the development of training data might have a higher relevance than the choice and modelling of classifiers [25]. To improve classifiers, data augmentation methods have been designed to artificially create training data with specific transformations [434]. Current research in data augmentation focuses on deep learning algorithms, which are state of the art for many classification tasks, as they still often suffer from a strong variance regarding the given problem if not enough data is provided. The artificial creation of training data serves as a kind of regularization and thus, simpler solutions are preferred [159, 518]. In addition, imbalance in datasets can be addressed [342, 519] and the safety of classifiers can be increased by making them resistant to the deception by skillful changes in the input sequences [292]. Data augmentation can also help to mitigate the "big data wall" problem, which relates to the fact that smaller companies, research groups and organizations are usually unable to acquire the same volume of data as large corporations [81].

[1] Minor typographical errors in the original publication have been corrected for this dissertation.

Regardless of deep learning, research into artificial data creation can benefit natural language processing (NLP) applications across several domains where training data is scarce or labeling is costly. For example, to enhance the situational awareness of emergency managers, a part of crisis informatics deals with the rapid recognition and subsequent classification of messages and pictures during disasters and emergencies [10, 198]. Due to the scarcity of financial and personnel resources, emergency services lose valuable time dealing with complicated identification tasks, which eventually can cost lives [167, 359]. This problem of scarcity also applies to small- and medium-sized enterprises (SMEs) when requiring a high quality and volume of labeled data for commercial tasks such as brand analysis or news classification [421]. In NLP, there is the difficulty establishing universal rules for transformations of textual data that can be carried out automatically and still maintain the quality of the labeling, which is especially sensitive in domains such as sentiment analysis [280]. Longpre, Wang, and DuBois [265] suggest that current pre-training or transfer learning methods in NLP already cover the goals of data augmentation. They argue that augmentation methods that only perturb the input data and do not provide new linguistic patterns fail to increase the classification quality of pre-trained models.

Thus, we propose a sophisticated generation-based method that overcomes these problems by incorporating new linguistic patterns (i.e., a high grammatical variety) which prove to be useful in combination with pre-trained models. This method does not simply create very similar instances, but very novel ones. Our approach uses two sub methods, whereof one is context-conditional by incorporating parts of the instances (e.g., first words or title) in the generation process and hence suited for long texts, while the other is context independent and suited for short texts. Although there is no clear distinction between long and short texts, we are guided by the 280 character limit (i.e., the length of a message in Twitter), at which most standard NLP data sets would be categorized as small. Thus, we seek to answer three research questions: How can we utilize text generation approaches of data augmentation that achieve a high novelty in the data while preserving the label quality to improve pre-trained machine learning classifiers **(RQ1)**? In which way is the incorporation of contexts of long text instances in classification problems helpful when using text generation as data augmentation method **(RQ1.1)**? How is it possible to achieve a quality improvement for classification tasks with short texts when augmenting with text generation **(RQ1.2)**?

Contributing to the domain of small-data analytics, our results indicate additive accuracy gains of up to 15.53% and 3.56% within a constructed low-data regime, compared to the no augmentation baseline and another data augmentation technique. As the current track of these constructed regimes is not universally applicable, we also show major improvements in several real world low-data tasks (up to +4.84 F1-score). Since we are evaluating the method from many perspectives (in total 11 datasets), we also observe situations where the method might not be suitable. We discuss empirical (i.e., insights into the domain-specific application of small data analytics), practical (i.e., new data augmentation methods based on the GPT-2 language model) and theoretical (i.e., a textual data augmentation basis which is beneficial for pre-trained classification models) implications for the successful application of our approach on different types of datasets.

The paper is structured as follows: After introducing related work on data augmentation, NLP and text generation approaches (Sect. 9.2), the paper presents both the concept and implementation of a novel text generation data augmentation algorithm (Sect. 9.3). Furthermore, it presents the method and findings of three rounds of evaluation (Sect. 9.4) before discussing the implications, limitations and potentials for future research (Sect. 9.5).

9.2 Related Work

9.2.1 Foundations of Data Augmentation

Data augmentation is a machine learning technique that artificially enlarges the amount of training data by means of label preserving transformations [434]. First variations of data augmentation can be identified in the well-known LeNet by LeCun et al. [237]. Using random distortions of training pictures, the MNIST-dataset was ninefold enlarged, so that a better detection of handwritten digits became feasible. A relevant term of data augmentation is label preservation, describing transformations of training data that preserve class information [81]. This means that this kind of transformations modifies texts of a given class to other texts that are as well related to this class. In data augmentation research, this is of high relevance because the absence of it would result in the generation of incorrectly labeled data. For the most part, an entity replacement within a sentence is sufficient for label preservation in sentiment analysis. However, the random addition of words may result in an alteration of the sentiment. Many researchers loosen the label preservation term. Then, transformations that break the preservation are legitimate as long as the label is adjusted simultaneously. Furthermore, transformations that preserve the right class with a high probability, but not with certainty, may exist. In this understanding, Shorten and Khoshgoftaar [401] designate the probability that the correct label is assigned after a transformation as the safety of a data augmentation method. For example, this uncertainty, if known, could be directly integrated in the label. If unknown, methods like label smoothing can model a general uncertainty.

In NLP, data augmentation is considered a difficult task [188] since textual transformations that preserve the label are difficult to define [209, 481]. Thus many methods have been tried out in research so far. Among them are methods for swapping [481], deleting [162, 336], inducing spelling mistakes [33, 81], paraphrasing [224], and replacing of synonyms [214, 496, 526], close embeddings [17, 473] and words predicted by a language model [106, 183, 209] on word-level. On a broader level, methods which change the dependency tree [375, 504], perform round-trip-translation [218, 391], or interpolate the input instances [63, 521] are used. Further studies have dealt with text generation approaches for data augmentation. While Rizos, Hemker, and Schuller [365] and Sun and He [428] are using recurrent neural networks and generative adversarial networks for short-text augmentation, Qiu et al. [336] sample instances from a variational autoencoder without length restrictions. Furthermore, Wang and Li [467] and Anaby-Tavor et al. [18] use the GPT-2 model for text generation. A more detailed analysis, taxonomy and listing of data augmentation techniques can be found in the data augmentation survey by Bayer, Kaufhold, and Reuter [29].

However, challenging many research directions in this area, Longpre, Wang, and DuBois [265] hypothesize that textual data augmentation would only be helpful if the generated data contains new linguistic patterns that are relevant to the task and have not yet been seen in pre-training.

9.2.2 Research Gap

Our data augmentation method is inspired by the text generation methods from Rizos, Hemker, and Schuller [365], Sun and He [428] and Qiu et al. [336] while also considering the limitations outlined by Longpre, Wang, and DuBois [265] and seeks to tackle three primary research gaps:

1. Considering short and long texts while maintaining coherence and achieving high novelty;
2. preserving the labels and quality of the augmentation method;
3. overcoming the challenge of limited usefulness of textual data augmentation in combination with pre-trained models.

First, in contrast to the works of Rizos, Hemker, and Schuller [365] and Qiu et al.[336], we consider short as well as long texts as input data instances to our augmentation method, which is covered explicitly by research questions 1.1 and 1.2. Additionally, and in relation to the main research question, our method is characterized by substantial label preservation in combination with the novelty and coherence of the data. At first, the generation is enriched with a special fine-tuning and prefix addition. Then, a document embedding filter is applied so that instances not associated with the actual class are omitted. Thus, the generation capabilities can be used in full extent while tailoring them to the class data. Furthermore, our experiments are based on the GPT-2-Model by Radford et al. [340] that achieves very good results in text generation.

Second, when it comes to usage of the GPT-2 model, Wang and Lillis [467] describe no measures for label and quality preservation in their GPT-2 augmentation. Anaby-Tavor et al. [18] indicate that the GPT-2 model will be further trained and improperly generated instances will be removed. In contrast to our method, the model of Anaby-Tavor et al. [18] is limited exclusively on sentences as instances and cannot generate coherent text. Furthermore, it uses other safety mechanisms for label preservation. For instance, they use a filter mechanism based on a classifier, that was trained on the class data. This can severely reduce the diversity of the data augmentation method.

Third, the method proposed in this paper is intended to overcome the issue that textual data augmentation can be of no or small value when used in combination with pre-trained classifiers [265]. In contrast to the study by Longpre, Wang, and DuBois [265], we use the ULMFit model by Howard and Ruder [156]. Nevertheless, the model is also pre-trained beforehand and fine-tuned on each task dataset. As a specialty, we also fine-tune the encoder with the augmented data for the baseline, making sure that at least the encoder has seen all linguistic patterns before.

9.3 Concept and Implementation

9.3.1 Conceptual Design

The text generation process can be based on any language model with good text generation capabilities. Language models indicate a probability distribution of sequences of words:

$$P_\Theta(w_t | w_{t-k}, .., w_{t-1}) \; \forall t \tag{9.1}$$

The model P_Θ predicts the probability that the current word is w_t given the predecessor words (context) $w_{t-1}, ..., w_{t-k}$. This enables the P_Θ to generate texts. A phrase prefix can be used as context to make the model follow a certain topic by completing exactly this part, abstracting from exact specifics as sampling methods. In addition, a temperature parameter can be introduced to adjust the randomness in the generation of the texts by scaling the logits in the softmax. In order to enable the sensible use of a language model for data augmentation, it has to be ensured that the procedure mainly generates texts which are similar to the training data and, in addition, reflect the respective class (label preservation or safety). In the following, our augmentation method is described, which comprises the specification of three steps for modeling this behavior.

In a first step, the pre-trained model P_Θ is further trained with the training data X_c of the class c that should be enriched. On the one hand, this enables the model to learn the words, spelling and form of the training data. On the other hand, a bias is generated with regard to the selected class. This means that, for the generation of data, the selected class can be retained more explicitly. In the following, we differentiate between the contextual data augmentation process that is suitable for longer texts and a context independent process for shorter training instances.

In order to further strengthen the safety and label preservation, special "start of text"-tokens are added to each training data in the fine-tuning input. In the text generation phase, these tokens are used as generation prefixes, signaling the model to generate texts similar to the specific training data. This ensures that the augmented examples are different to each other but remain based on the actual data. If the training data consists of longer texts, i.e. instances containing more than 280 characters, this token can be selected context-based, for example, by appending the first words or the title of each instance (e.g. "$<$ |startoftext| $>$ $(w_1...w_k)_i$" where $(w_1...w_k)_i$ is the beginning sequence of the instance i). This results in a high diversity of the generated data. Although, if the texts of the dataset are short and no context tokens can be used for appendage, the context independent variant is chosen, where the number of the occurrence of the instance in the training set is concatenated (e.g. "$<$ |startoftext| $>$ |i|" where i is the occurrence). As the language model is fine-tuned on the training data, it can be assumed that it learns to associate the unique token with the respective instance. Thereby, the model is able to recognize the prefix and completes it on the basis of memorization. Ultimately, this implies a strengthened label preservation. However, so that the data is not completely reproduced from memory, uncertainty is introduced in the sampling by adjusting the temperature parameter.

Filtering the generated data is the final heuristic to increase label preservation. For this purpose, document embeddings for each instance of the generated texts and training data of a class are created. The embeddings reflect the content of the respective instances. If in this latent space a data instance from the generated data X_{gen} is too far away from the actual training data X_c of the class to be augmented, it can be assumed that the content differs semantically and/or syntactically, which is why such data is discarded:

$$X_{\text{filtered}} = \{x_i \in X_{\text{gen}} \mid \text{dist}(\text{Emb}(x_i), \text{Centroid}(\text{Emb}(X_c))) < \delta\} \quad (9.2)$$

The large generative model is able to interpolate textual content in a sensible and non-trivial way. These capabilities are very promising for data augmentation by creating highly diverse samples that are coherent and contain new linguistic and semantic patterns with regards to the actual data. However, only through the application of the safety steps can the model generate class-related data that does not represent the wrong label.

9.3.2 Implementation

Figure 9.1 shows and summarizes the three different steps of the safety enhancement, sorted according to algorithmic order. The class safety of the procedure can be significantly increased by this, although it cannot be completely ruled out that the correct label is obtained. For the implementation, we use GPT-2 by Radford et al. [340] with 355 million parameters. We used GPT-2 as it is well suited for small-data analytics due to its diverse generation capabilities coming from the size. The model is enriched with the three different extensions discussed in the conceptual design section.

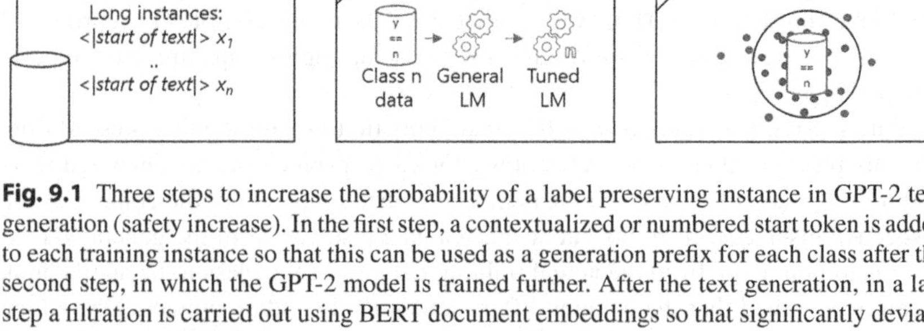

Fig. 9.1 Three steps to increase the probability of a label preserving instance in GPT-2 text generation (safety increase). In the first step, a contextualized or numbered start token is added to each training instance so that this can be used as a generation prefix for each class after the second step, in which the GPT-2 model is trained further. After the text generation, in a last step a filtration is carried out using BERT document embeddings so that significantly deviant instances are not included

In the first steps, the GPT-2 model is imported and the specific class data is extracted. Subsequently, all instances of this class are given a prefix token ("< |startoftext| > |{num}|") and suffix token ("< |endoftext| >"). If the training data consists of longer instances with an embeddable context, the "|{num}|"-field is removed. In all other cases "{num}" is replaced by the position of the current data instance. Afterwards, the model is fine-tuned with this data several hundred or thousand epochs, dependent on the dataset size, so that the loss of the model is greatly reduced. This should sufficiently ensure that the model prioritizes the training data in the generation.

Thereafter, texts are generated for each class instance. If the training instances exceed a certain number of words they are considered as long and the token "< |startoftext| >" and a specific context of the document (e.g. title or the first words) are added to the generation, else "< |startoftext| >" is used in combination with the index of the respective instance. Temperatures between 0.7 and 0.9 should be set in the generation step [489], whereby a higher number represents greater randomness/creativity. In the last step of the procedure, the generated data is filtered. This is done by using Sentence-BERT [356] to create document embeddings of the data. Generated instances that are, according to a manually set threshold, too far away from the centroid of the correct data, are deleted from the result set. To minimize this interaction, a predefined value (e.g. 0.3) is set and the algorithm displays the 10 furthest instances that are still within this threshold. Depending on how many instances are wrong, the threshold is moved further and the process is started again. For example, if there is one false instance in the set of 10 generated instances, the threshold is increased slightly. If there are only true instances, the threshold is decreased. This is done until a meaningful parameter is found.

The algorithms for long and short instances are given in Appendix A.2.3 of the Electronic Supplementary Material.

9.4 Evaluation

9.4.1 Selection of Application Domains

To conduct our evaluation, we selected the three cases of sentiment analysis, news classification and crisis informatics. First, sentiment analysis is, according to Medhat, Hassan, and Korashy [280], the analysis of opinions, attitudes and emotions toward individuals, events or topics. It is a very common NLP task and used in a broad variety of applications, as, for example, the decision making process of organisations and individuals is increasingly dependent on public opinions [253]. As it is part of several textual machine learning benchmarks and highly used in research, it is a sensible task for the experiments of this paper. To place this experiment in the setting of small-data analytics, an artificially downsampled version will be used.

Second, due to the constantly growing number of news items and their information sources, it is becoming more and more complicated keeping track of topics and finding specific articles [56]. Classifiers are used to automatically divide news into predefined classes [217]. However, news are highly dynamic, so that the source domains are constantly shifting and new classes emerge. As an extreme example in this context, one can compare the news landscape before and after COVID-19 occurred. Such shifts and newly emerging topics have as result that new data has to be labeled all the time, leading to classifiers with small data bases. While this is already sufficient to form a focus in this study, we are also interested in exploring classification as well as data augmentation for large texts, as this receives little attention in research.

Third, the research field of crisis informatics draws on computing and social science perspectives to study the ways in which ICT enables, constrains, and mediates human practices related to crisis and disaster [411]. Besides the topics of crisis communication, community interaction, and inter-organizational collaboration [360], crisis informatics examines the application of machine learning to reduce the information overload of irrelevant information, extract useful information from social media (e.g., eyewitness reports, multimedia files), and enhance information quality for both an improved situational awareness and decision making of emergency services [192]. Despite the considerable volumes of social big data disseminated during large-scale emergencies, there is a class imbalance since only a small number of social media posts contribute to situational awareness [10]. Furthermore, emergency services such as fire or police departments often lack the financial and personnel resources to engage in comprehensive dataset labeling tasks [167, 359]. In contrast, there might be a lack of available raw data in small-scale and uncommon types of emergencies, qualifying crisis informatics as an interesting application field for data augmentation and small-data analytics.

9.4.2 Model and Datasets

In accordance with the research questions, the previously conceptualized and implemented data augmentation methods are evaluated in this chapter based on a constructed low-data regime with the SST-2 dataset (Results I) and real-world low-data regimes regarding topic classification of long (Results II) and short texts (Results III). We use the ULMFit model by Howard and Ruder [156] that consists of a pre-trained encoder coupled with a linear pooling network and a softmax output. The encoder is fine-tuned for each task on all the available task-specific data (including the augmented instances). Then, the whole network is trained on a supervised task.

For the evaluation of the context independent method with short texts, we focused on sentiment analysis and the classification of crisis Twitter data. Sentiment analysis will be performed with subsampled SST-2 [410] datasets to simulate a low-data regime on a standardized dataset, similar to Hu et al. [160] and Kumar, Choudhary, and Cho [223]. As these constructed conditions are restricted in their real-world applicability, we perform further evaluations with real-world low-data regimes.

The crisis classification tasks have very limited resources, as described by Kaufhold, Bayer, and Reuter [198]. The first three datasets from Olteanu, Vieweg, and Castillo [313] are labeled according to whether or not they are informative on the specific topics that are related to the Boston Bombings, the Bohol Earthquake, and the West Texas Explosions in 2013. The other two datasets from Schulz, Guckelsberger, and Janssen [388] consist of city-specific Twitter posts that are labeled as incident-related or not.

For the evaluation of the contextual method with long texts, we gathered news articles for topic classification from 2019 and 2020. For the contextual start token we used the respective titles. Topic classification in the news context also faces the problem of few data instances, because the news and dependent topics are often very dynamic and research data is limited. In our case, for every topic, expert groups of two people decided whether the inspected article is relevant to the topic. Furthermore, there is a labeling guideline for every topic so that disagreements are excluded if possible (see Appendix A.2.2 of the Electronic Supplementary Material). The topics include three economic issues: layoff, management change (MC) and mergers and acquisitions (M&A) as well as two crisis issues: flood and wildfire.

Before augmenting the data, a fifth of every set was split into a hold-out set.

9.4.3 Evaluation Settings and Pre-Evaluation of Hyperparameters

All data augmentation methods are compared against a baseline with the help of 10-fold training executions. Additionally, the sentiment analysis results are compared with the EDA data augmentation method from Wei and Zou [481]. For the sentiment analysis task we augment both classes, while for the others we augment the minority class.

The text generation process offers various possibilities for hyperparameter optimization, which we evaluated with different datasets to avoid overfitting. Ablation studies are shown on the sentiment dataset (Sect. 9.4.4.1). During generation, the model has a temperature parameter which, the larger it is chosen, the more creative the texts will be and new linguistic patterns occur. However, a too high value can lead to instances that are not topic related. A value that is too low can mean that the model repeats itself very often, while a value too high can result in a loss of the actual theme in the texts [202]. According to the author of the implementation used in this paper, the most suitable value is between 0.7 and 0.9 [489]. In an evaluation with the management change topic, 0.7 was most suitable for the existing case of application, compared to 0.8 and 0.9.

The filtering of generated documents is of importance in this process, as GPT-2 can generate novel instances that may have no relation to the actual class. The parameter of this filtering was chosen individually so that the ten most distant documents would still be labeled accordingly to the class. In the ablation studies of Sect. 9.4.4.1 we show that this filtering is necessary to achieve the high results of this method.

Another aspect that can be added to hyperparameter optimization concerns the number of documents to be generated per training instance. As this is a very important factor we consider it in the evaluations of Sect. 9.4.4.1. More details on the hyperparameters of the model can be found in the appendices.

9.4.4 Results I: Sentiment Analysis (context independent method)

The high quality of the augmented data from a human perspective is also reflected in the quantitative evaluation results, that are presented in Table 9.1. The proposed data augmentation method has almost in every case better results than the baseline and the EDA method by Wei and Zou [481]. Particularly it gains the best improvements the less data is available (additive up to 15.53% and 3.56% compared to the baseline and EDA). However, even with the most data, the augmentation method achieves additive performance improvements in the best run of 0.49% and 1.22% compared to the baseline and the EDA method.

The method has the highest improvements if less data is available because the prior knowledge of the GPT-2 model is most effective there. The model also produces well written instances which is not the case with the EDA method that sometimes fails to improve the baseline. Furthermore, the rationale of low-performing data augmentation methods of Longpre, Wang, and DuBois [265] comes into play. In

Table 9.1 Accuracy of the non-contextual text generation process, EDA [481] and the baseline with regard to different SST-2 subsamples (10 runs)

Dataset	Run	Baseline	EDA	Text Gen
SST-2 100	AVG (SD)	0.5581 (0.0463)	0.6934 (0.0124)	**0.7134** (0.0207)
	Best	0.6226	0.7139	**0.7495**
SST-2 300	AVG (SD)	0.7241 (0.0119)	0.7217 (0.0047)	**0.7402** (0.0067)
	Best	0.7417	0.7295	**0.7534**
SST-2 500	AVG (SD)	0.7505 (0.0077)	0.7534 (0.0074)	**0.7598** (0.0126)
	Best	0.7651	0.7671	**0.7754**
SST-2 700	AVG (SD)	**0.7646** (0.0054)	0.7578 (0.0038)	0.7627 (0.0066)
	Best	0.7705	0.7632	**0.7754**

contrary to the EDA method, the proposed augmentation algorithm enriches the training data with new linguistic patterns that have not already been seen by the encoder.

9.4.4.1 Ablation Studies

Further, we want to show an excerpt of the relevant results from our ablation evaluations (Fig. 9.2). First, we test different augmentation sizes that are shown on the SST-2 100 dataset. We limit this evaluation to a maximum of 10 augmentation samples per instance, as higher numbers demand more computing time. It is evident that the higher the size, the better the results are. A human inspection indicates, however, that higher numbers might not be as beneficial since the repetition within the samples per instance increases. The human inspection process is detailed in Section A.2.5 in the Appendix of the Electronic Supplementary Material.

Furthermore, we also removed the steps of the augmentation process to see the contributions of each. In a first testing case, we did not include the number of the instance in the fine-tuning and generation phase (indicated by "w/o n." in Fig. 9.2). The decrease in the average accuracy by 5.42 points shows that this component is highly important for the whole augmentation process. This also applies to the last step of the augmentation method (indicated by "w/o f." in Fig. 9.2). Without the manual filtering the average accuracy is reduced by 2.64 points. This insight was already noticed when the filtering parameter was chosen in every task and some instances seemed to be unrelated to the class.

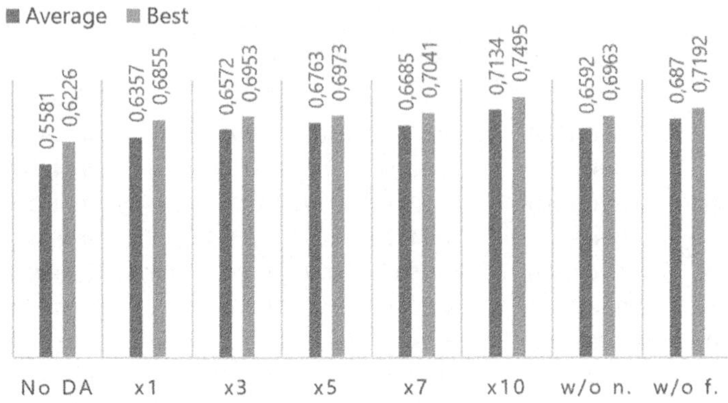

Fig. 9.2 Evaluation of different augmentation sizes and the omission of the numbering token (w/o n.) and the filtering (w/o f.) step on the SST-2 100 dataset (10 runs)

In summary, this indicates that all the steps of the augmentation process need to be included to reach the highest scores. The next evaluation studies are based on this best combination.

Furthermore, we conduct an error analysis by inspecting the generated instances of our technique and comparing them to the EDA method. For our method, it is not clear from which original instance the generation originates, as it could be an interpolation of more than one instance. Nevertheless, we try to find the closest original instances by measuring the resemblance by Levenshtein distance.

We select some insightful examples, which are presented in Fig. 9.3. The first example shows that EDA is able to keep the label, but substitutes a word that should not be substituted (*oscar wilde play* → *academy award wilde play*). On the other hand, our method is able to find suitable words for the replacement and expands the original instance in a meaningful way without distorting the label or the general content. Nevertheless, while we did not see this case in our analysis, it might be possible that the language model is not able to infer the label during fine-tuning and that the augmentation would change the label. A similar case can be seen regarding the Dublin task of Sect. 9.4.6, where we assume that the model is not able to infer the label due to a very high diversity in the instances and the label space. While the model does not incorrectly change the label, we see that more than 50% of the generated instances are "????????????". The more data is generated, the more such instances are created and the content-rich instances are repeated.

Original:	smart, sassy interpretation of the oscar wilde play.
Our Method →	harp, sassy interpretation of the oscar wilde play, with an unexpected twist.
EDA →	smart sassy interpretation of the academy award wilde play
Original:	RT @BBCBreaking: 12 confirmed dead, approximately 200 injured in #West fertiliser plant explosion in Texas, say state officials
Our Method →	Searchers Find 12 Bodies After Texas Explosion http://.. (CNN) - Hundreds believed injured in Texas fertilizer plant explosion, medical examiner says.
EDA →	rt bbcbreaking: 12 confirmed dead approximately 200 injured in west fertiliser plant explosion in texas atomic number say state official

Paraphrasing attempt   Novelty attempt

Fig. 9.3 Two instances and their transformations by EDA and our method. The first is from the SST-2 task 9.4.4 and the second from the West Texas Explosion task 9.4.6. Text passages where each augmentation method attempted to paraphrase the original instance are highlighted in blue; attempted interpolations or introduction of novelties are highlighted in green

The second example displayed in Fig. 9.3 shows that our method brings a high variation with sensible content but also reformulates "say state officials" to "medical examiner says" which might not be correct. For the task at hand, it is not decisive but it might be with regard to very specific tasks. Moreover, EDA sometimes changes the instance so that it is grammatically incorrect, as it can also be seen in this example. This can be a problem, for example, when using language models that do not expect a specific expression, such as "atomic number", in the context of this instance. Furthermore and even worse, it might happen that the method deletes an essential word, like "not", in the sentence "This movie was not bad", creating an instance with a wrong label when used for sentiment classification.

These examples show that the method proposed in this paper is able to create high-quality and diverse instances. The EDA method instead sometimes creates instances that are wrong or not fitting into the context due to its random character, which is especially critical when using pre-trained language models. Section A.2.5 in the Appendix of the Electronic Supplementary Material provides further examples and analyses of our data augmentation method. It is shown, for example, that the model sometimes completely replicates instances of the training data. This property is not necessarily bad, as it can at least be seen as a sophisticated oversampling method that clones the very important data.

9.4.5 Results II: News Classification (context dependent method)

A qualitative inspection of the data generated for the news dataset shows that high-quality, coherent, and diverse texts have been generated for the different article titles. Nearly all instances had a clear reference to the actual class. In the filtering step, primarily instances where the GPT-2 model frequently repeated words were sorted out.

The classification results of our contextual approach are presented in Table 9.2 in comparison to the baseline. Looking at the overall evaluation across all topics, it can be seen that the classifiers also achieve very good results through the contextual data augmentation method. Especially for the economic tasks, relative increases

Table 9.2 Accuracy and F1 scores of the contextual text generation process and the baseline with regard to the five news article topics (10 runs)

Dataset	Run	Accuracy	F1
Layoff	AVG (SD)	0.8350 (0.015)	0.7695 (0.012)
	Best	0.8545	0.7905
with DA	AVG (SD)	**0.8706** (0.009)	**0.8179** (0.011)
	Best	**0.8848**	**0.8354**
MC	AVG (SD)	0.8760 (0.005)	0.7217 (0.021)
	Best	0.8809	0.7627
with DA	AVG (SD)	**0.8853** (0.015)	**0.7559** (0.031)
	Best	**0.9077**	**0.8052**
M&A	AVG (SD)	0.8926 (0.005)	0.6953 (0.011)
	Best	**0.8999**	0.7075
with DA	AVG (SD)	**0.8975** (0.003)	**0.7095** (0.011)
	Best	**0.8999**	**0.7266**
Flood	AVG (SD)	**0.8462** (0.007)	0.8779 (0.007)
	Best	0.8540	0.8867
with DA	AVG (SD)	0.8408 (0.010)	**0.8804** (0.006)
	Best	**0.8594**	**0.8931**
Wildfire	AVG (SD)	0.9287 (0.016)	0.9253 (0.017)
	Best	**0.9419**	**0.9395**
With DA	AVG (SD)	**0.9312** (0.007)	**0.9297** (0.006)
	Best	0.9395	0.9375

in the maximum F1 value of over 4% for the MC and layoff topic and 2% for the M&A task can be recorded. The M&A evaluation illustrates the importance of the F1 measure. While good improvements in accuracy were also achieved for management change and layoff, there is virtually no change in the M&A topic. However, the more significant F1 measure increases substantially. A small improvement is also evident for the flood topic (+0.64 with the maximum F1 value). For the wildfire topic, the text generation approach does not achieve better results. The values are almost the same as for the baseline. However, the standard deviations of the results from the baseline are more than twice as high as with the data augmentation method.

It can be ascertained that the contextual text generation approach is very well suited for the present topic classification tasks. A possible bias regarding the economic topics could be attributed to the pre-trained GPT-2 model. The model was trained with documents from outgoing Reddit links. There may be comparatively few crisis data among the approximately 8 million documents so that the model is less able to represent this topic area. Another possible explanation for the smaller values of the crisis topics is the already very high classification quality of the two tasks. The flood topic, which was chosen because of the poorer classification quality compared to the wildfire topic, still exceeds the three economic topics in the F1-measure. The wildfire topic is by far the best task for the classifiers (about +5% F1-measure compared to the flood task). An improvement of the two topics may not be possible anymore because the dataset can contain errors as some labeling decisions are difficult and subjective.

9.4.6 Results III: Crisis Informatics (context independent method)

In the last section, we stated that GPT-2 might be less usable for crisis data. Since the usage of machine learning in crisis situations is very promising and getting good models is an ongoing issue due to little data and the challenge of domain adaption, we further examine this consideration by focusing only on crisis data for the second evaluation.

Inspecting the newly generated data, we see that the model often produces identical outputs for different runs. The generated data of the context dependent models of the second evaluation is clearly more diverse. However, the context independent method also seems to have the potential to perform very well, as Table 9.3 shows. It is especially beneficial for the classification tasks of Olteanu et al. (2015) (first three datasets). The averages and best runs outperform the baseline additively by 2.1% to 3.8% and 1.5% to 2.5% in the F1-measure. For the tasks of Schulz, Guckelsberger,

Table 9.3 Accuracy and F1 scores of the non-contextual text generation process and the baseline with regard to the five crisis Twitter topics (10 runs)

Dataset	Run	Accuracy	F1
Boston	AVG (SD)	0.7886 (0.019)	0.7344 (0.030)
Bombings	Best	0.8062	0.7720
with DA	AVG (SD)	**0.8003** (0.021)	**0.7588** (0.024)
	Best	**0.8311**	**0.7979**
Bohol	AVG (SD)	0.9097 (0.014)	0.8857 (0.021)
Earthquake	Best	0.9302	0.9126
with DA	AVG (SD)	**0.9238** (0.011)	**0.9062** (0.015)
	Best	**0.9399**	**0.9277**
West Texas	AVG (SD)	0.8486 (0.020)	0.8340 (0.025)
Explosion	Best	0.8804	0.8765
with DA	AVG (SD)	**0.8755** (0.010)	**0.8721** (0.011)
	Best	**0.9004**	**0.8970**
Dublin	AVG (SD)	**0.9893** (0.002)	**0.9199** (0.015)
	Best	**0.9912**	**0.9351**
with DA	AVG (SD)	0.9858 (0.002)	0.8945 (0.014)
	Best	0.9878	0.9116
New York	AVG (SD)	0.9302 (0.016)	0.8428 (0.027)
City	Best	**0.9463**	**0.8701**
with DA	AVG (SD)	**0.9346** (0.003)	**0.8472** (0.007)
	Best	0.9385	0.8555

and Janssen [388], however, the method does not provide any substantial improvements regarding the direct scores. For the Dublin dataset, the augmentation method even seems to have a negative effect. Yet, a decrease in the F1 standard deviation is achieved for every task.

Although we can confirm the assumption of the sentiment analysis that the context independent variant of the augmentation method creates less diverse instances, it analogously has a very positive impact on the classification quality when applied to real-world low-data regimes. Especially the narrowly defined problems by Olteanu, Vieweg, and Castillo [313] are well suited, leading to the consideration that the difference of the results lies in the nature of the problems. While the first three tasks are bond to a special crisis event, the two other tasks are just incident related with no other focus than the respective city. It may be that these two tasks are too broadly defined with very different instances so that the model was not able to properly fine-tune to generate sensible instances. Apart from that, the gain in robustness is clearly visible because a decrease in the F1 standard deviation is achieved on each task. The evaluation also shows that our data augmentation methods can achieve good results not only on economic topics.

9.5 Discussion and Conclusion

While there are numerous beneficial data augmentation methods in computer vision, textual transformations are more difficult to define [209, 481] and often result in mixed results [265]. In order to address these issues, we presented two data augmentation methods for long and short texts based on text generation techniques to enhance the knowledge base on small-data analytics. Our results on 11 datasets, which are listed in an aggregated form in Table 9.4, contribute to answering the following research questions.

Table 9.4 Average and maximum F1 performance deltas of the data augmentation methods in comparison to their respective baseline counterparts across all datasets

Dataset	Delta Avg.	Delta Max
SST-2 (100) – Acc.	+15.53%	+12.69%
SST-2 (300) – Acc.	+1.61%	+1.17%
SST-2 (500) – Acc.	+0.93%	+1.03%
SST-2 (700) – Acc.	−0.19%	+0.49%
Layoff – F1	+4.84%	+4.49%
MC – F1	+3.42%	+4.25%
M&A – F1	+1.42%	+1.91%
Flood – F1	+0.25%	+0.64%
Wildfire – F1	+0.44%	-0.20%
Boston Bombings – F1	+2.44%	+2.59%
Bohol Earthquake – F1	+2.05%	+1.51%
West Texas Explosion – F1	+3.81%	+2.05%
Dublin – F1	-2.54%	-2.35%
New York City – F1	+0.44%	-1.46%

How can we utilize text generation approaches of data augmentation that achieve a high novelty in the data while preserving the label quality to improve pre-trained machine learning classifiers (RQ1)? We proposed two data augmentation methods that are based on text-generating language models. We constructed three different steps (see Fig. 9.1) for ensuring a high label preservation within the transformations of these models. As a first step, the data was primed by a special token. This token can signal the model to generate training data for this class in the generation phase. In order to ensure that the model is familiar with the class data and the token, fine-tuning was carried out with the prepared data in the second step. After the generation, filtering based on the BERT document embeddings [356] formed the final step. In our experimental setup, we used a pre-trained encoder and fine-tuned it on the various tasks, including the augmented data, to face the challenge of creating a sophisticated data augmentation method that also performs well on pre-trained models. The two derived methods achieved very good results in the evaluation phase with several performance gains and reductions of the standard deviations.

When evaluating the utility of the algorithm, however, further criteria must be considered. While it is rather easy to embed the augmentation method into a classification process, the GPT-2 model needs some time to be executed. The generation of one example of a long dataset took about 10-30 seconds. Noise-inducing methods such as EDA take much less than a second to complete an instance, as they only perform simple operations such as changing the order of words, deleting some words, or generating misspellings [33, 81, 162, 336]. However, text generation methods such as ours are limited by the time required by the generation process. Nevertheless, compared to the time it takes a human to label a new instance, our method is still very advantageous. The time required can also be significantly reduced, for example, by using a different language model that is faster. Furthermore, the used GPT-2 model is mainly limited to English, making it less usable for multilingual tasks. However, this can also be mitigated by using another language model as the proposed method is suitable for different language models.

In which way is the incorporation of contexts of long text instances in classification problems helpful when using text generation as data augmentation method (RQ1.1)? When dealing with long texts, we decided to integrate a context-based token in the generation phase so that the generated texts are more explicit for the respective instance and highly diverse among all instances. A closer look into the generated samples from the evaluation phase confirmed this assumption. Furthermore, the generated instances seem to be very coherent and related to the task at hand, due to the strengths of the GPT-2 language model. More importantly, four of the five tasks could be improved by including the newly generated and filtered instances. This led to an additive increase in the average and maximum F1 value of up to 4.8% and 4.5% respectively. However, we noticed that the augmentation technique could not improve the classification results when the classifier already performs very well without additional data.

How is it possible to achieve a quality improvement for classification tasks with short texts when augmenting with text generation (RQ1.2)? For classification tasks with short texts, a context-based token integration is not possible, wherefore we included the number at which the respective instance occurred in the fine-tuning. On closer inspection of the newly created instances, several duplicates were found. This did not have negative implications for the evaluation, since repeating some training examples resembles the process of simple oversampling. It may even be interpreted as a more sophisticated version, where some examples are completely new and the others are oversampled from the most fitting data points. Accordingly, a great performance gain could be achieved in the constructed and real-world low-data regimes, leading to improvements of up to 15.53 and 3.81 points respectively. We noticed that this augmentation method was not suitable for two special real world tasks. We hypothesize that these two tasks are too broadly defined on the rationale that the GPT-2 model is not able to infer the right context just based on the fine-tuning of the data.

9.5.1 Empirical, Practical and Theoretical Contributions

Considering our findings, the study revealed practical, theoretical and empirical contributions:

New data augmentation methods based on the GPT-2 language model. The evaluation results of the data augmentation methods indicated that the GPT-2 model, in combination with three safety steps, can achieve a considerable improvement in the text classification tasks (see Table 9.1 to 9.3). In contrast to the similar approaches of Wang and Lillis [467] and Anaby-Tavor et al. [18] that utilized GPT-2 for text generation too, our method is more generally applicable and offers more safety steps. Wang and Lillis [467] describe no measures for label preservation in their approach, and Anaby-Tavor et al. [18] only enable data augmentation for instances consisting of one sentence. Furthermore, we include a filtering mechanism which includes the human expertise, strongly increasing the diversity without much supervision. The advantages of the text generation approach proposed here facilitate a wide use, qualifying it as basic element for further adaption in prospective classification applications.

A textual data augmentation basis which is beneficial for pre-trained classification models. Longpre, Wang, and DuBois [265] show that data augmentation might not be helpful when dealing with state-of-the-art pre-trained models. This seems logical from a theoretical perspective since pre-training and the transfer to new tasks also follow the goal of reducing the amount of necessary training data. In order to get an enrichment anyway, sophisticated augmentation methods are needed, which should provide unseen linguistic patterns that are relevant to the task [265]. We addressed this issue by leveraging the GPT-2 model that was trained with more than 8 million web pages. This gives the great potential to include new linguistic patterns in the generated data (example instances can be found in the Appendix A.2.5 of the Electronic Supplementary Material). For creating task relevant data, we derived three steps that increase the possibility of class related content. In the evaluation we showed that the proposed method is able to improve the pre-trained encoder model. In contrast to Longpre, Wang, and DuBois [265], we did not test the method on a transformer model. However, we trained the pre-trained ULMFit encoder for both testing cases (no augmentation and augmentation) with the augmented data so that for the encoder no data is unseen beforehand.

Empirical insights into the domain-specific application of small-data analytics. In this work, we gathered new empirical insights into the application of data augmentation in the research domains of sentiment analysis, news classification, and crisis informatics. In crisis informatics, various studies have examined the use of domain adaptation, transfer learning, active learning, and online learning to reduce the labeling effort [167, 198, 305]. However, few research has examined the application of textual data augmentation for crisis management [467], which we enhance by the evaluation and interpretation of seven augmented datasets. For sentiment analysis we constructed a low-data regime, like Kumar et al. [224] and Hu et al. [160]. Small-data analysis research is gaining popularity and there is a need to establish datasets that can be used, understood and compared by all types of researchers. We strengthen this research direction by basing our evaluation on this dataset. Furthermore, with the five news classification datasets we are exploring the important topic of long text classification, that is often neglected in research. News classification is mostly done with short descriptions of the articles only, as in the AG News dataset [526].

9.5.2 Limitations and Outlook

During the evaluation, a restriction was made with regard to the language model used. While applicable with other language models, it is not clear if the performance gain remains the same. The GPT-3 model by Brown et al. [47] seems to be the next sensible choice for increasing the results. However, the fine-tuning step that is likely to be necessary is currently not possible due to the high resource utilization of the model. Nonetheless, because of its size and linguistic expressiveness, it may be especially helpful to address the challenge stated by Longpre, Wang, and DuBois [265] in which pre-trained models might not gain any improvement from data augmentation. In relation to the study by Longpre, Wang, and DuBois [265], future research could test the proposed method with transformer models. It would also be interesting to see how smaller language models perform, that may be much faster. In addition, there might also be an option to fully automate the filtering step, which further increases the universal usability, even if the human effort is already very low now.

Despite many efforts in data augmentation, the big data wall problem addressed at the beginning is still of great relevance. However, if in the future, according to various assumptions, very large models, such as GPT-3 by Brown et al. [47], are better able to solve these problems, the high resource wall problem opens up, which only allows large companies to train and use these models.

Acknowledgements This work has been co-funded by the German Federal Ministry of Education and Research in the project CYWARN (13N15407) and by the LOEWE initiative (Hesse, Germany) within the emergenCITY center. The calculations for this research were conducted on the Lichtenberg high performance computer of the TU Darmstadt.

Design and Evaluation of Deep Learning Models for Real-Time Credibility Assessment in Twitter

10

Abstract

Social media has an enormous impact on modern life but is prone to the dissemination of false information. In several domains, such as crisis management or political communication, it is of utmost importance to detect false and to promote credible information. Although educational measures might help individuals to detect false information, the sheer volume of social big data, which sometimes need to be analysed under time-critical constraints, calls for automated and (near) real-time assessment methods. Hence, this paper reviews existing approaches before designing and evaluating three deep learning models (MLP, RNN, BERT) for real-time credibility assessment using the example of Twitter posts. While our BERT implementation achieved best results with an accuracy of up to 87.07% and an F1 score of 0.8764 when using metadata, text, and user features, MLP and RNN showed lower classification quality but better performance for real-time application. Furthermore, the paper contributes with a novel dataset for credibility assessment.

Original Publication *Marc-André Kaufhold, Markus Bayer, Daniel Hartung, and Christian Reuter. "Design and Evaluation of Deep Learning Models for Real-Time Credibility Assessment in Twitter". In:* Artificial Neural Networks and Machine Learning—ICANN 2021—30th International Conference on Artificial Neural Networks, Bratislava, Slovakia, September 14–17, 2021, Proceedings, Part V. *Ed. by*

Supplementary Information The online version contains supplementary material available at https://doi.org/10.1007/978-3-658-48778-2_10.

M. Bayer, *Deep Learning in Textual Low-Data Regimes for Cybersecurity*, Technology, Peace and Security I Technologie, Frieden und Sicherheit, https://doi.org/10.1007/978-3-658-48778-2_10

Igor Farkas, Paolo Masulli, Sebastian Otte, and Stefan Wermter. Vol. 12895. Lecture Notes in Computer Science. Springer, 2021, pp. 396–408. DOI: https://doi.org/10.1007/978-3-030-86383-8_32. URL: https://doi.org/10.1007/978-3-030-86383-8_32 [1]

Note Supplementary material containing descriptions of the dataset developed and further results can be found in Appendix A.3 of the Electronic Supplementary Material.

Material
Application and Experimental Code: https://github.com/PEASEC/Deep-Learning-Credibility-Assessment

10.1 Introduction

Social media is an integral part of modern everyday life as they allow the creation and exchange of user-generated content. Besides everyday life, social media is used by journalists for reporting, analysing, and collecting information, by organisations to monitor customer feedback and sentiment, but also by citizens and emergency services to gain situational awareness in conflicts and disasters [192]. On the contrary, social media is prone to the dissemination of (potentially) false information, including conspiracy theories, fake news, misinformation, or rumors [432]. While counter-measures such as gatekeeping information, increasing media literacy, or passing new laws seem to be promising approaches [191], the sheer volume of *big social data*, which sometimes needs to be analysed under time-critical constraints, calls for automated and (near) real-time credibility assessment methods. Thus, a multitude of different machine learning approaches were established to automatically distinguish false and credible information [332, 404, 458]. Despite their merits, when reviewing existing deep learning approaches for credibility assessment in social media, we found that most approaches provided binary or multi-class models but did not allow a steady (e.g., percentage) prediction of credibility. Furthermore, most approaches require extensive computations, thus lacking the ability for real-time application in social media, and, to our best knowledge, none of these approaches incorporated previous posts of the user into their analysis. Thus, the

[1] Minor typographical errors in the original publication have been corrected for this dissertation.

paper seeks to answer the following research question: **Which deep learning models and parameters are suitable for real-time credibility assessment in Twitter?**

By answering this research question, the paper makes several contributions. It (i) conducts a review of existing credibility assessment methods (Sect. 10.2), (ii) presents the design and fine-tuning of three deep learning models for credibility assessment, (iii) provides a novel dataset for credibility assessment in Twitter (Sect. 10.3), and (iv) evaluates the quality of the designed models, also examining the usefulness of incorporating previous user posts into credibility assessment (Sect. 10.4). The paper finishes with a discussion of the findings and implications and highlights possible limitations and potential for future work (Sect. 10.5).

10.2 Related Work

Since the study of *credibility* is highly interdisciplinary, there is no universal definition for it [117]. However, it can be understood as a measure which comprises both objective (e.g., useful, good, relevant, reliable, accurate) and subjective (e.g., a perception of the receiver) components. Credible information is characterized by trustworthiness (unbiased, true, good purpose) and expertise (competence, experience, knowledge) [119]. When estimating the credibility of information in social media, users are confronted with different types of harmful information [432] that can be distinguished by the *intention of the publisher* (i.e., intentional or non-intentional) and the *truth of content* (i.e., true or false) [117]. Both disinformation and misinformation are objectively false, but only disinformation (often referred to as fake news [235]) is published intentionally false. Moreover, rumors are statements that cannot be immediately verified as either true or false [335].

Amongst others, harmful information is disseminated to manipulate political elections and public opinions or to generate financial revenues [11]. Moreover, false information might affect the decision making of emergency services in conflicts or disasters, effectively contributing to the loss of lives. Countermeasures against harmful information comprise the gatekeeping of information by media, increasing the media literacy of citizens, passing new laws and regulations, or detecting harmful information via algorithmic detection approaches [191]. When reviewing literature on credibility assessment in social media (Table 10.1), we did not only find individual systems but also interesting survey papers comparing different machine learning approaches [332, 404, 458]. The approaches are primarily based on Twitter data, often attributed to different *domains*, including credibility, fake news, or rumors, and have a different *scope of analysis*.

Table 10.1 Comparison of ML credibility classifiers. Sometimes not all methods are listed (*). Abbrev.: Decision Tree (DT), Decision Rule (DR), Bayesian Network (BN), Bayes Classifier (BC), Support Vector Machine (SVM), Random Forrest (RF), Convolutional/Recursive Neural Network (C/R/NN), Naive Bayes (NB)

Ref.	Domain	Scope	Output	Realtime	Methods
[57]	Credibility	Event	binary	no	DT, DR, BN, SVM
[58]	Credibility	Event	binary	no	RF, LR, *
[50]	Fake News	Event	binary	no	RF, *
[166]	Credibility	Message	binary	untested	NN
[351]	Astroturfing	Mem	binary	no	DT, SVM
[139]	Credibility	Message	5 classes	untested	SVM-rank, *
[494]	Crisis Credibility	Event	binary	untested	SVM, BN, DT
[148]	Credibility	Event	tertiary	no	SVM, DT, NB, RF
[335]	Rumors	Event	steadily	no	BC
[149]	Fake News	Message, Source	binary	untested	NN, NB, DT, SVM, RF
[371]	Fake News	Event	binary	no	RNN, NN
[260]	Fake News	Message	4 classes	almost	RNN, CNN
[490]	False Information	Message	5 classes	yes	RNN, CNN

First, event-based approaches cluster social media messages into events to determine the credibility of the event [50, 57, 58, 118, 148, 494]. Second, propagation-based approaches analyse the caused engagement, such as mentions or retweets, of published messages [260, 351, 371]. Third, message-based approaches assess the credibility of individual messages, using metadata and textual features [139, 149, 166, 335, 490]. Especially *methods* based on neural networks achieved high classification performances, e.g., accuracies of 85.20% using NN [166] or 89.20% [371] using RNN and NN. Despite the variety of features involved, to our best knowledge, none of these approaches incorporated previously published messages of the user into their analysis. Further, only two of the approaches allow a near *real-time* application [260, 490], i.e., being able to classify tweets directly after their dissemination. This is the case because most approaches are event-based, requiring event detection before classification can take place, or rely on temporal features, such as the number of likes or retweets, which change over the course of time and could lead to a flawed credibility score at retrieval.

In terms of *output*, most approaches allow a binary (i.e., credible or incredible information) or multi-class credibility assessment [139, 260, 490], although some works outlined that they do not reproduce reality in a sufficient manner [58,

80]. Thus, the use of a steady regression seems promising [335] since it allows a percentage-based representation of credibility and better accounts for the subjective component of credibility [117]. Although multiple attempts have been made to establish standard datasets for credibility assessment [290, 403, 540], almost all publications used their own dataset, probably due to the methodological requirements of their approaches. The lack of standardized datasets is noticed by diverse authors, emphasizing the lack of comparability of the evaluation results of different approaches [332, 404, 458].

10.3 Concept and Implementation

Based on the outlined research gaps, we seek to implement neural network-based approaches for credibility assessment that (i) work with public Twitter data due to the ease of access, (ii) use regression to allow a steadily (i.e., percentage-based) assessment of credibility, and (iii) allow a near real-time application of the trained models. Furthermore, we intend to (iv) check if the analysis of previously published messages of a user positively impacts the performance of credibility assessment. We also (v) compose a novel dataset for credibility assessment.

10.3.1 Features and Model

In order to train our models, we reviewed the features used by previous approaches. The used features can be roughly categorized into four types: (i) **metadata features** (n=10) of a tweet provided by the Twitter API, such as hashtags, links, or mentions, (ii) computationally extracted **text features** (n=25) from the tweet's body, such as number of words, text length, or sentiment, (iii) **user features** (n=17) provided by the Twitter API, such as the number of followers or published tweets, and (iv) **timeline features** (n=140), including the maximum, minimum, arithmetic mean, and standard deviation (4 $*$) of both the metadata and text features (10 + 25) of the last 40 tweets of the user.

First, the baseline model is a simple **multilayer perceptron (MLP)** that consists of an input layer with 192 neurons for the features described before. These are projected into a hidden layer with 32 neurons with $tanh$ activations. Since the problem to solve is a regression task, a sigmoid function was selected for the output and the entire network is trained with mean square error (MSE). The layers are fully connected with a dropout rate of 0.3. Further hyperparameters of the learning process are a learning rate of 0.01, a batch size of 256, and the maximum number of epochs of 10,000, which is contained by early stopping on the development set. We choose ADAM [207] as optimizer.

Second, to extend the baseline, we embed the sentiment and textual content of a tweet with a **recurrent neural network (RNN)** and feed those tweet embeddings into the baseline model. As a first step, GloVe [330] pretrained Twitter embeddings (dimension of 50) are utilized to create word embeddings of each word in the tweet. These word embeddings are enriched by another dimension that represents the VADER [164] sentiment value. Every embedding is then processed by a RNN that produces a hidden state (tweet embedding) that serves as another input into the baseline MLP.

Third, another approach to extend the baseline is to use fine-tuned **BERT embeddings** [89]. We replaced user mentions, URLs, and emoticons with special tokens. Then we fine-tuned the base BERT model with its CLS token as output with the training data (batch size: 16, learning rate: $5 \cdot 10^{-5}$ and 3 epochs). The fine-tuned model is then used to produce the additional input for the baseline. The BERT connection to the baseline is regularized by a dropout connection with rate 0.3. As the dimensionality of the BERT embeddings is substantially higher, we increased the number of hidden neurons of the baseline to 128.

10.3.2 Automatic Dataset Composition

The task of credibility assessment requires much data from different topics and time frames to make the model invariant to these patterns. Accordingly, we searched for Twitter datasets that can be combined into a larger set. The **PHEME** [540] dataset contains 300 binarily annotated posts from which both the classes "true" and "false" are mapped to a credibility score of 1 and 0 respectively in our coding schema. In contrast, the **Twitter15** [259] and **Twitter16** [269] datasets are categorized into four classes. The classes "true" and "false" are mapped analogously to the PHEME dataset. For the tweets of the class "unverified", we decided for a more uncertain score of 0.3 that reflects a tendency towards dubious content. After the manual inspection of the last class "no rumor", we chose a score of 0.9 as instances of this class seem to be primarily true. This way, additional 2,308 instances were added to the corpus of this paper.

Then, we implemented an automatic coding scheme for the **FakeNewsNet** [403] dataset. It contains several topics to which a large number of tweets are assigned. For each topic, an associated headline was labeled as true or false. Since the assignment of tweets to the topics was carried out using keywords, posts may also be incorrectly assigned to a topic and a mapping from the headline label to the label of tweets in it is not possible. It also allows a topic that is annotated as "false" to contain posts that expose the topic as wrong, which is especially important in our use case. To compensate for this, we perform a temporal, keyword-based, similarity, and topic filtering, which is described in detail in the Appendix A.3.2 of the Electronic Supplementary Material. Thus, 1,378 credible and 729 implausible tweets were retrieved and mapped to a target score of 0.9 and 0.1, respectively.

We also used the **Twitter20** dataset (see Appendix A.3.1 of the Electronic Supplementary Material), where various German tweets of the COVID-19 pandemic were individually labeled. The following assumptions have to be true so that a translated version can be included into the dataset of our paper: (i) incorrect information in German and English are syntactically the same, (ii) the dataset contains only a few posts with misinformation or satire, and (iii) during the translation of the articles no linguistic properties (e.g., rhetorical stylistic devices) that are a characteristic of misinformation are lost. We used the Google Translate API and automatically corrected wrong @ and # placements, to preserve the general syntax of tweets. Since the assumptions do not necessarily have to apply, we have decided to create a "default" dataset (Fig. 10.1a) without and a "large" dataset (Fig. 10.1b) with these translated instances.

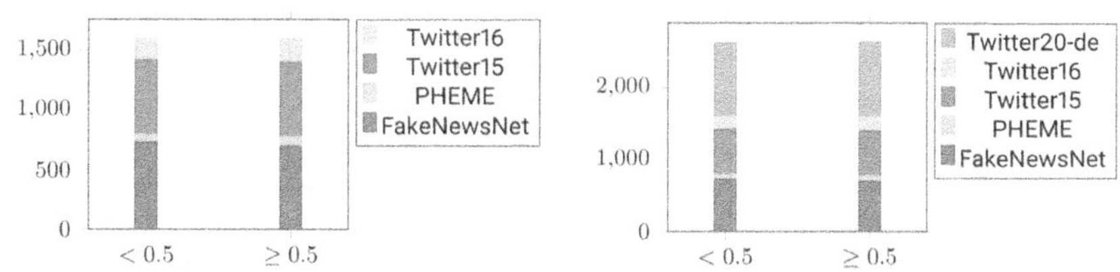

Fig. 10.1 Credibility Datasets. (a) Default dataset (N=3,178, whereof $n_{credible}$ =1,589). (b) Large dataset (N=5,225, whereof $n_{credible}$ =2,619)

10.4 Evaluation

For the evaluation of our models, we use metrics based on regression fitting to the dataset development, classification for comparisons with past and future work, and execution time for insights related to real-time application. For the classification, tweets with a score of less than 0.5 are classified as 0 and 1 otherwise. We split 80%, 10% and 10% of the posts into training, development and testing sets. We also ensure that all posts by a user who is represented more than once in the dataset are included in the training set so that no information of the other sets is already seen during training. For the implementation we used PyTorch, Huggingface Transformers and NLTK. The system used for the evaluation has an Intel i7-9750 with 6 cores and 2.6 GHz, a NVIDIA GeForce RTX 2070 graphics card with 8 GB Graphics memory and 32 GB of RAM.

10.4.1 Evaluation of Model Quality

For the evaluation of the performance of the different model architectures and feature combinations, we first tuned the hyperparameters (e.g., batch size and dropout-rate) on the development set of both datasets. We chose to proceed with the models that have the lowest MSE. The evaluation results regarding the development set can be found in Table A.5 in the Appendix A.3 of the Electronic Supplementary Material.

The testing set results on both datasets are shown in Table 10.2. It is clear to see that the adaption of the standard MLP model is very beneficial. Especially the BERT model can gain additive accuracy improvements of up to 21.63%. Looking at the feature constellations in the MLP network, it is evident that they are suitable for distinguishing credible and implausible posts without the addition of sentence or BERT embeddings (reaching up to 66.77% accuracy and 0.6513 F1 score). The Tweet, user and text feature constellation even reaches a slightly better MSE than the RNN basis model on the default dataset.

The more sophisticated BERT-based model, however, draws less benefit from the additional feature inputs. With the default set, a minimal improvement of less than 1 accuracy point is achieved, while the features for the large set even degrade the BERT model. Sometimes we noticed improvements from the timeline feature, but no significant results were found when the test set evaluation was performed. When inspecting relative changes with regard to both datasets, it becomes apparent that the BERT-based model has a greater positive impact on the first set. The additive accuracy improvements on this dataset are of up to 21.63% compared to just up to 14.12% on the large dataset.

From a dataset development perspective, one might think that the larger dataset contains more false annotated data, since the classifier scores are worse on this dataset. This can apply, e.g., if one of the assumptions given in Sect. 10.3.2 is incorrect and significant linguistic properties were lost during the translation of the Twitter20 dataset. Another consideration could be that the large dataset covers more different or domain-specific tweets; this makes classification more difficult but increases the generalizability and practicality of a classifier. When inspecting the translated posts in the Twitter20 dataset, we noticed some mistakes in the translated

Table 10.2 Results of the quality analysis per model on both datasets. Abbrev.: Tweet Features (Tw), User Features (Us), Text Features (TX), Advanced timline features (ATi)

Model	Features	MSE	Acc	Pre	Rec	F1
MLP(default)	Base	–	–	–	–	–
	Tw, Us, Tx	0.1367	0.6552	0.6323	0.6490	0.6405
	Tw, Us, Tx & ATi	0.1474	0.6677	0.6471	0.6556	0.6513
RNN(default)	Base	0.1380	0.7116	0.6879	0.7152	0.7116
	Tw, Us, Tx	0.1202	0.7367	0.7190	0.7285	0.7237
	Tw, Us, Tx & ATi	0.1199	0.7429	0.7226	0.7417	0.7320
BERT(default)	Base	0.0806	0.8621	0.8497	0.8609	0.8553
	Tw, Us, Tx	**0.0794**	**0.8715**	**0.8618**	**0.8675**	**0.8674**
	Tw, Us, Tx & ATi	0.0805	0.8621	0.8591	0.8477	0.8533
MLP(large)	Base	–	–	–	–	–
	Tw, Us, Tx	0.1803	0.6469	0.6734	0.6162	0.6435
	Tw, Us, Tx & ATi	0.1841	0.6412	0.6895	0.5572	0.6163
RNN(large)	Base	0.1720	0.7042	0.7266	0.6863	0.7059
	Tw, Us, Tx	0.1639	0.7118	0.7143	0.7380	0.7260
	Tw, Us, Tx & ATi	0.1603	0.7042	0.7538	0.6679	0.7002
BERT(large)	Base	0.1347	**0.7844**	0.7883	**0.7970**	**0.7927**
	Tw, Us, Tx	0.1339	0.7824	0.7897	0.7897	0.7897
	Tw, Us, Tx & ATi	**0.1319**	0.7824	**0.7962**	0.7786	0.7873

text. However, we tend to the second explanation as the actual content was preserved most of the time and we were still able to identify the credibility.

This consideration comes also into play when inspecting the stronger impact of the BERT-base model on the default dataset. BERT can have a major impact when it is applied to a dataset with fewer data instances as it can transfer knowledge from its previously learned tasks. The other algorithms can only get closer to the evaluation results if the dataset grows in its size, since they do not have this initial capacity.

Furthermore, the BERT model just slightly improves with the feature engineering process while the features seem more useful when applied to the other models. This might be due to the high capabilities of the pre-trained model. The incorporation of textual features might be redundant as the language model is able to identify some of these by itself. Some of the features might even be misleading and in this low-data regime unwanted statistics are more likely to appear during the training process leading to better scores for the other models that do not have the generalization capabilities of BERT. The RNN model builds upon GloVe embeddings which also impose a certain generalization that is reflected in the results. However, with this model we still expect a bias towards unwanted statistics.

10.4.2 Evaluation of Model Execution Time

To measure the execution time of our models, all tweets and previous posts of the large dataset were loaded to the RAM in order to reduce variances of HDD memory access. The individual models were executed using the whole dataset to measure the execution time of different steps, such as the model initialization, the processing time per tweet, and the processed tweets per second (see Table 10.3). For the RNN modell, there are two options to read the required embeddings: (i) a filesystem-based approach that reads and indexes the embedding file once (fs) and (ii) a memory-based approach, where the whole file is loaded into RAM. While the first approach consumes less memory and has a shorter initialization time, the second approach offers a faster access to the embeddings. For the first approach, the use of an SSD ($\approx 3,500\,{}^{\mathrm{MB}}/_{\mathrm{s}}$ reading speed) or HDD ($\approx 100\,{}^{\mathrm{MB}}/_{\mathrm{s}}$ reading speed) did not yield measurable differences in execution time.

Table 10.3 Results of the temporal analysis per model. The column *tweets/second* ignores the initialization time

	Configuration	Init.	Time/ Tweet	Tweets/ Second
Features	Text	203ms	914μs	1094
	Tweet	0s	144μs	6944
	User	60ms	130μs	7692
	Timeline	265ms	40,000μs	25
MLP	Basis	1,400ms	44μs	22,727
	Adv. Timeline	1,430ms	52,000μs	19.2
RNN	Basis (fs)	7,653ms	1,354μs	738
	Basis (ram)	32s	179μs	5586
	Adv. Timeline (fs)	7,653ms	107,000μs	9.3
	Adv. Timeline (ram)	32s	88,000μs	11.3
BERT	Basis (gpu)	3,706ms	7,495μs	133
	Basis (cpu)	3,720ms	150,000μs	6.6
	Adv. Timeline (gpu)	3,680ms	224,000μs	4.4
	Adv. Timeline (cpu)	3,695ms	>4s	$< 1/4$

The BERT model strongly benefits from GPU acceleration. For comparison, Table 10.3 highlights the execution times with (gpu) and without (cpu) acceleration by a graphics card. For other models, the use of a GPU did not yield measurable performance improvements. Generally spoken, complex models require longer execution times than simple models. The base RNN model is seven times faster than BERT using a GPU and more than 110 times faster than BERT without a GPU. Furthermore, the processing of timeline features, i.e., incorporating up to 40 previous posts of a tweet, requires significantly more time. While the RAM-based RNN model is able to classify up to 5.5k tweets per second, BERT processes up to 133 tweets per second with a GPU, but only 6.6 without a GPU. In that model, additional features show negligible impact on the overall execution time.

10.5 Discussion and Conclusions

Nowadays, social media is widely used for multiple purposes, such as relationship maintenance, journalism, customer interactions but also for crisis management. However, these activities can be severely impeded by the propagation of false information. Hence, it is important to promote credible and to counter implausible information. In this work, we reviewed existing approaches before designing and evaluating three neural network models capable of near real-time credibility assessment in Twitter to answer the following research question: **Which deep learning models and parameters are suitable for real-time credibility assessment in Twitter?**

Our findings indicate that our BERT-based model achieves the best results when using metadata, text, and user features, reaching an accuracy of 87.07% and F1 score of 0.8764 on the default dataset. In comparison to existing works, the results appear to be promising. While Helmstetter and Paulheim [149] reached an F1 score of 0.7699, Iftene et al. [166] achieved an accuracy of 85.20%. Although Ruchansky, Seo, and Liu [371] reached an accuracy of 89.20% and F1 score of 0.9840, their approach is propagation-based, thus having limited real-time capability, and classifies events instead of individual tweets. Similarly, Liu and Wu [260] reach an F1 score of 0.8980; however, their approach focuses on the detection of disinformation and also incorporates propagation-based features.

Furthermore, we compared the real-time capabilities of our three models. While our MLP baseline is capable of processing high volumes of data (>20k tweets/sec) with a low resource demand, the accuracy of up to 66.77% does not allow for a reliable classification. In contrast, our RNN model is still capable of processing high volumes of data (>5k tweets/sec when used in RAM) for classification while reaching more promising accuracies of up to 74.29%. Finally, BERT reached accuracies of 87.07% but was only able to process a considerably lower amount of data (>0,1k tweets/sec when used with a GPU). When including the previous posts of the user into computation, we did not achieve consistent improvements of classification performance; however, the real-time capability of all feature and model combinations was lost.

10.5.1 Practical and Theoretical Implications

We compared existing datasets and combined suitable ones into a **novel dataset to increase the amount of available data for model training (C1)**. Since available datasets are used for varying credibility classification tasks, several steps of transformation were required to convert them into a unified structure. Due to the combination of datasets, it comprises a richer number of users, topics, and message characteristics. Our future work will include the application of data augmentation techniques to increase the size and richness of the dataset.

We provided a **review of existing machine learning approaches for credibility assessment in Twitter (C2)**. In contrast to other works, we critically examined and compared approaches for credibility assessment in Twitter. Many of the reviewed approaches did not use a development set, relied on a small dataset, or conducted many hyperparameter optimizations, which entails the risk of overfitting. While difficult to compare, it seems that propagation-based models achieve the best results [371]; however, they lack the ability of real-time application. As the engagement based on tweets unfolds over time, propagation-based models seem promising when no time contraints are present.

In addition, our work contributes with **insights into the real-time capability of neural networks for credibility assessment (C3)**. Comparing our models in terms of real-time capability, their usefulness seems to be dependent on many factors. While our MLP baseline shows excellent execution times for large-volume data processing, the lack of classification performance disqualifies its real-world applicability. In contrast, our RNN model still offers suitable execution times and maintains a better classification performance. Finally, despite achieving the best classification results, BERT offers limited realtime capability when used for large-scale data analysis unless considerable GPU power is used for processing. In the end, we would still advise to do further research regarding the BERT model, as it has the best generalization capabilities. The credibility research shows that simple algorithms tend to be very biased towards the topics and domains in the dataset and often bahave more like a topic classifier.

10.5.2 Limitations and Outlook

While this work is subject limitations, they also offer potentials for future research. First, although BERT achieved the best classification results, it was also the slowest classifier. Variations of BERT, such as DistilBERT [379], provide smaller models or shared weightings within the model to achive a lower memory usage and faster execution time. Thus, future work could examine if they achieve comparable classification results for credibility assessment. Second, the classifier is limited by merely using a dataset based on textual Twitter data. Although it can be used for other social media, it might perform worse due to different linguistic features. Thus, the exploration of a cross-platform dataset, supported by active learning and data augmentation techniques, could be worthwhile for future research [198]. Furthermore, pictures displaying text messages (requiring optical character recognition techniques) or external sources could be incorporated into the credibility assessment concept. Third, novel but similar publications emerged during the implementation of our study. For instance, Tian et al. [440] contributed with a rumor detection algorithm that achieves an F1 score of 0.862 but does not provide a steady regression of findings. A further work used ALBERT to reduce the memory usage of BERT, reaching an F1 score of 0.795 compared to 0.71 of the original BERT model [439]. Furthermore, additional research was conducted to detect fake news spreaders by analyzing their previous posts [97, 348].

Acknowledgements This work has been co-funded by the German Federal Ministry of Education and Research in the project CYWARN (13N15407) and by the BMBF and the Hessen State Ministry for Higher Education, Research and Arts within the SecUrban mission of the National Research Center for Applied Cybersecurity ATHENE.

11 CySecBERT: A Domain-Adapted Language Model for the Cybersecurity Domain

Abstract

The field of cybersecurity is evolving fast. Security professionals are in need of intelligence on past, current and—ideally—on upcoming threats, because attacks are becoming more advanced and are increasingly targeting larger and more complex systems. Since the processing and analysis of such large amounts of information cannot be addressed manually, cybersecurity experts rely on machine learning techniques. In the textual domain, pre-trained language models like BERT have proven to be helpful as they provide a good baseline for further fine-tuning. However, due to the domain-knowledge and the many technical terms in cybersecurity, general language models might miss the gist of textual information. For this reason, we create a high-quality dataset and present a language model specifically tailored to the cybersecurity domain which can serve as a basic building block for cybersecurity systems. The model is compared on 15 tasks: Domain-dependent extrinsic tasks for measuring the performance on specific problems, intrinsic tasks for measuring the performance of the internal representations of the model as well as general tasks from the SuperGLUE benchmark. The results of the intrinsic tasks show that our model improves the internal representation space of domain words compared to the other models. The extrinsic, domain-dependent tasks, consisting of sequence tagging and classification, show that the model performs best in cybersecurity scenarios. In addition, we pay special attention to the choice of hyperparameters against catastrophic forgetting, as pre-trained models tend to forget the original knowledge during further training.

M. Bayer, *Deep Learning in Textual Low-Data Regimes for Cybersecurity*, Technology, Peace and Security I Technologie, Frieden und Sicherheit, https://doi.org/10.1007/978-3-658-48778-2_11

Original Publication *Markus Bayer, Philipp Kuehn, Ramin Shanehsaz, and Christian Reuter. "CySecBERT: for the Cybersecurity Domain". In:* ACM[1] Trans[2]. Priv. Secur. *27.2 (Apr. 2024).* ISSN: *2471-2566.* DOI: https://doi.org/10.1145/3652594. URL: https://doi.org/10.1145/3652594 [3]

Material

Model: https://huggingface.co/markusbayer/CySecBERT
Dataset: https://github.com/PEASEC/cybersecurity_dataset
Experimental Code: https://github.com/PEASEC/CySecBERT

11.1 Introduction

Cybersecurity has become an increasingly critical concern in today's digital world, as the number of cyber attacks continues to rise and their consequences become more severe [485]. The COVID-19 pandamic[4], the Russian war of aggression against Ukraine [103] and the resulting shifts in diplomatic and geopolitical dynamics have only intensified this development. As a result, there is a growing need for effective cybersecurity measures to prevent, detect, and respond to these threats. One of the key components of effective cybersecurity is the ability to analyze and understand the vast amounts of data generated by various sources, such as logs, network traffic, and threat intelligence reports. However, much of their analysis is still performed manually by security experts, which is a time-consuming and labor-intensive process.

To address these challenges, recent advancements in Natural Language Processing (NLP) and machine learning have shown promise in automating the analysis of cybersecurity-related data. For instance, cyber threat intelligence (CTI) [153], an important domain of cybersecurity, involves the collection and analysis of information about emerging threats and vulnerabilities. Information thereby is often disseminated in the form of indicators of compromise (IOCs) or unstructured natural language text in blog posts and news articles [413, 459]. Applying NLP techniques to

[1] https://github.com/PEASEC/cybersecurity_dataset

[2] https://huggingface.co/markusbayer/CySecBERT

[3] Minor typographical errors in the original publication have been corrected for this dissertation.

[4] https://enterprise.verizon.com/en-gb/resources/articles/analyzingcovid-19-data-breach-landscape/

CTI can help automate the extraction and understanding of relevant, evidence-based knowledge, thus significantly reducing the manual workload for experts [445].

With regard to underlying NLP mechanisms, word embedding methods that use a sparse vector space to represent words are prominent examples [283]. In this context, language models such as Bidirectional Encoder Representations from Transformers (BERT) [89] have become the standard basis models in all machine learning tasks that involve natural language as input. These models are already pre-trained on a general level and can be adapted to the task at hand by so-called fine-tuning. However, research has shown that the full potential of such models cannot be realized when applied to domain-specific tasks [14, 34, 140, 238]. This is intuitive because these models intend to cover as many domains as possible and, especially in normal-sized models, specific domain knowledge is lost due to capacity constraints or because the knowledge is not even included in the training data. To gain domain-specific knowledge, pre-trained models can be further trained on domain-specific corpora to achieve better results in this particular domain [238]. However, it must be ensured that catastrophic forgetting does not occur, which means that the model forgets its original knowledge.

Models trained on general domain corpora, such as Wikipedia, and without further training often reach their limits when applied to domain-specific tasks such as cybersecurity [301]. Their limitations can be explained by two primary reasons:

1. Unfamiliarity with domain-specific terminology: General domain models may not have encountered specific vocabulary of the cybersecurity domain, such as names and designations of new vulnerabilities or unique threat actor groups. This lack of exposure can lead to reduced performance when analyzing cybersecurity texts, as the model may fail to recognize crucial information.
2. Semantic ambiguity across domains: General domain models may fail to disambiguate words with different meanings in different contexts. For example, the word *virus* might be interpreted by a general model as referring to a biological disease rather than to a type of malware, which is the more relevant meaning in the context of cybersecurity [301].

In this paper, we propose CYSECBERT, a language model based on BERT [89] for analyzing cybersecurity texts. Our aim is to enable state of the art NLP for the security domain and to provide a model which is highly suitable for practical cybersecurity use cases and a solid base for further research in this field. By evaluating our resulting model on different tasks (*i.e.*, intrinsic and extrinsic tasks) we ensure that it indeed enriches the cybersecurity domain. In a preliminary evaluation phase, we try to identify appropriate hyperparameters to minimize the problem of forgetting previously trained knowledge, which is then verified using a standard NLP benchmark. In this study, we pre-train a model on a thoroughly chosen cybersecurity corpus consisting of various datasets, such as scientific papers, Twitter, webpages, and the national vulnerability database. A well-performing model for this use case may supersede high manual workload for researchers and experts. Although there are well performing models for various very specific purposes in the cybersecurity domain [240, 344], the importance of a general cybersecurity model that can serve as a basis for all kinds of tasks is undeniable. The following contributions are made by this paper:

- A pre-trained, general purpose cybersecurity language model based on BERT, called CySecBERT **(C1)**.
- Experiments for hyperparameter tuning in light of catastrophic forgetting **(C2)**.
- An evaluation of CySecBERT based on several tasks tailored to the cybersecurity domain, including intrinsic and extrinsic tasks, as well as general benchmark, to measure if and to which degree the model forgets past knowledge **(C3)**.
- A comparison of our model to a related cybersecurity model and to the original BERT model, as well as a discussion about its shortcomings and potential improvements **(C4)**.

11.2 Related Work

This subsection provides an overview of relevant work on the topic of BERT models. We thereby outline models adapted to different domains that have emerged following the publication of BERT. Moreover, we summarise work that already proposes BERT-like language models for the cybersecurity domain. Finally, we specify the research gap that we intend to fill in the scope of our research.

11.2.1 BERT Models in Different Domains

In various publications, researchers have demonstrated that it is possible to achieve good classification quality on domain-specific text corpora with pre-trained models such as BERT. Of particular interest for our research is the method of domain-adaptive pre-training (DAPT) [140], which describes the process of training an already pre-trained language model on a domain-specific, domain-dependent dataset and which is conducted in the same way as the pre-training. Hence, this technique differs from classical fine-tuning in that the model is not specialized for just one task, but serves as a building block for many tasks in the field. It has been implemented in several other domains since the introduction of BERT [34, 238, 347]. A prominent example is BIOBERT, introduced by Lee et al. [238], where BERT was adapted to a biomedical corpus. BIOBERT was initialized with weights from Devlin et al. [89]'s BERT model and then pre-trained once again, this time with a large biomedical dataset, where the dataset was more than five times larger than BERT's. Evaluating the resulting model with a subsequent fine-tuning process of three different biomedical text mining tasks, which are Named Entity Recognition (NER), Relation Extraction (RE), and Question Answering (QA), Lee et al. [238] were able to largely outperform BERT and previous state-of-the-art models on these aforementioned tasks. Models that address other domains present similar approaches. SCIBERT [34] for example focuses on scientific publications whereas DA- ROBERTA, introduced by Krieger et al. [216] covers media bias. Gururangan et al. [140] underpin our method of additional pre-training on BERT by yielding good results in their application of this approach on ROBERTA [263], a variant of BERT which uses the same transformer-based architecture. In contrast to studies such as BIOBERT by Lee et al. [238], in which only a single domain at a time is considered, Gururangan et al. [140] covered several different domains.

Similarly, researchers have also explored BERT models for the cybersecurity domain. For example, Ranade et al. [347] propose a BERT model for this domain called CyBERT. Although their paper states that fine-tuning on BERT took place, in fact, they further pre-trained BERT for the cybersecurity domain. Fine-tuning is performed on top of this pre-trained cybersecurity model and is primarily used for application. In general, however, their research goal is similar to ours.

Moreover, there are further cybersecurity BERT models, which, however, are fine-tuned instead of subjected to continued pre-training, as is required for true DAPT. Hence, this makes them less suitable for other task of the cybersecurity domain. MALBERT [344] is a BERT-based model from the cybersecurity domain focusing on the detection of malicious software. Another security-related work is CATBERT, introduced by Lee, Saxe, and Harang [240]. They replaced some transformer blocks with adapters and fine-tuned the BERT model for the detection of phishing emails. Mendsaikhan et al. [281] introduced a BERT-based Natural Language filter for identifying and classifying cyber threat-related information from publicly available information sources with high accuracy.

An overview of the approaches with their domains and how they are trained can be found in Table 11.1. These works are related to ours because the approach of adapting BERT to a specific domain is similar to our methodological framework and differs mainly in the specific target domain. All in all, the different pre-trained

BERT approaches can be of use for our work as an orientation and for comparisons of our results to theirs w.r.t. performance. Notwithstanding the fact that BERT has achieved great results in various domains, the full potential for the cybersecurity domain has yet to be exploited.

Table 11.1 Overview over relevant existing BERT models for special domains. The method explains if the model was only fine-tuned (FT) or also pre-trained (PT)

Model / Paper	Domain / Use Case	Method	Model Base
BIOBERT [238]	biomedical	PT (+ FT)	BERT
SCIBERT [34]	scientific	PT (+ FT)	BERT
[140]	Papers (bio. + CS), news, reviews	PT (+ FT)	ROBERTA
MALBERT [344]	malware	FT	BERT, ROBERTA, DistilBERT
CATBERT [240]	phishing	FT	DistilBERT
EXBERT [510]	exploit prediction	FT	BERT
[281]	CTI	FT	BERT
CyBERT [347]	cybersecurity	PT	BERT

11.2.2 Research Gap

The previously established research objective has led us to develop a model with the aim of achieving satisfactory performance for cybersecurity textual material in various tasks. BERT has already been transferred to different domains, resulting in domain-specific models (BIOBERT [238], SCIBERT [34]) and has been applied even to specific domains in the cybersecurity field, producing models like MALBERT [344] or CATBERT [240]. Besides, there are also BERT models such as from Mendsaikhan et al. [281] that are not specific to a particular cybersecurity domain and that can handle various different texts, but are then fine-tuned for a particular task, as in the case of the BERT model from Mendsaikhan et al. [281] CTI classification. In contrast to these models and associated research, the proposed CySecBERT is able to handle many different sources and text forms, and is therefore the basic building block for all cybersecurity-related tasks.

As outlined in the introduction, there are a multitude of research problems in the field of cybersecurity, based on the essential aspect of information extraction. A solid method to address this can improve research in the cybersecurity field at a stroke. Furthermore, the outcome enables extensibility and the application of additional layers on top of the model, for instance CRF [415], (Bi)LSTM, or both combined [182].

Ranade et al. [347] also address the delineated research gap to some extent. Unfortunately, no comparison was made with the results of BERT as the baseline, but only a presentation of their model's outcome was given. We compare our CYSECBERT with theirs, which is varied in the model training and the corpus [347]. Furthermore, in delimitation to their work, we evaluate a whole span of different cybersecurity tasks, ranging from classification to NER and clustering tasks and we include the results of BERT for comparison. We also take into account the phenomenon of catastrophic forgetting, where the pre-trained model forgets its already acquired knowledge in the new training phase. This issue has not been targeted in other works of this area. The similarity in in the work of Ranade et al. [347] and our work results from the nature of the research task, which all the more underlines the importance of the approach. We understand that the attention given to this research gap is important and encouraging at the same time. Multiple works addressing a similar objective can be complementary and thus accelerate filling the gap in research. Nonetheless, our work distinguishes itself from the work of Ranade et al. [347] at several points including the evaluation step, the applied data, and overall by the extent of our work.

11.3 Methodology

This section provides a brief background on domain adaptive pre-training, including the planned training process, presents the dataset which is used to adapt our proposed language model to the cybersecurity domain, the architecture of the model, and a pre-evaluation phase.

11.3.1 Domain Adaptive Pre-Training

DAPT of language models to a specific domain is a common method to achieve an advanced domain-specific language model (see Sect. 11.2). It has shown to increase model performance in several ways. Not only is model performance on downstream tasks increased, hence receiving better evaluation results, but training time is also reduced for such tasks due to smaller datasets for the training process to achieve similar performance. These prospects lead us to expect that the cybersecurity domain will benefit greatly from a domain-adapted, pre-trained language model for every possible task, *e.g.*, NER and relevance classification, to name a few.

We aim to adapt BERT to the cybersecurity domain based on domain specific text corpora (see Section Sect. 11.3.2) [140]. Our DAPT pipeline is built with *Huggingface*[5] and *Weights and Biases*[6]. The final domain-adapted pre-trained model is based on `bert-base-uncased`. Likewise, the text corpus is tokenized using the `bert-base-uncased` model. The training itself is done on the Lichtenberg Cluster[7].

During the training phase, we try to mitigate the problem of catastrophic forgetting [277] by reducing the learning rate, the training steps, and the size of the dataset compared to BERT pre-training. In this way, the susceptibility to catastrophic forgetting should be greatly reduced because the new learning process is subordinated. Nevertheless, we test whether the problem also occurs with the created model by evaluating it on a non-cybersecurity task. While we expect no improvements, we want to analyse to what extent the old knowledge is altered.

Table 11.2 Statistics of the number of tokens and entries based on the training dataset

#Tokens	Min[8]	Max[8]	Sum	Median	Mean	Entries
Blogs	44	0.1M	169M	710	1.1k	151k
arXiv	533	0.7M	167M	8.2k	9.9k	16k
NVD	5	1.9k	12M	58	71	171k
Twitter	1	500	179M	39	45	4M
Total	1	0.7M	528M	40	122	4.3M

11.3.2 Text Corpus

When creating the text corpus, we paid great attention to the quality of the data, as this quality transfers to the model [28]. The text corpus is composed of four different subcorpora: (i) blog data, (ii) arXiv data, (iii) National Vulnerability Database (NVD data, and (iv) Twitter data. Our decision for this selection is based on the kind of information that is used by security professionals, and on the fact that these sources

[5] https://huggingface.co/

[6] https://wandb.ai/

[7] http://www.hhlr.tu-darmstadt.de/

[8] Minimal or maximal token per entry.

are commonly used in recent publications regarding machine learning. Furthermore, those sub-corpora vary considerably in their structure. While the NVD contains short, objective and precise language with semi-structured information, and Twitter consists of short posts with objective, subjective, emotional, on- as well as off-topic, etc. content. arXiv encompasses long papers with highly educational language, and blog posts are typically longer articles with less formal language.

The blog posts build a solid foundation for different writing styles and practical information in information security, including vulnerability and exploit information, threat notifications [252], and foundational knowledge. We initially aimed for a set of 41 different website domains to be crawled. During the automated crawling process, three of them either blocked our requests, or contained so much advertisements and cookie information that we decided to omit those sources. This resulted in 38 different blogs which we crawled, like troyhunt.com, darkreading.com, schneier.com, and krebsonsecurity.com, to name a few. Besides, we filtered duplicates and instances shorter than 300 characters[9] and filtered for English texts using fasttext [185]. This process resulted in a total of over 151k blog posts, based on the initial list of 165k web pages.

Secondly, we use arXiv papers from the category *Cryptography and Security*[10] [238]. Due to errors during the text extraction process, we ignored papers with lower length than 3000 characters, resulting in over 16k papers.

Thirdly, we use vulnerability descriptions of the NVD [95, 219]. Experts curate those texts[11], which is why they need no further processing. Hence, we do neither filter nor pre-process these information.

Finally, we use Twitter as information source [65, 363, 374], where we crawled datasets containing the following keywords based on information we received from CERT members:

- *infosec OR security OR threat OR vulnerability OR cyber OR cybersec OR infrasec OR netsec OR hacking OR siem OR soc OR offsec OR osing OR bugbounty*

Additionally, we crawled dedicated datasets of data breaches, like the Microsoft Exchange Server Data Breach, for example. Of all the tweets collected, we only used

[9] A randomly selected and manually inspected sample of corner case blog posts, *e.g.*, shorter or longer posts, showed that short posts contained mostly advertisements or cookie notifications.

[10] For text extraction we used opendetex for papers in tex format or textract for papers in pdf format.

[11] https://www.cve.org/ResourcesSupport/FAQs#pc_cve_recordscve_record_descriptions_created

those that were annotated as English by Twitter. While the posts include retweets, we did not gather replies. Overall, we managed to crawl nearly 4M tweets with over 179M tokens in total.

We would like to emphasise that all collected entries have been minimally edited at most, as we want to have the most natural form in which the model will later function. A summary of all datasets is depicted in Tab. 11.2. To get an overview of the words and tokens contained in the dataset, we created word clouds for the dataset, which can be seen in Fig. 11.1. The first image represents individual words, the second highlights bigrams, and the third illustrates the tokenized dataset. Across all word clouds, a prominent association with the security domain is evident.

Fig. 11.1 Visualization of the training corpus through word clouds: (a) single words, (b) bigrams, and (c) tokenized input for the model

11.3.3 Architecture

As explained before, we adapt the domain of the BERT model with further pre-training. The architecture is based on the transformer architecture, introduced by Vaswani et al. [453]. Transformers leverage the self-attention mechanism which enables the model to weigh the significance of each word in a sequence relative to the others, thereby capturing long-range dependencies and contextual information. They consist of encoders and decoders, but BERT only employs the encoder portion, i. e. it only encodes the text into an internal representation. During pre-training, BERT utilizes masked language modelling by randomly masking a certain percentage of words in the input sequence. The model is then tasked with predicting the masked words based on their surrounding context. This encourages BERT to develop a deep understanding of the language structure, as it has to infer the masked words using the available context.

From this follows the mathematical representation of the key components: Given an input text which is transformed into a sequence of tokens $\mathrm{X} = \{x_1, x_2, ..., x_\mathrm{n}\}$. From them, the token embeddings, i. e. fixed-size vector representations (with positional encodings) $\mathrm{E} = \{e_1, e_2, ..., e_\mathrm{n}\}$ are obtained. For each token, we compute the Query (Q), Key (K), and Value (V) vectors by projecting the embeddings using learned weight matrices WQ, WK, and WV:

$$\mathrm{Q_i} = e_\mathrm{i} * WQ \qquad \mathrm{K_i} = e_\mathrm{i} * WK \qquad \mathrm{V_i} = e_\mathrm{i} * WV \tag{11.1}$$

Next, we calculate the attention scores for each token pair (i, j) (with d being the embedding dimension):

$$score(\mathrm{i, j}) = \frac{\mathrm{Q_i} * \mathrm{K_j^T}}{\sqrt{\mathrm{d}}} \tag{11.2}$$

Then, we apply the softmax function to the scores to obtain attention weights:

$$weights(\mathrm{i, j}) = e^{score(\mathrm{i, j})} / \sum_{\mathrm{k}=1}^{\mathrm{n}} e^{score(\mathrm{i, k})} \tag{11.3}$$

Finally, we compute the self-attention output for each token by taking a weighted sum of the Value vectors:

$$output_\mathrm{i} = \sum_{\mathrm{j}=1}^{\mathrm{n}} (weights(\mathrm{i, j}) * \mathrm{V_j}) \tag{11.4}$$

BERT uses multiple self-attention heads to capture different aspects of the context. Each head computes its own self-attention output $output_{h,i}$, which are concatenated and projected using another learned weight matrix WO:

$$multiheadoutput_\mathrm{i} = [output_{1,\mathrm{i}} \oplus output_{2,\mathrm{i}} \oplus ... \oplus output_{\mathrm{H,i}}] * WO \tag{11.5}$$

Then, BERT applies a position-wise fully connected feed-forward network to the multi-head attention output, with W1, b1, W2, and b2 as learned weight matrices:

$$nnoutput_{\mathrm{i}} = ReLU(multiheadoutput_{\mathrm{i}} * W1 + \mathrm{b1}) * W2 + \mathrm{b2} \tag{11.6}$$

Each encoder layer in BERT consists of the multi-head self-attention mechanism, followed by layer normalization, the position-wise feed-forward network, and another layer normalization:

$$outputmha_{\mathrm{i}} = LayerNorm(multiheadoutput_{\mathrm{i}} + e_{\mathrm{i}}) \tag{11.7}$$

$$outputnn_{\mathrm{i}} = LayerNorm(nnoutput_{\mathrm{i}} + outputmha_{\mathrm{i}}) \tag{11.8}$$

This process is repeated for each encoder layer in the model, as depicted in Fig. 11.2.

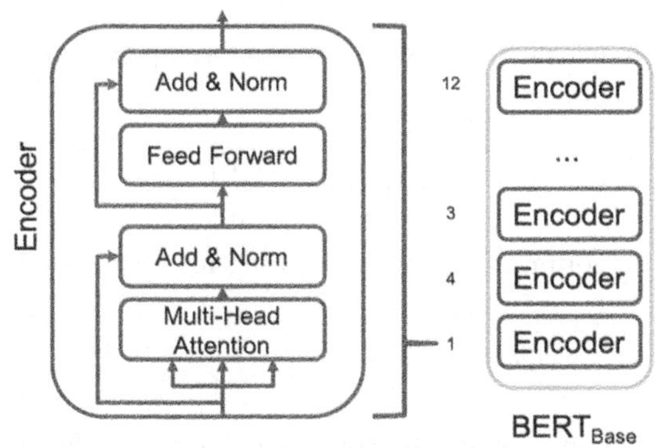

Fig. 11.2 Visualization of the BERT encoder stack [104]

While these mathematical representations describe the core components of the BERT architecture, including self-attention, multi-head attention, and the position-wise feed-forward network within the encoder layers, further details can be extracted from the work by Devlin et al. [89].

11.3.4 Preliminary Evaluation: Catastrophic Forgetting

Catastrophic forgetting describes the phenomenon that an already trained model tends to catastrophically forget the previously trained knowledge when trained on further data [201]. If catastrophic forgetting occurs, our model might lose its ability to understand or interpret nuances and contexts that may not be explicitly covered in the cybersecurity domain but are still relevant for processing cybersecurity-related texts. (1) In real-world applications, data often contains a mix of domain-specific and general language. For instance, cybersecurity texts may include general language expressions, analogies, or references that require a broader understanding of language beyond just technical terms. (2) There are certain tasks where the model has to handle in domain and out of the domain instances. Like, for example, when the task is in a intersection of cybersecurity and another domain (like law, policy, and management). A model exclusively trained on cybersecurity might struggle with texts that blend these disciplines. A model that retains its general language capabilities alongside its domain-specific knowledge is more robust and flexible [423]. Finally, the work of Rongali et al. [369] shows that avoiding catastrophic forgetting in domain-specific BERT models even improves the performance on the domain tasks. This illustrates the benefits of a well-balanced model that can handle both domain-specific and general language data effectively.

Research [89, 426] suggests that the second training should be more lightweight than the first, so that the already trained knowledge is not overshadowed. In this regard, the three most important hyperparameters are the learning rate, which determines how much the new data updates the weights of the model, the epochs, i.e. how long the model is trained and how many updates are made, and finally the size of the dataset. The work of Sun et al. [426], for example, addresses the problem of catastrophic forgetting by examining different learning rates, demonstrating that a lower learning rate is best. Besides, the authors of the BERT paper Devlin et al. [89] recommend training with only a few epochs. Accordingly, further training can often lead to overfitting of the data and, as already discussed, to catastrophic forgetting. This remark can also be transferred to the size of the datasets for DAPT, as they are very large and correspondingly many training updates are carried out. The difficulty is to find the right configuration of these parameteres to avoid catastrophic forgetting and at the same time to enable enough learning that the model achieves high quality in the new domain. Therefore, we test different configurations of the hyperparameters' learning rate, dataset size, and number of epochs in this preliminary evaluation. Since training a model already requires a considerable amount of time and resources (about 4 days on 4 Nvidia V100 GPUs), we cannot perform an extensive hyperparameter search, such as the bio-inspired search by Ibor et al.

[165]. Hence, we narrow down the range of configurations by taking into account the research findings and by orienting our approach towards the original BERT training process. The authors of BERT trained the model with a learning rate of 1×10^{-4} for about 40 epochs. We therefore propose the following configurations:

1. Learning Rate: 1×10^{-4}, 2×10^{-5}, 5×10^{-5}
2. Epochs: 20, 30, 40
3. Data: 5%, 10%, 15%

For the other hyperparameters, we followed the original BERT approach, *i.e.*, we used a weight decay of 0.01, a dropout rate of 0.1, 10000 warm-up steps, and ADAM as the optimisation algorithm [207]. We trained all models with a batch size of 64 on 4 Nvidia Tesla V100 GPUs.

To be able to measure the degree of forgetting, as well as the performance in the security domain, we evaluate the different models on the SuperGLUE task BoolQ and the security task MSExchange (a more detailed description of the datasets follows in Sect. 11.4.1).

The results are presented in Table 11.3. As can be seen, almost every configuration performs well in both tasks. The learning rate of 2×10^{-5} appears to be better in both tasks, which is in line with the results of Sun et al. [426]. Similarly, 30 epochs perform better on both tasks than the higher and lower configurations. However, with regard to the size of the dataset, the results are not so clear. While a dataset size of 10% of the BERT training dataset size performs best on the MSExchange dataset, a dataset size of 5% of the original BERT training dataset size seems to perform

Table 11.3 Pre-evaluation of different CySecBERTconfigurations considering the catastrophic forgetting task and the performance in the security domain with the boolq task (accuracy) and the MSExchange task (F1), respectively

Configuration	BoolQ	MSExchange
LR 2×10^{-5}, Epochs 30, Data 10%	0.6752	**0.8869**
LR 1×10^{-4}, Epochs 30, Data 10%	0.6722	0.8620
LR 5×10^{-5}, Epochs 30, Data 10%	0.6700	0.8544
LR 2×10^{-5}, **Epochs 20**, Data 10%	0.6660	0.8860
LR 2×10^{-5}, **Epochs 40**, Data 10%	0.6614	0.8846
LR 2×10^{-5}, Epochs 30, **Data 5%**	**0.6785**	0.8816
LR 2×10^{-5}, Epochs 30, **Data 15%**	0.6716	0.8496

best on the BoolQ task. This is not surprising, as the degree of forgetting is lower with less new training data. Since the model with 10% still performs very well on the BoolQ task and because we weight the performance in the security domain more heavily, we decided to use the model with 10% of the original BERT training dataset. The training loss of this CYSECBERT model can be seen in Fig. 11.3. Accordingly, the loss decreases logarithmically and improves only very slowly after 300k steps.

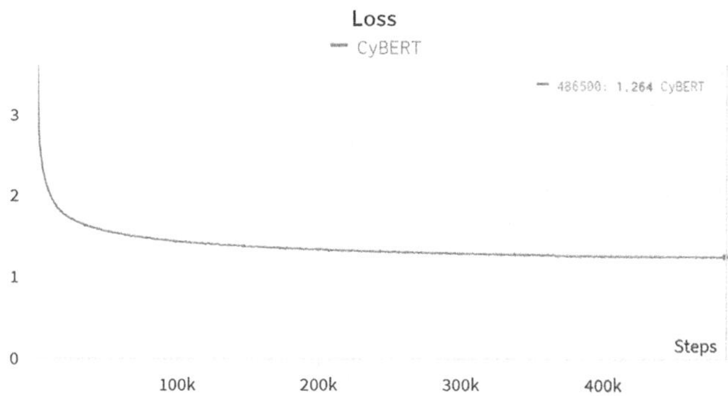

Fig. 11.3 Training loss of the CYSECBERT model

11.4 Evaluation

In this section, the evaluation process and the corresponding results are presented in detail. We cover a short description of the evaluation tasks in Sect.11.4.1, followed by the presentation and interpretation of the results in Sect. 11.4.2.

11.4.1 Experiments and Tasks

Since our goal is to publish a model that is highly usable for the cybersecurity domain, we evaluate it against the current standard method of the domain (BERT) and against another cybersecurity language model (CyBERT from Ranade et al. [347]). We use different types of tasks, *i.e.* intrinsic and extrinsic evaluation tasks, which are reasonably chosen for the field of cybersecurity. While the extrinsic tasks measure how well the trained model performs on downstream tasks, *i.e.* measure real-world application, the intrinsic tasks measure the model itself without any kind of additional classifier, *e.g.* by measuring the representations of the model and by demonstrating an overall fit to the domain.

As intrinsic tasks, we apply a word similarity task using parts of the dataset by Mumtaz et al. [301] and a clustering evaluation with a Twitter dataset. The clustering dataset is based on a random sample of Log4j Twitter posts. For this task, posts are converted into latent representations with different BERT models. The latent representation consists of the concatenation of the last four layers of the model output and of the mean values across all words in the post. A KMeans clustering algorithm with k-values from 5 to 9 is executed on the gathered and transformed posts. The evaluation scores are measured with the Silhouette Coefficient (the higher the better). This is an internal clustering metric that analyses the clusters created and does not require gold labels. This is important because there can be many solutions and gold labels can be misleading in the case of clustering [30]. The cybersecurity word similarity dataset consists of words with their equivalents, all from the field of cybersecurity. Moreover, the dataset is an extension of the public cybersecurity word similarity dataset from Mumtaz et al. [301] and contains over 300 word pairs. For this extension of the original dataset, a cybersecurity expert followed the process of Mumtaz et al. [301] and added more cybersecurity words that can be considered similar. Normally, the word similarity evaluation is based on the cosine similarity of the word embeddings when static embeddings are used. The problem is that BERT is context-dependent, which means that a clear word embedding cannot be given without context and the standard method of measuring word similarity is no longer suitable. For this reason, we have developed a new method for evaluating word similarity, in which the model predicts whether two given words are similar. Similar to the works on zero-shot learning, we create a meaningful cloze task consisting of a sentence with a masked word that the model fills in, which is implicitly the answer to the similarity question. The task is written in the following way:

"*Are* $word_1$ and $word_2$ *similar? [MASK]*", where **[MASK]** can be either "Yes" or "No", which represent the masked words that the model has to fill.

Example: "*Are* ***virus*** *and* ***malware*** *similar?* **[MASK]**"

This trick of zero-shot learning means that no explicit classification model needs to be trained and the task remains intrinsic. Besides, we aim to show that the evaluated model does not only predict the similarity of each word to all other words, but also detects when two words are not similar. We therefore randomly take word paris from the dataset that are not similar and add them to the evaluation. This dataset is then used for evaluation and an F1 score is calculated for the model's predictions.

As part of the extrinsic tasks, we employ two cybersecurity classification tasks from Riebe et al. [363] and Bayer, Frey, and Reuter [27]. In the first task, the classifier has to decide whether a Twitter post is related to the field of cybersecurity, and in the second task, it has to predict whether a post might be relevant to experts in the field during a major cybersecurity event. Furthermore, we use the sequence tagging dataset by Kuehn et al [219]. Sequence tagging is the task of finding specific words in a text and it requires that each word in a text is tagged, often as IOB: Either I-nside, o-utside, or b-eginning, referring to the specific words being searched for. Specifically, the dataset consists of several NER tasks (recognition of named entities) for predicting relevant details of NVD descriptions. We then chose the task of tagging the software name (SN), software version (SV), and the attack complexity. We decided for the employment of these tasks because of their different performances in Kuehn et al. [219]'s work, in order to analyse how well the models perform on varying levels of difficulties of the tasks.

Furthermore, we evaluate CySecBERT and BERT on the SuperGLUE benchmark [463]. This is a common NLP benchmark which we utilize to identify signs of catastrophic forgetting [277]. We assume that our cybersecurity model is not able to achieve a better or even equivalent result, as a certain degree of forgetting is acceptable and necessary for learning. Nevertheless, we expect the performance not to be too poor, as this would otherwise indicate that some basic knowledge would have been forgotten during the domain training phase.

Neural networks consist of many random hyperparameters that have to be fixed before each training and that are not transferable to other learning processes. As shown by Reimers and Gurevych [355], the frequent execution of a training process can be used to avoid incorrect inferences arising from a randomly better choice of hyperparameters. Accordingly, and following the experiments of Sanh et al. [379] and Liu et al. [263] all extrinsic experiments were performed five times and the mean values as well as the standard deviation are given in the respective tables.

11.4.2 Results

As stated in Sect. 11.4.1, we aim to evaluate CySecBERT, as well as BERT and CyBERT [347] on different tasks which are mainly located in the cybersecurity domain, incorporating both intrinsic and extrinsic evaluation tasks. Additionally, we run the SuperGLUE task to test our model for catastrophic forgetting.

11.4.2.1 Intrinsic Tasks

To measure the representation quality of the model, we evaluate it by applying two intrinsic tasks from cybersecurity: clustering and word similarity.

CLUSTERING The results of the clustering task are given in Table 11.4. While the CyBERT model of Ranade et al. [347] only performs better than the baseline when forming 5-7 clusters, our approach performs better in every constellation, according to the Silhouette Score. In fact, CYSECBERT outperforms Ranade et al. [347]'s cybersecurity model by a considerable margin for each number of clusters, with consistent improvements ranging from +0.002 to +0.059 points. Our model shows the highest improvement when 9 clusters are formed. On the one hand, these results demonstrate that we can obtain more coherent clusters thanks to our trained language model. On the other hand, from a more general perspective, the results show that the model is more able to represent the given instances in a meaningful latent space.

Table 11.4 Silhouette Score of the first intrinsic task, clustering the data of a Log4J dataset. The best values are marked

# Clusters	BERT	CyBERT [347]	CYSECBERT
5	0.114	0.141	**0.143**
6	0.115	0.124	**0.150**
7	0.118	0.133	**0.167**
8	0.125	0.117	**0.163**
9	0.130	0.113	**0.172**

Nevertheless, even better results could be expected if we used an approach like SentenceBERT by Reimers and Gurevych [356] for our model, as they have proven to be much more suitable for representing complete documents, like tweets in our case.

WORD SIMILARITY The word similarity task results are displayed in Table 11.5. The baseline model has the worst performance with a F1-score of 0.44. This is to be expected, as most domain-specific words were not or only very rarely included in the standard BERT training. Our CYSECBERTmodel is clearly superior to the other two approaches, which is remarkable as it confirms the previous intrinsic results. However, we would like to point out that this task is different from other word similarity tasks as it does not reflect word similarities directly through the word representations, but by questioning the model in a cloze fashion (see Sect. 11.4.1).

Table 11.5 Overview of the results of the word similarity task, where the scores are indicated by the F1 score

Name	Word Similarity
BERT	0.4382
CyBERT [347]	0.4861
CYSECBERT	**0.6382**

11.4.2.2 Extrinsic Tasks

After we have already demonstrated that the model produces meaningful representations of cybersecurity-related vocabulary and data, we want to test whether our model is also comparably more suitable for real-world applications, i.e. for extrinsic tasks, the so-called downstream tasks of machine learning. The tasks that we have chosen for the cybersecurity domain are (i) NER (ii) general relevance classification, and (iii) CTI classification.

NER The results of the NER tasks are shown in Table 11.6. The table shows that the results of the original work of Kuehn et al. [219] are worse than those of the methods evaluated in our study. While the basic BERT model and the CyBERT model of Ranade et al. [347] are more similar to each other, e.g. in terms of software naming, our model consistently outperforms both. Only in the tagging of the software version does the CyBERT model of Ranade et al. [347] perform significantly better than the baseline BERT model, whereas our model nevertheless improves this result. Hence, it could be speculated that the CyBERT training data from Ranade et al. [347] contained entries of software versions at a higher frequency than the normal BERT data. However, the CyBERT model deteriorates the results for software names,

Table 11.6 NER score based on tagged software versions (SV), software names (SN), and attack complexities (AC) of NVD descriptions. The results are given as F1 scores and the best values are marked. *Showing the reported results of the work. **No comparable results available

Name	SV	SN	AC
[219]*	0.8735 (–)	0.8584 (–)	– (–)**
BERT	0.9247 (0.0064)	0.8837 (0.0037)	0.3323 (0.0135)
CyBERT [347]	0.9298 (0.0019)	0.8834 (0.0029)	0.3336 (0.0214)
CYSECBERT	**0.9302** (0.0066)	**0.8871** (0.0025)	**0.3472** (0.0116)

which could indicate that either a large number of software names are missing in high frequency in their dataset, or that they have been neglected due to errors in the training process. The highest improvements of our model can be seen in attack complexity, outperforming the CyBERT model of Ranade et al. [347] by with 0.0136 points. Nevertheless, we perceive the results of this particular task as overall not very satisfactory. Reasons therefore have already been discussed by Kuehn et al. [219] and are mainly related to the problem of too little data in this task.

RELEVANCE CLASSIFICATION (CYSECALERT) In the first classification task of our experiments, the models are trained to predict whether a Twitter post is related to the cybersecurity domain (see Table 11.7). This can be considered a general cybersecurity task, as the model only has to identify cybersecurity-related words. The original work by Riebe et al. [363] has the worst results, possibly due to the use of a classical machine learning method. It is particularly interesting to see that the Ranade et al. [347] model performs worse than the basic BERT model. Our model significantly improves the baseline and the CyBERT model by 0.0104 and 0.0236 points in the F1 score, respectively. All models have a relatively low standard deviation, indicating that the fine-tuning process is stable across all runs.

SPECIALIZED CTI CLASSIFICATION (MS EXCHANGE) The second classification task is about identifying specialized CTI information where very specific words are needed to classify the instances. The results of this task are also presented in Table 11.7. Surprisingly, unlike in the previous tasks, the CyBERT model of Ranade et al. [347] is able to improve the baseline, showing that while it does not contribute improvements in the more general tasks, it could be beneficial in more specific tasks. There is a high improvement of our model compared to the baseline (+0.027), which we expected since this task focuses on very domain-dependent language and specific vocabulary. Moreover, although the CyBERT model of Ranade et al. [347] is advantageous for this task, our model still improves the results significantly by +0.0103 F1 points. Besides, it is observable that our model has a significantly lower

Table 11.7 Classification results of the MS Exchange and CySecAlert dataset, given as F1 scores. The best values are marked. *Showing the reported results of the work

Name	MS Exchange	CySecAlert
[363]*	–	0.8051 (–)
[27]*	0.8536 (0.0007)	–
BERT	0.8599 (0.0193)	0.8779 (0.0084)
CyBERT [347]	0.8766 (0.0153)	0.8647 (0.0095)
CYSECBERT	**0.8869** (0.0026)	**0.8883** (0.0064)

Table 11.8 Classification results of the few-shot experiments of the MS Exchange dataset. The best values are marked

Name	Model	Accuracy	F1
Baseline	BERT	0.4965 (0.0190)	0.5870 (0.1881)
ADAPET	BERT	0.6589 (0.0135)	0.6254 (0.0432)
[27]	CySecBERT	**0.7913** (0.0056)	**0.8063** (0.0027)

standard deviation than the other two models, which again indicates a very stable training process. Table 11.7 also contains the result of the work by Bayer, Frey, and Reuter [27], which is very similar compared to the BERT baseline, but has the best standard deviation.

Furthermore, we also included the few-shot learning task of this dataset in Table 11.8, where only 32 instances are given as training data. CySecBERT is an integral part of the few-shot approach of Bayer, Frey, and Reuter [27]. The results show a significant improvement over the baseline and over the state-of-the-art few-shot learning method ADAPET by Tam et al. [431]. This demonstrates how the CySecBERT model is incorporated and applied in further research, contributing to the advancements in other disciplines. Details of the approach and further results can be extracted from the work of Bayer, Frey, and Reuter [27].

11.4.2.3 Catastrophic Forgetting

In the final part of our evaluation, we revisit the problem of catastrophic forgetting. To this end, we evaluate our model with the SuperGLUE benchmark to test if the model degrades the results considerably, which would indicate that the model has forgotten the initial knowledge acquired in the BERT training phase. The results of this task and a comparison to the BERT model is displayed in Table 11.9. As expected, we can see that our model reduces almost every task outcome. Nevertheless, the worsening of results are not an indication for catastrophic forgetting, as the differences are still relatively small, with a mean drop of about –0.05 points. This shows that although the model has lost some of its knowledge, most of it is still present. Besides, it is particularly interesting that the performance in the cb task has even increased and the result of the boolq task has remained almost the same.

In addition, we have again included the results of the CyBERT model by Ranade et al. [347], which, interestingly, performs almost the same as the BERT model. One could interpret this as an indication that the model did not suffer catastrophic forgetting during training. However, taking the results of the extrinsic and intrinsic

Table 11.9 Results of the SuperGLUE benchmark, each indicated in the evaluation metric proposed in the benchmark

Name	record	rte	wic	wsc	boolq	cb	copa	multirc	total
BERT	0.6416	0.5949	0.6476	0.5538	0.6760	0.3704	0.606	0.4067	0.6010
CyBERT [347]	0.6346	0.6173	0.5980	0.5337	0.6887	0.5676	0.615	0.4146	0.6065
CySecBERT	0.6137	0.5545	0.5887	0.5404	0.6752	0.5551	0.486	0.3915	0.5468

domain tasks into account, we can conclude that their model might have learnt very little overall, as it is seldom better, sometimes worse and most often on par with BERT. This can be attributed to the training process of Ranade et al. [347], who trained their model only on 17,000 documents with one epoch, whereas our model is trained on 4,300,000 documents with 30 epochs.

11.4.2.4 Conclusion

In the evaluation, we have shown that the model developed in this work is very well adapted to the cybersecurity context. The tasks have demonstrated that the CySecBERT model is able to outperform the BERT baseline and the CyBERT model by Ranade et al. [347] consistently across all cybersecurity tasks. We evaluated these models on intrinsic cybersecurity tasks where we summarized how accurate the models represent documents and words in latent space. Based on these tasks, the fundamental quality of the language model has been assessed. In addition, we evaluated the three models using extrinsic cybersecurity tasks that demonstrate the practicality of the model in most real-world application contexts. Our model improves the results of these tasks by up to 0.027 F1 points compared to the other two models and it achieves its highest improvement on an in-depth cybersecurity task where very specific language differences have to be considered. Moreover, we analysed the phenomenon of catastrophic forgetting by evaluating our model on standard NLP tasks. Although we observed a deterioration in performance in these tasks, it is only within the expected range of decline. Besides, we concluded that the CyBERT model of Ranade et al. [347] may have been trained on a much too small corpus with too few training steps (CyBERT: 17,000 documents and one epoch versus our CySecBERT: 4,300,000 documents and 30 epochs). We can say with confidence that our model is capable of handling a wide range of cybersecurity tasks while retaining the original language modelling knowledge.

11.5 Discussion, Conclusion, and Outlook

In this work we propose a novel state-of-the-art cybersecurity language model based on BERT [89]. We performed DAPT on this model with a sensibly chosen cybersecurity corpus. The corpus consists of a variety of source data structures, such as blogs, scientific paper, as well as Twitter data. The data and the sources were selected to be appropriate for cybersecurity research and practice. Furthermore, the size of the dataset and the structure of the training process were chosen to prevent catastrophic forgetting on the one hand, and, on the other hand, to enable enough learning for the model to contribute to the general field and to the specific niches of cybersecurity. We explored this through a pre-evaluation phase by tuning the hyperparameters in terms of catastrophic forgetting and domain quality. We then highlighted the performance of our model with a thorough main evaluation of various tasks and drew a comparison to the BERT baseline as well as the current state of the art of cybersecurity language models. First, we evaluated the models on two intrinsic tasks, where we demonstrate that the quality of our model improves in terms of the learned representation space, i.e. how well the cybersecurity-specific instances (words and texts) can be distinguished from each other. Table 11.4 and 11.5 show the substantial performance increases by our model. Second, we evaluated the model together with the other two models for cybersecurity-specific classification and NER tasks to exemplify the usefulness and practicality of the model in application contexts. Our model outperforms the other models in every task, which can be derived from Table 11.6 and 11.7. The greatest improvement is observed in the special CTI classification dataset, suggesting that the model is particularly beneficial when dealing with very specific cybersecurity language that is not contained in the training dataset of the standard BERT model. Our evaluation concludes with a focus on catastrophic forgetting by which we assessed the performance of our model against a general NLP benchmark. While these results (Table 11.9) point out that our model does indeed degrade them overall, they also show that, as intended, there is no catastrophic forgetting and that the final model has combined much of its original knowledge with the new knowledge about cybersecurity.

While we are aware that the current state of the art on research in language modeling and NLP generally tends to focus on larger language models like GPT-3 by Brown et al. [47], we have chosen the BERT model on purpose. Most of the cybersecurity research and especially practice does not have the necessary resources to apply large language models. In most cases, the BERT model can still be considered the standard model in ML application contexts such as the cybersecurity domain. In this way, our work is most beneficial to the research landscape and to practice.

11.5.1 Practical and Theoretical Implications

Our work contributes to research and pratice through a novel, state-of-the-art cybersecurity model called CYSECBERT, which we have published. We also publish the associated dataset so that it can contribute to further research. Thus, our work has several implications for practice and research:

A novel, state-of-the-art language model for cybersecurity that is useful for various tasks. With our research surrounding the model, we have aimed to find a solution to increase the performance of machine learning in as many cybersecurity language tasks as possible. Our model provides utility for a large number of tasks, as can be estimated based on the success in extrinsic task scores as well as inferred from intrinsic task scores. Thereby, they show that the representation space is better for the domain-dependent language with our model. With the release of our model, we are paving the way for better cybersecurity tools, as practitioners can easily use the new model in existing pipelines, for example, in alert aggregation [231], in detection of phishing websites [495, 507], or even in malware detection [377]. More advanced tools will then also be the result of new research derived from the model and have the potential to improve the results in various tasks by incorporating further research ideas, as has already been implemented as part of the research of Bayer, Frey, and Reuter [27], for example, where the CYSECBERT model has been integrated. This can be pursued on a smaller scale, where the model is not the focus but serves as a foundation on which further techniques such as data augmentation, meaningful data selection, few-shot learning or specific applications are built. Yet, it can also be done on a larger scale, where the model is the subject of research, for example by analysing its results in explainable AI approaches.

In our evaluation, we demonstrate that the CYSECBERT model can serve as a substitute for other models in the field of cybersecurity by providing new state-of-the-art performances. However, the CYSECBERT model will not be suitable as a replacement for every type of cybersecurity model. We expect that further work in this area will adopt our methodology and will train other models that may be even more specialized regarding a specific cybersecurity topic or that may have a much larger neuron size.

Evaluation of catastrophic forgetting in terms of learning rate, dataset size, and number of epochs. When training the CYSECBERT model, we paid special attention to catastrophic forgetting phenomena where the trained model loses its valuable initial knowledge. To ensure that the level of catastrophic forgetting is kept to a minimum, we first performed a preliminary evaluation of the hyperparameters' learning rate, dataset size, and number of epochs, as research in this area suggests that the second training process should not overshadow the first training [89, 426].

We narrowed down the space of possible hyperparameter constellations by these considerations and evaluated seven fully trained CySecBERT models on a standard NLP and on a cybersecurity task to measure the degree of catastrophic forgetting and the quality in the cybersecurity domain. Our results are in line with research and show that the training process should not be too heavyweight, e.g. by heavy updates due to a high learning rate, but also not too lightweight, so that domain knowledge is not acquired, e.g. due to a smaller dataset. In the final section of our main evaluations, we evaluated the CySecBERT model on the SuperGLUE benchmark, which consists of standard NLP tasks, and compared it with the BERT model and the CyBERT model of Ranade et al. [347]. While the results show that our model performs worse than the BERT model, this is to be expected since some level of catastrophic forgetting will always occur when further training the model towards a certain domain. However, it is interesting to note that the results of the CyBERT model of Ranade et al. [347] are very similar to the BERT model, which is due to the small training process which the CyBERT model originates from. While the CyBERT model is trained on only 17,000 documents with one epoch, our model is trained on about 4,300,000 documents with 30 epochs.

It is not clear whether these results, especially with regard to the hyperparameters, are generalizable to other language models, especially to very large language models such as GPT-3 and 4. For an evaluation with a broader hyperparameter search, as for example executed with a bio-inspired, evolutionary method from Ibor et al. [165], or for a focus on these very large language models, we expect research to be primarily concerned with catastrophic forgetting and with having the necessary extensive resources.

A cybersecurity dataset containing most relevant sources for the training process. The dataset was created with consideration given to including a variety of sources and textual types. This ensures that the model can be applied to a broad range of cybersecurity tasks. However, we anticipate future work analysing the published dataset, for example, to enhance its quality based on specific criteria or to identify any unintended biases. The dataset can also serve as a basis for training other language models. Although we have deliberately chosen this size of the dataset for BERT training to prevent catastrophic forgetting, it might be useful to expand it, which can easily be done by collecting more data from the sources that we have already selected. It might also make sense to utilise larger language models that might achieve even greater performance or that might even use neural architecture search to find a suitable architecture for special use cases, as proposed by Shang et al. [396] or Okunoye and Ibor [310].

11.5.2 Ethical Considerations

We would like to emphasise that we did not explicitly focus on and analyse social biases in the data or the resulting model. While this may not be so damaging for most application contexts, there are certainly applications that depend heavily on these biases, and including any kind of discrimination can have serious consequences. As authors, we would like to express our warnings regarding the use of the model in such contexts. Nonetheless, we aim for an open source mentality, observing the great impact it can have, and therefore transfer the thinking to the user of the model, drawing on the many previous discussions in the open source community.

Acknowledgements This research work has been funded by the German Federal Ministry of Education and Research and the Hessian Ministry of Higher Education, Research, Science and the Arts within their joint support of the National Research Center for Applied Cybersecurity ATHENE and by the German Federal Ministry of Education and Research in the projects CYWARN (13N15407) and CYLENCE (13N16636). The calculations for this research were conducted on the Lichtenberg high performance computer of the TU Darmstadt.

Multi-Level Fine-Tuning, Data Augmentation, and Few-Shot Learning for Specialized Cyber Threat Intelligence

12

Abstract

Gathering cyber threat intelligence from open sources is becoming increasingly important for maintaining and achieving a high level of security as systems become larger and more complex. However, these open sources are often subject to information overload. It is therefore useful to apply machine learning models that condense the amount of information to what is necessary. Yet, previous studies and applications have shown that existing classifiers are not able to extract specific information about emerging cybersecurity events due to their low generalization ability. Therefore, we propose a system to overcome this problem by training a new classifier for each new incident. Since this requires a lot of labeled data using standard training methods, we combine three different low-data regime techniques—transfer learning, data augmentation, and few-shot learning—to train a high-quality classifier from very few labeled instances. We evaluate our approach using a novel dataset derived from the Microsoft Exchange Server data breach of 2021 which was labeled by three experts. Our findings reveal an increase in F1 score of more than 21 points compared to standard training methods and more than 18 points compared to a state-of-the-art method in few-shot learning. Furthermore, the classifier trained with this method and 32 instances is only less than 5 F1 score points worse than a classifier trained with 1800 instances.

Supplementary Information The online version contains supplementary material available at https://doi.org/10.1007/978-3-658-48778-2_12.

M. Bayer, *Deep Learning in Textual Low-Data Regimes for Cybersecurity*, Technology, Peace and Security I Technologie, Frieden und Sicherheit, https://doi.org/10.1007/978-3-658-48778-2_12

Original Publication *Markus Bayer, Tobias Frey, and Christian Reuter. "Multi-level fine-tuning, data augmentation, and few-shot learning for specialized cyber threat intelligence". In:* Comput. Secur. *134 (2023), p. 103430.* DOI: https://doi.org/10.1016/J.COSE.2023.103430. URL: https://doi.org/10.1016/j.cose.2023.103430 [1]

Note Supplementary material containing the annotation guidelines for the dataset can be found in Appendix A.4 of the Electronic Supplementary Material.

Material
Experimental Code: https://github.com/PEASEC/Specialized-Cyber-Threat-Intelligence Dataset: https://github.com/PEASEC/msexchange-server-cti-dataset

12.1 Introduction

Social media is where cutting-edge and critical cyber threat information is disseminated, which is highly relevant to researchers, security providers, security operation centers, urban infrastructures, and computer emergency response teams (CERTs), among others [291, 367]. While there have been several research works on general cyber threat event detection [93, 110], the aim of this work is to enable fine-grained and potentially individualized collection of cybersecurity information in open data sources such as Twitter.

A major challenge in gathering cybersecurity-related information, also called Cyber Threat Intelligence (CTI) [279], which needs to be specialized, i.e. customizable, is that information in this area is very dynamic and varies greatly from past events (in terms of specific names, different attack vectors, specific attack paths, affected functions, etc.) [61]. As a result, supervised machine learning yields poor results because these dynamics cannot be captured in the learning process. Alternatively, new classifiers could be trained for each cyber threat event so that the new features are taken into account. However, since machine learning usually requires a large amount of data for normal training, this would result in having to label a dataset for each cyber threat event, which is unrealistic considering the effort involved and the need for fast and up-to-date information. Against this background, the concept of active learning systems takes a first step towards label reduction for supervised machine learning for cyber threat events [363]. Active learning supports the labeling process, so that only the instances with the highest learning value need to be labeled

[1] Minor typographical errors in the original publication have been corrected for this dissertation.

for machine learning. However, despite this method, too much data is still needed to train a useful classifier. The endeavor sought in this work takes an even stronger stance on labeling reduction by proposing a system consisting of few-shot learning, transfer learning, and data augmentation, which are all techniques to reduce the amount of manual labeling required for a high-quality classifier. With few-shot learning, it is sufficient if the model is already trained with very few instances, as opposed to hundreds or thousands in the case of active or normal learning [47]. This includes special learning techniques as well as transfer learning, where knowledge from a previous task is transferred to the new one. Data augmentation is used to create artificial instances from the training data using label-preserving transformations [29].

The concept of few-shot learning is extended in this work through the use of multi-level transfer learning. The different levels start with a model that has been trained on a large general dataset and thus has a basic prior knowledge. During the next steps, this model is approximated more and more to the actual task domain. In this way, it can be ensured that the model is given a basic cybersecurity reference in order to be able to counter the dynamics in the task, in addition to being familiar with the task. This is particularly relevant for urban infrastructures, which require high resilience against cyberattacks, as well as for CERTs, as they need to collect and communicate information in the most reliable and targeted way possible [362]. The data augmentation strategy is inspired by the work of Bayer et al. [28] and follows the example of Yoo et al. [512] by utilizing the large generation model GPT-3 to generate new instances based on the few existing labeled ones.

Our paper includes several contributions relevant for the cybersecurity and machine learning community:

- A novel pipeline combining transfer learning, data augmentation, and few-shot learning for rapid development of effective specialized CTI classifiers.
- Novel techniques of data augmentation and few-shot learning to deal with a small number of training instances.
- A new specialized CTI dataset annotated by three experts and based on the 2021 Microsoft Exchange Server data breach.

The code and dataset of this study are freely available[2].

The remainder of the paper is structured as follows: After introducing related work on transfer learning, data augmentation, few-shot learning, and cyber threat

[2] Code: https://github.com/PEASEC/Specialized-Cyber-Threat-Intelligence
Dataset: https://github.com/PEASEC/msexchange-server-cti-dataset/

event detection and intelligence (Sect. 12.2), we explain the concept of our method (Sect. 12.3). It is subdivided in three components which are described in detail. In Sect. 12.4 the evaluation is presented and findings are given in detail. The last section (Sect. 12.5) contains a discussion of the implications, limitations, and potentials for future research.

12.2 Related Work

12.2.1 Cyber Threat Event Detection and Intelligence

Cyber threat event detection can be defined as the process of automatic scraping of the webspace and Open Source Intelligence (OSINT) to detect possible cybersecurity events [363, 374, 381]. Social media platforms like Twitter are part of OSINT and provide a great space to share and discuss cybersecurity vulnerabilities, for example. While vulnerability databases such as the National Vulnerability Database (NVD) are often of high quality and much higher credibility of vulnerability information, Twitter posts can be more up-to-date and rich [15]. There are some automated systems and research that already scrape Twitter and other OSINT sources to detect cyber events. Some examples are the *CySecAlert* system by Riebe et al. [363] or *SONAR* by Sceller et al. [381], which collect cybersecurity relevant tweets from Twitter, filter them, and present them in a manageable dashboard.

CTI on the other hand describes the process of collecting additional information after the first detection of a cyber threat event. The process helps deliver the context of the vulnerabilities found to assist CERTs and cybersecurity organizations make sound decisions and find quick solutions [5, 445, 459]. CTI is currently mostly accomplished by manually collecting information on different platforms [5]. It relies heavily on manual tasks and is therefore labour intensive [459]. However, there are already some threat intelligence platforms, such as Facebook ThreatExchange or CrowdStrike, that are able to automatically detect, monitor, and analyze cyber threat occurrences [445]. A manageable dashboard is also provided by the Cyber Threat Observatory [193], which aggregates cybersecurity information from various sources, including social media, security advisories, indicators of compromise (IoCs) and CVEs. For an overview of different CTI platforms and tools, see the work of Kuehn et al. [221].

However, these systems need too much time to adapt to a newly discovered threat that is, for example, propagated on Twitter, and cannot be extensively customised. It is possible that this has not yet been addressed because current machine learning systems are generally too rigid and cannot be easily generalized to new situations. Our work aims to solve this problem by providing a novel pipeline that allows the rapid training of new specialized CTI classifiers through significantly reduced labeling requirements. This is achieved with novel techniques in the field of transfer learning, data augmentation and few-shot learning.

Once the CTI information has been collected, there are several methods and research approaches that can be used to analyze this information and provide useful insights. TTPDrill by Husari et al. [163], for example, extracts and constructs attack patterns from threat and blog articles. IoCMiner by Niakanlahiji et al.[307] is able to extract IoCs from Twitter data. Similarly, GoodFATR by Caballero et al. [51] collects IoCs from Twitter and five other sources (including Telegram and blogs). They also give a good overview of different IoC extraction works. Many of these works use or can be complemented by machine learning. While our approach aims to provide a specialized CTI collection in new cyber threat events, it could also be used to enhance the quality of these works.

12.2.2 Transfer Learning

Transfer learning describes the process of transferring knowledge gained from training a neural network from one task to another related task [323, 444]. This technique is now one of the standard learning methods for machine learning, especially in the field of natural language processing (NLP). It is particularly powerful for tasks where there is not enough training data or it is difficult to manually adjust the data for training. In these cases, it is possible to use a pre-trained neural network that was trained to solve a related task or with more easily accessible data. Afterwards the neural network is fine-tuned with the task-specific data to fit the wanted task. One of the most frequently used pre-trained models is BERT by Devlin et al. [89]. BERT (short for Bidirectional Encoder Representations from Transformers) is a pre-trained deep bidirectional transformer for language understanding. In essence, it is trained by predicting words in a sentence given the other words, also called masked language modeling. It has a lot of widely used descendants trained for many different tasks, such as BioBERT [238], SciBERT [34], and CamemBERT [276]. While BERT is already a considerably large model, nowadays far larger models, like GPT-3 from [47], are trained. Compared to BERT's base model with 110 million parameters, GPT-3 has 175 billion parameters, however, GPT-3 is not publicly available and cannot be easily fine-tuned due to its size.

Transfer learning can be an important step to overcome the high labeling requirements through knowledge transfer. Unlike other work in this field, we do not just train a pre-trained model for the actual task, but propose to train a model further and further towards the actual task through several fine-tuning steps. Other work may not have addressed this because it requires multiple datasets that are more and less specific to the task at hand. Moreover, this technique can only be used to handle a small number of tasks. In our case, however, this is exactly what we want, as we need a basic model that can be easily adapted to the different cybersecurity events and thus only very few labeled instances are needed.

12.2.3 Data Augmentation

Data augmentation is the concept for artificially enlarging the training datasets for machine learning by transforming the existing ones. Originated and heavily used in computer vision, it is now also increasingly being explored on textual data [29]. NLP data augmentation techniques can be applied to the raw text or also on the numerical representations. Ranging from small transformations, i.e. flipping characters [33] or inducing adversarial noise [181], to interpolated [427] or even newly created instances [18], data augmentation can have great effects. Nevertheless, as Longpre, Wang, and DuBois [265] point out, the success of data augmentation in NLP is often not perceivable when fine-tuning large pre-trained models. A data augmentation technique needs to incorporate new linguistic patterns as otherwise the changes are too small and already captured by the pre-training phase of the model. For example, simple synonym replacement methods have not been shown to be beneficial with pre-trained models, as these synonyms are already mapped to nearly the same vector for their numerical representation [297]. On the other hand, there are generation models that can integrate new linguistic patterns, for example, through their own training data during pre-training, as for example shown by Yoo et al. [512] with the GPT-3 model. The challenge with using these models is to make the generations truly label preserving. This is, for example, done by Anaby-Tavor et al. [18], Abonizio and Junior [4] and Bayer et al. [28]. The models are conditioned by fine-tuning on the label-induced training data (or just the class data) and are then tasked to complete a text given the label conditioned beginning (prompt). As this is oftentimes not sufficient to achieve a high label preservation, a filter mechanism is used that removes artificial instances that are unlikely to fit the class. For example, Anaby-Tavor et al. [18] use a classifier trained on the data to predict whether the new instance can be assigned to the expected label.

In this work, we take advantage of recent research directions by combining the strategy of using GPT-3 to generate training data and then filtering out the instances that are not close enough to the respective class. In this way, we can create instances with a high degree of novelty, i.e. instances with linguistic patterns that were not previously included in the training data, but also preserve the label. If we use this for our goal of reducing the data needed for specialized CTI, we try to generate instances that have these novel linguistic patterns at best in the cybersecurity domain, but are still very close to the original data, especially since the data must be very specific.

For an overview of the data augmentation methods that could also be used in this study, we advise the reader to have a look at the survey from Bayer, Kaufhold, and Reuter [29].

12.2.4 Few-Shot Learning

Few-shot learning describes the training of effective classifiers on the basis of a small number of examples. While there are several strands of research on few-shot learning [44], in this study we focus on the use of pre-trained language models. At the latest, the large language model GPT-3 by Brown et al. [47] paved the way for using these kinds of models, as it reaches astounding performance even without task-specific training data. However, as GPT-3 is too large for most companies and research institutes, the research field adapted smaller language models to reach similar or even better few-shot performances [431].

Pre-trained language models can be especially beneficial for few-shot settings when the instances are reformulated in a cloze-style way. Cloze tests [435] are tests where some words in the text are missing and have to be completed. For few-shot learning, instances are rephrased, often into questions, so that the text contains the label (or a word that can be mapped to the label), generally within the answer to the question. The label, known (training) or not known (testing and inference), is masked out, so that the language model can fill it with the right word and a label can be inferred. Using the language model directly is more effective for few-shot learning than the classical way of training a classifier head on top of it, as there are no more randomly initialized parameters that have to be learned [123].

A pattern describes the transformation of the input instance to the cloze-like text. The verbalizer maps the predicted words for the mask to the label. An example for a pattern and a verbalizer can be seen in Fig. 12.1.

Gao, Fisch, and Chen [123] show that the choice of template and verbalizer has a major impact on the resulting performance. Since domain knowledge is often necessary for these, the authors propose a method to automatically find meaningful templates and verbalizer. For this purpose, they use a language model and the existing training instances to predict the words for the verbalizer and template. Zhang et al. [523] take a different perspective on automatic template generation with the DART method by making the template differentiable. They use special tokens in the template that are mapped into trainable parameters. These template parameters

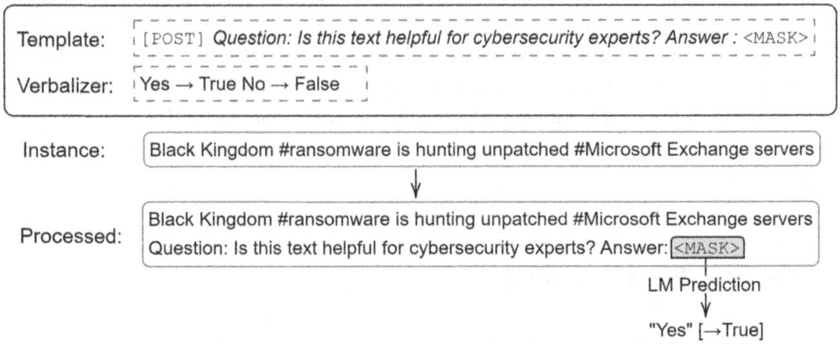

Fig. 12.1 Example of a template and a verbalizer and how they are applied on an instance

are then optimized together with the target label. PERFECT by Mahabadi et al. [271] leverages task-specific adapters to replace template tokens. Adapters make it possible to train only the newly added parameters, which are able to transform the hidden states, while freezing all other parameters.

Schick and Schütze [383] propose a semi-supervised few-shot learning technique, called PET. They take several manually designed templates and use the training data to train on each one a pre-trained language model. They take these models to generate pseudo-labels for unlabeled data. A classifier is then trained on the resulting dataset. Tam et al. [431] adapt the PET method to not be dependent on additional training data and can even improve the performances. Contrary to the preceding PET technique, the word probabilities are computed not only for the verbalizer words, like "yes" and "no", but also for all other words. In the training, incorrect class tokens are explicitly penalized and correct tokens are encouraged. Furthermore, ADAPET [431] introduces a label conditioning step in which the model is tasked to predict other tokens in the sentence given the label.

In our pipeline, we incorporate the ADAPET technique into the proposed multi-level fine-tuning technique and adopt it for the task of specialized CTI. Together with the novel technique of data augmentation, we can create a system that enables specialized CTI by reducing the amount of data required for high-quality classifiers.

12.2.5 Research Gap

Our study addresses several research gaps which are highly relevant for researchers as well as practitioners. Most importantly, our research paves the way for fine-grained and specialized CTI. Current research addresses CTI from a very coarse perspective, by building classifiers, like Riebe et al. [363], that are able to find general cybersecurity information. As a result, only a small amount of data reduction can be achieved in these information-overloaded situations. On the other hand, specialized classifiers are not designed to generalize well to new situations. Our work fills this gap by introducing a pipeline for specialized CTI, where new cyber threat events are encountered with the very fast creation of new classifiers. By addressing this fine-grained information gathering challenge, we create a novel dataset combined with a sophisticated labeling guideline for CTI. Furthermore, with our pipeline we address research gaps of machine learning low-data regimes. Our data augmentation strategy is the first to explore the generation capabilities of large language models with constraining them through filtering mechanisms. We combine the works of Yoo et al. [512] and Bayer et al. [28] by using GPT-3 with a human-in-the-loop filtering mechanism. We extend the few-shot learning research by proposing a multi-level fine-tuning approach. In the process, the model learns a very broad knowledge in the first levels, which in the later levels becomes more and more directed to the specific CTI task.

12.3 Concept

12.3.1 Dataset Creation

The goal of dataset creation is to extract specific CTI information during a significant cyber threat event. In this work, we focus on Twitter as a data source because it provides a wide range of vulnerability information, unlike the NVD, which only provides brief information that is not as up-to-date, does not provide direct mitigation advice, and does not include exploit information [15]. However, we are also aware of the disadvantages of Twitter, as the information may not be as credible, for example (more on this in the limitations in Sect. 12.5.2).

This dataset is subsequently binary-labeled according to the relevance of the information for CTI and for cybersecurity experts. We focused on the Microsoft Exchange Server data breach of 2021, where four zero-day exploits were discovered. While the first report of a vulnerability was already made in January of that year, in March various attackers were found to be exploiting the vulnerabilities and a proof of concept was released.

We used the Twitter APIv2 to gather 50,000 tweets in March that fulfill the query *"Microsoft Exchange" OR "MS Exchange" OR "CVE-2021-26855" OR "CVE-2021-26857" OR "CVE-2021-26858" OR "CVE-2021-27065"*. From this, we filtered out the tweets that were not in English, resulting in 39474 tweets. The used query is intended to replicate the process of filtering Tweets by security experts in the event of the incident. Examination of the resulting posts shows that only a subset of them are really relevant to an expert. While they contain a lot of relevant and up-to-date information such as references to patches and remedies, proof of concepts, code, IoCs and attacker names, they also contain a lot of irrelevant information aimed at the general public or, for example, only spam, podcasts and news aggregations. For these tweets, we drew a random sample of 3001 posts for labeling and resolved the links shortened by Twitter, as the full URLs could be an important indicator in the context of CTI.

For the labeling process, we have created a codebook that contains guidelines describing what content is relevant and what is not, closely following the staff of CERTs who work with this type of data. Our goal was to collect precise information that would yield maximum benefit for them. This does not include, for example, information that is primarily intended for a wider audience or information that is not current. For instance, valuable insights can be gained through details about exploits, IoCs, and the vulnerabilities themselves (including remediation, impact, solutions, etc.). Figure 12.2 presents one case that is relevant and another that is not, both from the process of labeling. The relevant example contains information about patching the Microsoft software, while the example labeled as not relevant does not provide any in-depth information or technical insights that would be immediately useful to a cybersecurity professional.

New episode alert! In this Juniper Threat Labs podcast I interview @MounirHahad for his take on the MS Exchange ProxyLogon vulnerability CVE-2021-26855. https://bit.ly/3lU1347 (https://t.co/6pniZgqTju)

#Microsoft Releases Patches for Older Versions of #Exchange Server https://www.zdnet.com/article/microsoft-exchange-attacks-now-microsoft-rushes-out-a-patch-for-these-unsupported-exchange-servers-too/ (https://t.co/U2jwnLMA9n)

Fig. 12.2 Two examples of the labeling procedure, the upper one not relevant and the lower one relevant

The labeling of the data was performed by three cybersecurity experts guided by the codebook. The guidelines, which gave annotators clear guidance on when to mark a post as relevant or irrelevant, were iteratively updated by the annotation leader. A first draft of this was developed with the help of the CTI concept [279]:

> "Threat intelligence is referred to as the task of gathering evidence-based knowledge, including context, mechanisms, indicators, implications, and actionable advice, about an existing or emerging menace or hazard to assets that can be used to inform decisions regarding the subject's response to that menace or hazard."

After an initial sifting of the tweets and again after the first labeling of 750 tweets, the process was refined by the annotation leader. The full guidelines can be found in the Appendix A.4 of the Electronic Supplementary Material.

The first round of annotation of 750 tweets was conducted by the annotation leader, who updated the guidelines after gathering several insights. He and the other two cybersecurity experts then annotated the 750 tweets again. After this round, all three experts discussed the cases they were not sure about and corrected them if necessary. Regarding the intercoder reliability the Kappa Scores were calculated (see Table 12.1). Subsequently, each annotator tagged 750 different examples, resulting in a total of 3001 commented Twitter posts for the complete dataset (the labels of the 750 instances of the first round were determined by majority vote).

Table 12.1 Intercoder reliability calculated with the Kappa Score

Coder	Score
C1 and C2	0.8763
C2 and C3	0.7446
C1 and C3	0.8709

The dataset was then split into a full and few-shot training set and development set. The splits (train, dev) consist of 1800 and 600 instances for the full set and 32 and 8 instances for the few-shot set, respectively. The test set is the same in both cases and consists of 601 instances. For a complete overview of the dataset splits and class distribution, see Table 12.2.

Table 12.2 Split of the dataset with count of relevant and not relevant labels in the datasets

Split	Count	Relevant	Not Relevant
Train (full)	1800	949	851
Train	32	16	16
Dev (full)	600	273	327
Dev	8	4	4
Test	601	304	297
Total	3001	1526	1475

12.3.2 Approach

Our system for dynamic, specialized cyber threat event detection consists of three components, all of which help to boost performance with little data. We explain the three components in detail in the following:

MULTI- LEVEL FINE- TUNING: In light of the success of large pre-trained models such as BERT, we propose to further tune such models on several levels of domain-dependent data (see Fig. 12.3). The levels begin from a broader view and are narrowed down to the actual task. In our case, we first take a pre-trained BERT model (which can be seen as the 0th level of fine-tuning), train it with masked language modeling on cybersecurity data. We then tune the resulting model for classification on the CySecAlert dataset [363] in which Twitter posts are generally assigned to the cybersecurity domain. Finally, we train it on the few training examples of the specialized cyber threat dataset. The rationale behind this is that the model gains more and more knowledge as it is tuned to more and more fitting tasks. The 0th level is about gaining general knowledge of text. In the first level, the dataset consists of papers, blogs, web pages, and also Twitter data, from which the model gains knowledge about cybersecurity language and also how Twitter data is written in this domain. In the second level, the model should gain a general understanding of the relevance of cybersecurity information. Finally, in the third level, the model is tuned to the actual task data to which it can transfer the knowledge of the previous levels.

GPT- 3 DATA AUGMENTATION: With data augmentation we can create new instances from existing ones, which can be particularly advantageous when the amount of data is small. We propose a data augmentation strategy based on text generation with GPT-3 [47], which is inspired by the method from Yoo et al. [512] and Bayer et al. [28]. GPT-3 can be tasked to complete a given text, also called

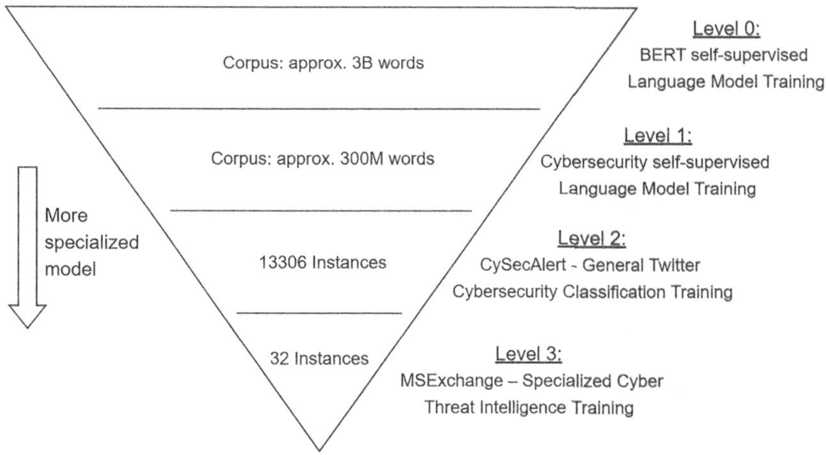

Fig. 12.3 Multi-level fine-tuning process that shows the model becoming more specialized as it is guided to the actual task with less data

a prompt. We utilize this mechanism so that the generation model creates new instances based on the training data of one class. Specifically, this means that we are concatenating all instances of one class with a class specific priming token. For the class of cyber threat information we prepend every positive instance with "cybersecurity →". For the irrelevant class we chose "other →" as priming token. In both cases the priming token is also appended at the end so that the model generates the instance(s) after it. Dependent on how many remaining generation tokens the model has after the prompt, it may generate more than one instance by picking up the priming token. After the creation of the instances we perform the human-in-the-loop filtering step proposed by Bayer et al. [28]. The training examples and generated instances are mapped into an embedding space. There, the generated instances that deviate the most from the training data are discarded. The distance from which this happens is determined by an expert.

FEW- SHOT LEARNING: We make use of the existing ADAPET [431] few-shot learning technique and adapt it to our case. With ADAPET, in contrast to normal use, no classification head is trained on the language models. The instances are transformed to cloze-style phrases and then the language model itself is used to predict the blank word in the phrases. The predicted word is subsequently transformed with a verbalizer to one of the labels. The cloze-style phrases are automatically formed with templates. For our task we use the following template:

"[POST] Question : Is this text helpful for cybersecurity experts? Answer: <MASK>. [SEP]"

The verbalizer maps the two possible words "yes" and "no" to the labels representing relevant and not relevant. As explained in Sect. 12.2.4 there also exist methods for automatically determining the pattern and verbalizer. We believe that these techniques are not necessary in our case, as we can integrate the expert knowledge regarding the task, which facilitates the learning process.

12.4 Evaluation

12.4.1 Dataset, Models and Evaluation Settings

Following the research goal of specialized CTI for security professionals, we constructed a setting, consisting of models and datasets, representing the real conditions. For the dataset, we labeled data from the 2021 Microsoft Exchange Server data breach. The specifics of the dataset can be found in Sect. 12.3.1. The labeled dataset, including few-shot and normal-shot splits, is freely available.

In our main evaluation we have different settings regarding the dataset and models. The *baseline* and initial model of our evaluation is the bert-base-uncased model by Devlin et al. [89]. For the baseline, this model is fine-tuned on the few-shot dataset representing the standard training strategy without any few-shot or data augmentation methods. For the *best case*, on the other hand, we train the bert-base-uncased model with the full dataset of 1800 instances. This is called the best case because we consider this amount of data to be the best case in the event of a new cybersecurity attack. In addition, we also train a model with ADAPET, as we consider this to be the current state of the art in few-shot research. In preliminary tests, we found that ADAPET performed best on the few-shot split with ALBERT [230] compared to DART and the PERFECT variant. To be consistent with our evaluation settings as opposed to the evaluation settings of ADAPET, we use the bert-base-uncased model, instead of the albert-xxlarge-v2 model by Tam et al. [431]. The evaluation settings of our procedure are divided into the three components mentioned. For the data augmentation technique we use GPT-3 (DaVinci) as text generation model, which is prompted with the specifics explained in Sect. 12.3.2. The multi-level fine-tuning process starts with the bert-base-uncased model, which is further pre-trained on a cybersecurity dataset, which is then fine-tuned with the ADAPET few-shot method on the CySecAlert dataset. This resulting model is finally trained on the few-shot split and evaluated on the test set of the Microsoft Exchange dataset. Furthermore, in addition to the CySecAlert fine-tuning process, we also use the ADAPET few-shot method for the fine-tuning of the Microsoft Exchange Server dataset. The mentioned components are also inspected within an ablation study, showing their individual contribution to the overall pipeline.

The evaluation performance is measured in accuracy and with the F1-score. For every evaluation setting, we perform five runs to rule out random factors. The results are given with the minimum, maximum, mean, and standard deviation.

12.4.2 Hyperparameters

As already mentioned, we are using bert-base-uncased as base model for our experiments. The evaluations are performed on a NVIDIA A100 with 40 GB GPU memory. The training runs on the CySecAlert and Microsoft Exchange dataset are performed with 5 epochs each. Furthermore, we use a batch size of 48, 100 warmup steps with a warmup ratio of 0.06, a learning rate of 0.00001, and weight decay of 0.001. As optimization algorithm, we use the Adam algorithm. For the data augmentation technique we use the GPT-3 text-davinci-002, which has 175 billion parameters. The filtering is performed with SBERT with the all-mpnet-base-v2 model.

12.4.3 Evaluation

The first section of our evaluation is about the data augmentation process, as we manually inspect the instances generated by GPT-3 and compare our method to two other data augmentation techniques.

After this, the main evaluation follows where we compare our methods to a baseline, state-of-the-art and best case experiment. Finally, we inspect our method by doing ablation studies, testing how each component evaluates.

12.4.3.1 Data Augmentation

Due to our human-in-the-loop approach, we already saw that the generated instances are of very high quality. An excerpt of the generated data is given in Table 12.3. For research purposes, we were also interested in the most likely original instances that the model used for generating specific instances. This is why we tried to find the training instance with the closest resemblance to the generated one. We measured the resemblance by generating sentence embeddings with SBERT [356] and comparing them with the cosine distance. These counterparts are also given in Table 12.3. These examples show that the data augmentation method is capable of many

Table 12.3 Generated data instances and their most similar original counterparts. The instances created are displayed first and the most similar ones second. URLs are removed from the text

Relevant	RT If you're running Microsoft Exchange Server on premises, you need to take these urgent security steps now. The zero-day exploits may have already caused a breach of your data. #infosec #cybersecurity #HAFNIUM http://..
	If you are hosting #MicrosoftExchange on premises you need to take these urgent security steps right now. The zero-day exploits may have already caused a breach of your data. #infosec #HAFNIUM http://..
	RT Black Kingdom ransomware is exploiting the Microsoft Exchange Server ProxyLogon vulnerabilities to encrypt servers. http://..
	Please take Information Security seriously. #CyberAttack can bring your reputation down. Another #ransomware operation known as 'Black Kingdom' is exploiting the Microsoft Exchange Server ProxyLogon vulnerabilities to #encrypt servers. http://..
	RT @SecureList: The ProxyLogon vulnerability in Microsoft Exchange Server is being actively exploited in the wild to install ransomware. http://..
	RT Just as predicted, the Microsoft Exchange exploit chain #ProxyLogon now confirmed being used to install ransomware #DEARCRY http://..
	RT RT @hackerfantastic: Microsoft Exchange Server Remote Code Execution CVE-2021-26855 Exploit.
	#BugBounty #RCE #infosec http://..
	RT Thousands of US companies have been hacked by Chinese hackers using This RCE.
	Microsoft Exchange Server Remote Code Execution CVE-2021-26855 Exploit.
	#BugBounty #RCE #infosec http://..
	RT @ryan_a_h: Microsoft just released their quarterly updates which include a patch for the Exchange zero-day. You can find more information here: http://..
	If you are hosting #MicrosoftExchange on premises you need to take these urgent security steps right now. The zero-day exploits may have already caused a breach of your data. #infosec #HAFNIUM http://..

(Continued)

Table 12.3 (Continued)

Not Relevant	Microsoft Exchange Server Attack Escalation Prompts #Patching Panic
	#cybersecurity #vulnerabilities http://..
	#MicrosoftExchange Server Attack Escalation Prompts #Patching Panic
	#cybersecurity #vulnerabilities http://..
	#Technology #Cybersecurity Microsoft Exchange Hackers Also Breached European Banking Authority #AiUpNow #techy http://..
	#Technology #TechNews Microsoft Exchange Hackers Also Breached European Banking Authority #Cybersecurity #AiUpNow #techy via http://..
	RT Microsoft Exchange Server has been hacked – here's what you need to know http://..
	RT Here's what we know so far about the massive Microsoft Exchange hack http://..
	Microsoft Exchange Server Flaws Expose Millions of Emails to Attack http://..
	RT Here's what we know so far about the massive Microsoft Exchange hack http://..
	Protected: Microsoft Exchange Server Attacks Escalate to Government, Healthcare and Financial Institutions http://..
	The Microsoft Exchange hacks: How they started and where we are http://..

different transformations. The first example demonstrates that the model sometimes replaces one or few words with synonyms (*hosting* $\rightarrow$ *running*) or adds context words (*#cybersecurity*). While in the second example, one can see that the model is able to paraphrase parts of the original instance (*Another #ransomware operation known as 'Black Kingdom' is exploiting the […]* $\rightarrow$ *Black Kingdom ransomware is exploiting the […]*), in the third example the entire instance is paraphrased (*Just as predicted, the Microsoft Exchange exploit chain #ProxyLogon now confirmed being used to install ransomware* $\rightarrow$ *The ProxyLogon vulnerability in Microsoft Exchange Server is being actively exploited in the wild to install ransomware*). The fourth shown instance is an example of the method stripping away parts, while still preserving the label (~~*Thousands of US companies have been hacked by Chinese hackers using This RCE.*~~ *Microsoft Exchange Server Remote Code Execution CVE-2021-26855 Exploit.*). For some generated instances, like the fifth example, we were not able to find similar instances. The instances might be entirely new based on the interpolation of the given instances and the knowledge of the underlying model.

Regarding the irrelevant class, we see that many generated instances are duplicates of the training instances, differing at most by very small changes, such as removing the hashtag in the first example (*#MicrosoftExchange Server Attack → Microsoft Exchange Server Attack*) or swapping the position of words in the second example (*#Technology #TechNews Microsoft [...] Authority #Cybersecurity #AiUp-Now #techy → #Technology #Cybersecurity Microsoft [...] Authority #AiUp-Now #tech*). While the third example, again, shows an instance where the content is paraphrased, the last two generated texts have no clear counterpart.

We also quantitatively evaluated our data augmentation strategy by evaluating the entire pipeline using our data augmentation method compared to two other popular methods in the field. One of these augmentation techniques was proposed by Wei and Zou [481], called Easy Data Augmentation (EDA), and consists of several text transformations: Replacing a word with a synonym or randomly inserting a synonym as well as randomly swapping and deleting words. The other data augmentation technique is Round-trip translation (often referred to as Backtranslation [391]), as for example in [105], where the instance to be transformed is first translated into another language (German in our case) and then back into the original language.

The results of this experiment can be found in Fig. 12.4. Here we can see that the data augmentation proposed in this work is clearly superior for our task, achieving +1.33 and +3.59 F1 points over EDA and backtranslation, respectively.

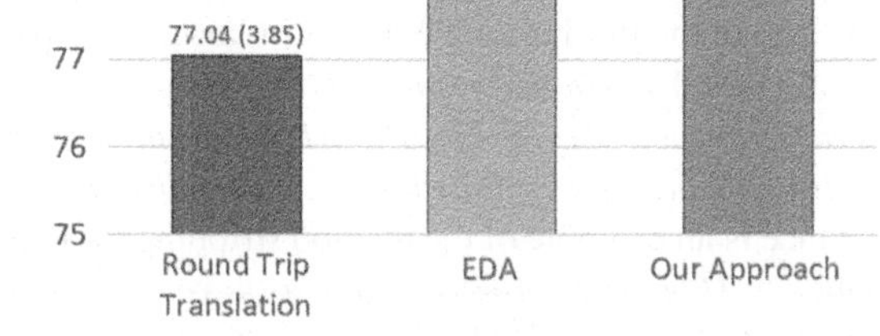

Fig. 12.4 Evaluation results of the data augmentation experiment. Showing the mean F1 results of 5 runs and the standard deviation in brackets

12.4.3.2 Main Experiments

In our main experiments, we test the whole pipeline proposed in Sect. 12.3. As a quick reminder, our method includes the multi-level fine-tuning with bert-base-uncased on cybersecurity data, the CySecAlert dataset and the actual few-shot learning task with 32 instances, as well as the GPT-3-based data augmentation technique and ADAPET for few-shot training. For a sensible comparison, we first follow the standard training procedure by fine-tuning a bert-base-uncased model with a classifier head on the few-shot training instances (baseline). Furthermore, we test a bert-base-uncased model with the ADAPET method, as it can be regarded as the state-of-the-art method for performing few-shot learning. We also perform a best case evaluation in which we train the bert-base-uncased model on the full training dataset (1800 instances) to see how a classifier would perform with enough data. A more detailed analysis of the approach itself can be found in the ablation studies in Sect. 12.4.3.3.

The results of the pipeline experiments are shown in Table 12.4. It is observable that the baseline is not able to learn any meaningful classification strategy with the few-shot dataset, reaching an accuracy of about 50% and F1 score of 58.70. ADAPET reaches a significantly higher accuracy with an additive improvement of about 15 points in accuracy and a F1 score of 62.57. This is, nevertheless, far from a good classification quality as the best case classifier reaches a F1 score of 85.35. With an F1 score of 80.63, our approach proposed in this paper could even almost keep up with the best case classifier. Particularly noteworthy at this point is that the best case classifier is trained with 1800 instances, while our approach only has access to 32 instances. Furthermore, our approach improves the current state of the art with 18.09 points in F1. A look at the violin plots in Fig. 12.5 shows that both the best case and our approach have a very good standard deviation, which means that both are robust to random changes.

Table 12.4 Detailed evaluation results of the main experiments. The values on the left show the minimum, in the middle the mean, in brackets the standard deviation, and on the right the maximum value

Name	Model	Accuracy (SD)	F1
Best Case	BERT	84.69/85.36(0.07)/86.02	84.87/85.35(0.47)/85.81
Baseline	BERT	46.26/49.65(1.90)/50.58	25.06/58.70(18.81)/67.18
ADAPET	BERT	64.89/65.89(1.35)/68.05	59.30/62.54(4.32)/69.81
Our Approach	CySecBERT	78.54/79.13(0.56)/80.03	80.42/80.63(0.27)/81.07

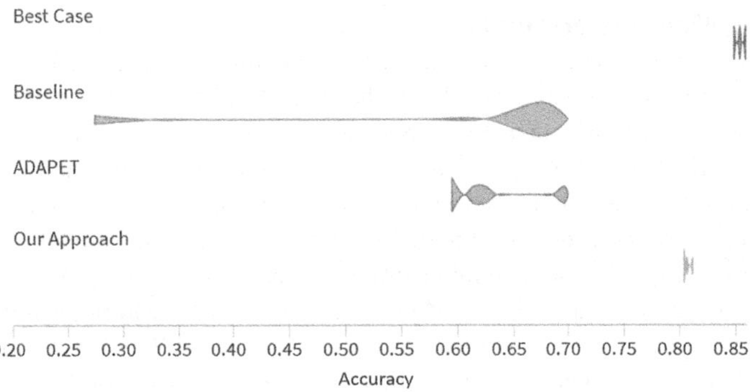

Fig. 12.5 Violin plots showing the accuracy differences of the main experimentation setting

The evaluation results show that our approach is able to identify cyber threat information from which we can deduce that a new classifier can be trained for upcoming cybersecurity incidents with limited data.

12.4.3.3 Ablation Studies

Finally, we want to give a more detailed insight into our method by showing how each component contributes to the resulting score. For this purpose, we conducted three further experiments in which we omitted one component in each case and evaluated the other two components. When multi-level fine-tuning is not used, we evaluate the BERT base model with the auxiliary data of the augmentation method and ADAPET for the learning objective. Without ADAPET, we train the cybersecurity pre-trained model on the CySecAlert dataset and the final task (with augmented data) with a classifier head. In the last experiment, the augmented data is simply omitted, while training the model in the multi-level fine-tuning process with ADAPET.

Upon examination of the results, presented in Table 12.5, it becomes clear that leaving out a component worsens the overall results. The highest loss is reached when the multi-level fine-tuning component is left out, showing how important it is. This behavior could be due to the many specific cybersecurity words trained by the general language modeling of cybersecurity data and to fine-tuning by a very related task that already gives the model an idea of how to distinguish between relevant and irrelevant content. Furthermore, we can clearly observe that leaving out ADAPET greatly worsens the results. When compared with the results of the main evaluation presented in Table 12.4, ADAPET even improves the values significantly more than compared to the baseline. This shows that ADAPET needs a strong base model

Table 12.5 Detailed evaluation results of the ablation experiments. The values on the left show the minimum, in the middle the mean, in brackets the standard deviation, and on the right the maximum value

Name	F1
Our Approach	80.42/ 80.63(0.27) /81.07
→ w/o Augmentation	78.48/ 80.33(1.27) /81.49
→ w/o Multi-Level Fine-Tuning	63.95/ 66.16(1.67) /67.43
→ w/o ADAPET	65.33/ 71.33(3.62) /75.08

to be highly beneficial. The smallest improvement is made with the augmented data. Although the data appeared to be of high quality (see Sect. 12.4.3.1), it did not significantly improve the classifier. Nevertheless, a small increase can be reached and the classifier training got more robust through the additional training data (smallest standard deviation).

12.5 Conclusion and Discussion

CTI, the collection of evidence-based knowledge of cybersecurity threats, is highly relevant for identifying and remediating security incidents. Professionals, security providers, CERTs, as well as many others in the cybersecurity realm can gain important information about the incidents, such as how severe they may be, which software and systems are affected, how to be protected, and if exploits exist. The challenges lie in the information overload and the high dynamics associated with every new threat event. To counteract the flood of information or to collect certain types of information, it is necessary to train a classifier. However, a trained classifier cannot generalise to new vulnerability events due to high dynamics (new names of vulnerabilities, paths, etc.) and new requirements (focus on exploitation, mitigation, consequences, etc.). To the best of our knowledge, this is the first work to address this problem by proposing a framework that enables rapid training of new, high-performance classifiers for specialized CTI. It consists of several components that allow the end user to label only a few data instances (tested here with 32 instances) to obtain a classifier that is comparable to one trained with 1800 instances. We also constructed a dataset labeled by three cybersecurity experts showing that this method indeed overcomes the problem of information overload and addresses high dynamics by being easily adaptable to new incidents.

12.5.1 Practical, Theoretical, and Empirical Contributions

Considering our findings, the study revealed (P) practical, (T) theoretical, and (E) empirical contributions:

(P) A novel pipeline for detecting specialized cyber threat information. Our work provides an approach to detect specific cyber threat information by addressing the problem that a trained classifier does not generalise well to new cybersecurity events and is not able to adapt to different requirements. Furthermore, it is aligned with the circumstances of such events. These circumstances include that information has to be gathered fast in the early stages of the events and that security institutions and experts do not have the time and capacity to label many instances. Therefore, we combine few-shot learning with multi-level fine-tuning and data augmentation to produce classifiers that only need few instances to perform with high quality. For few-shot learning we utilize ADAPET by Tam et al. [431] combined with the multi-level fine-tuning process. For data augmentation we use GPT-3 to create instances with novel linguistic patterns. Our pipeline reaches a F1-score of 80.63 on a specialized cyber threat dataset, which is 21.93 points above the score of a classical learning scheme. Other work, such as the cyber threat event detection systems of Riebe et al. [363] or Sceller et al. [381], allow for coarse-grained information gathering. To the best of our knowledge, our system is the first to provide rapid detection of specialized cyber threat information, by needing only very few data instances to create high-quality classifiers. This way, in the event of a current cybersecurity incident, experts can quickly create a classifier tailored to their specific needs and gather important information.

Moreover, our general approach to learning with very few examples can also be used for these detection systems or other cybersecurity problems, such as in IoC extraction. This leads over to the theoretical contributions of our work.

(T) New few-shot learning technique based on multi-level fine-tuning. We propose a novel few-shot learning approach for creating classifiers of high quality with a smaller amount of training data. The idea behind this approach is to fine-tune a machine learning model in several levels where enough data is available (see Figure 12.3). In our study we first further trained a BERT model on a general cybersecurity corpus. This model was then trained on a general Twitter cybersecurity relevance dataset. From this point, the model has a fundamental understanding of cybersecurity texts and is also able to distinguish cybersecurity-related content from irrelevant content. With this pre-trained knowledge, the model only needs few data instances to be able to differentiate specific cybersecurity content. As shown in this study, this new technique can also be combined with other techniques like ADAPET or data augmentation to further reduce the amount of needed training data.

However, we show that this multi-level fine-tuning approach has the greatest impact on classification quality of all techniques (+14.47 F1, see Table 12.5). The multi-level fine-tuning approach significantly advances research in few-shot learning, as it allows for a much higher model quality and at the same time can be combined with previous few-shot studies, such as ADAPET [431], DART [523], or PERFECT [271].

(T) New insights on data augmentation with large pre-trained language models. In our study, we also implemented a data augmentation technique that combines the works of Yoo et al. [512] and Bayer et al. [28]. As in the former, we used the large language model GPT-3 with a prompting strategy and filtered the generated instances with a human-in-the-loop technique, as in the latter. The idea is that GPT-3 can create instances with a very high degree of novelty, resulting in some very valuable instances. However, this novelty comes with the problem of poor label preservation, as the instances may be too far away from the class. For this reason, we also introduced this filtering strategy where the original labeled data of a class is compared with the generated data and those that are too far away from the original data are discarded. The boundary is determined by an expert who examines those instances close to a predefined boundary. As shown in Sect. 12.4.3.1 and Table 12.3, this procedure generates instances with very different transformation patterns, including word substitution, paraphrasing, and partial removal. It even leads to instances that are entirely novel. Furthermore, we included a quantitative evaluation in Sect. 12.4.3.1 comparing the data augmentation strategy against two of the most common NLP data augmentation strategies. It shows that our method is clearly superior for the task of specialized CTI in our pipeline.

However, in Sect. 12.4.3.3, we showed that omitting this method from the overall pipeline only slightly reduces the resulting score. This means that the model learns very little from the augmented data when multi-level fine-tuning and ADAPET are already used. Nevertheless, the evaluation results show a reduction in the standard deviation, which indicates that the model has become more robust with the artificial data.

(E) A specialized CTI dataset for further research purposes. In this study we created a CTI dataset based on the 2021 Microsoft Exchange Server data breach. The dataset was constructed by three experts. The guidelines have been revised several times in an attempt to flesh out the concept of cyber threat analysis as much as possible. Along with the code and the dataset, the guidelines are available in the repository. All annotators reached a good intercoder reliability showing that the guidelines and the general annotation process was successful. Further research can benefit from this dataset as it is, to our knowledge, the first to contain a relevance coding regarding CTI in Twitter in relation to a specific cybersecurity event.

12.5.2 Limitations and Outlook

In this work, we focus on Twitter as a data source, as it can be very up-to-date and rich, since the cybersecurity community is very active when it comes to vulnerabilities in Twitter. However, we would like to point out that using Twitter as a data source also brings some disadvantages. On Twitter, for example, anyone can share information, which can lead to a lot of speculative or even false information. A system that relies only on Twitter may not be as reliable and could easily be fooled. This is why we look forward to studies making the proposed approach more robust. In addition, we believe that other data sources should also be examined for vulnerabilities. We plan to integrate the component implemented in this work into an already developed, manageable dashboard that additionally includes a credibility component and aggregates information from many different sources.

In terms of the overall concept, we look forward to research studies testing the performance of this approach in other domains and on further cyber threat events. For example, it would be interesting to see if the same improvements can be achieved in medical or crisis domains, where data is also scarce. On a smaller scale, we also look forward to work applying our methodology to other cybersecurity events. Our pipeline was only evaluated with the MS Exchange data breach, but can be generalized to other CTI-related incidents as this was a priority in our development process. Moreover, our experiments are limited to the BERT base model. It would be interesting to see if the improvements are as high when a larger model like RoBERTa [263] is used. Likewise, one could also test other language models for the data augmentation technique. Especially interesting would be to test if open source models, like GPT-NeoX-20B [37], reach a good augmentation performance.

A part of our experiments was to fine-tune the model on the CySecAlert dataset of Riebe et al. [363]. The authors of this work propose an active learning component to achieve high classification scores with less data. With a view to future research, it might be sensible to also include active learning into the concept of our approach to further increase the classification quality. In practice, our approach would in the worst case lead to users labeling very similar examples, resulting in poor execution of data augmentation and poor classification quality, which can happen quickly when labeling such a small amount of data. Therefore, an active learning system could help to collect very different examples. Otherwise, experts can also be trained to label diverse examples.

Acknowledgements This work has been co-funded by the German Federal Ministry of Education and Research in the project CYWARN (13N15407) and funded by the Deutsche Forschungsgemeinschaft (DFG, German Research Foundation)—SFB 1119 (CROSSING)–-236615297, as well as the German Federal Ministry of Education and Research and the Hessian Ministry of Higher Education, Research, Science and the Arts within their joint support of the National Research Center for Applied Cybersecurity ATHENE. Calculations for this research were conducted on the Lichtenberg high performance computer of the TU Darmstadt.

XAI-Attack: Utilizing Explainable AI to Find Incorrectly Learned Patterns for Black-Box Adversarial Example Creation 13

Abstract

Adversarial examples, capable of misleading machine learning models into making erroneous predictions, pose significant risks in safety-critical domains such as crisis informatics, medicine, and autonomous driving. To counter this, we introduce a novel textual adversarial example method that identifies falsely learned word indicators by leveraging explainable AI methods as importance functions on incorrectly predicted instances, thus revealing and understanding the weaknesses of a model. Coupled with adversarial training, this approach guides models to adopt complex decision rules when necessary and simpler ones otherwise, enhancing their robustness. To evaluate the effectiveness of our approach, we conduct a human and a transfer evaluation and propose a novel adversarial training evaluation setting for better robustness assessment. While outperforming current adversarial example and training methods, the results also show our method's potential in facilitating the development of more resilient transformer models by detecting and rectifying biases and patterns in training data, showing baseline improvements of up to 23 percentage points in accuracy on adversarial tasks. The code of our approach is freely available for further exploration and use.

Original Publication *Markus Bayer, Markus Neiczer, Maximilian Sam-singer, Björn Buchhold, and Christian Reuter. "XAI-Attack: Utilizing Explainable AI to Find Incorrectly Learned Patterns for Black-Box Adversarial Example Creation". In:* 2024 Joint International Conference on Computational Linguistics, Language Resources and Evaluation (LREC-COLING 2024). *LREC-COLING '24. 2024*[1]

[1] Minor typographical errors in the original publication have been corrected for this dissertation.

M. Bayer, *Deep Learning in Textual Low-Data Regimes for Cybersecurity*, Technology, Peace and Security I Technologie, Frieden und Sicherheit, https://doi.org/10.1007/978-3-658-48778-2_13

Material
Experimental Code: https://github.com/PEASEC/XAI-Attack/

13.1 Introduction

Adversarial examples are specially crafted inputs to machine learning models that aim to trick them into making incorrect predictions. These inputs can be created by deliberately modifying existing examples to fool the algorithm [132]. It is evident that such deceptions can have serious consequences if they are framed as attacks. One example is crisis informatics, where deep learning models are used by response teams to gather and analyse information about, e.g., a natural disaster or terrorist attack. An attacker could construct examples that have nothing to do with the incident, but are designed in such a way that the model recognises them as relevant, leading to poorer insights and lower confidence in the information gathered.

Beyond that, the exploration of adversarial examples is very important, as it can show where a model has learned spurious correlations or shortcuts. Many machine learning practices result in the preference of shortcut solutions [126]. On the one hand, this can be desirable, as the models may be less prone to overfitting and generalize better [350]. On the other hand, this can lead to solutions that are too simple, which, for example, result from patterns and biases in the training dataset that correlate with classes in the data but are not actually responsible for the class. The research field of adversarial examples not only enables us to recognise them, but also offers methods to correct them. A commonly used strategy is adversarial training, where the model is re-trained with adversarial examples. However, current methods suffer from small or very specific robustness gains, partly due to their narrow design and partly due to ineffective evaluation methods.

Therefore, this paper considers an optimal adversarial example as one that has a significant learning factor. Instead of following current methods, we aim to uncover the model's incorrectly learned patterns by analyzing its erroneous predictions. As second novelty, we also propose to use feature/token attribution explainable AI (XAI) methods (e.g. LIME or SHAP) as importance functions to highlight the model's incorrectly learned indicators. Incorporating XAI into research on adversarial examples offers the opportunity to make more sophisticated importance calculations and to make them more flexible by providing a framework that can easily replace them. Our method coupled with adversarial training allows for a more robust model by highlighting and erasing patterns of the training data identified by the model which are either not truly indicative or not solely responsible for class determination. Hence, our approach promotes models to favor decision boundaries that are intricate when necessary, and straightforward when appropriate.

Unfortunately, there is currently no optimal evaluation method to ascertain a model's true robustness. The greatest challenge in adversarial training evaluation is that biases in training data often echo in the test data [126], leading to poor results. Therefore, we propose an out-of-distribution evaluation method that specifically addresses attack robustness, which, in combination with human and transfer evaluations, shows the performance of XAI-Attack with transformers.

Our work contributes the following aspects:

- **(C1)** A novel method for creating adversarial examples based on identifying indicators for wrong predictions.
- **(C2)** Proposal to utilize XAI methods as importance functions for adversarial example creation.
- **(C3)** A novel out-of-distribution evaluation setting for adversarial training that enables a more accurate assessment of robustness.

The code of this study is freely available[2].

13.2 Related Work

There is no unique formal definition of adversarial examples in the literature. In this work, we follow the definition of Goodfellow, Shlens, and Szegedy [132], which states that an adversarial example is an instance that has been intentionally curated from existing examples to fool the machine learning model. We add for clarification that the new instance should be semantically similar to its original instance. An adversarial example is deemed semantically similar to its original instance if, despite any textual variations, the underlying meaning pertaining to its label remains unchanged. Some works, such as [472], also imply that the adversarial examples should only be modified by a small change, ideally imperceptible to us humans, a prerequisite which we drop explicitly in our definition, as, for example, Brown et al. [46] or Ebrahimi et al. [99]. Adversarial examples can be created in black-box and white-box form, focusing on the model to be tricked, also known as the victim model. A white-box attack is one in which the internals of the model are completely transparent to the attacker [36]. In black-box attacks, the attacker can only query the victim model for an instance and get the prediction. Depending on the victim model, it outputs the prediction as soft labels, i.e. the softmax outputs, or as hard labels, i.e. only the class labels.

[2] https://github.com/PEASEC/XAI-Attack/

13.2.1 Feature Space

While adversarial examples are intensively studied in computer vision [225, 326, 430], they are more difficult to create for textual data due to the discrete nature and semantic coherence of text [124, 179]. This is also the reason for adversarial example methods being divided into data and feature space. In terms of feature space, they are often designed to maximise loss by adding noise in a white-box attack scenario. To solve the search for the right perturbation, several different maximisation methods and variants have been proposed in research [29]: PGD [77, 131], FreeAT [395], YOPO [520], FreeLB [538], VAT [292, 293], SMART [181], ALUM [258], and more. These feature space methods are often integrated directly into the training process, with SMART achieving the highest adversarial training scores in many tasks [29].

13.2.2 Data Space

However, our method acts in the data space. This includes, e.g., the work of Ebrahimi et al. [99], in which the letters of the input texts are flipped in such a way that the loss increases. For this, the gradients of the method and accordingly a white-box scenario are needed. Jin et al. [184] propose the black-box method TextFooler, which works at the word level and replaces important words with synonyms that are chosen based on embedding similarity and the highest change in prediction confidence. BERT-Attack [247] and BERT-based adversarial examples (BAE) [124] can be seen as variants of this approach, where BERT is used to create a list of substitution words. BERT-Attack differs from BAE in that it differentiates between words and subwords. BAE, conversely, proposes a replace and insert operation and additionally uses a Universal Sentence Encoder to filter the generated tokens to ensure high semantic similarity to the original text. The textbugger method of Ye et al. [509] also finds the important words and then creates either word-level changes with GLoVe embeddings or character-level changes based on rules. The authors propose a black-box and white-box variant. There are also methods for sentence-level adversarial examples, e.g. Iyyer et al. [173], which paraphrase a sentence with certain syntactic structures, the Entailment Preserving Transformations method of Thorne and Vlachos [438], which transforms the sentences based on templates, or the SSAE network by Li et al. [243], which consists of two auto-encoders to preserve syntax as well as semantics and to insert pertubations into the latent space. The research field also includes specialised methods such as the work of Qaraei and Babbar [334] for extreme multilabel text classification scenarios, Song et al. [414]

for text retrieval or Wan et al. [461] and Ebrahimi, Lowd, and Dou [98] for neural machine translation.

13.2.3 Robustness Evaluation

Research indicates that numerous methods exist for enhancing the robustness against adversarial examples, such as employing adversarial example generation alongside adversarial training, or detecting and filtering malicious inputs [133, 477]. Assessing adversarial robustness is crucial for determining the effectiveness of a defense method and, ultimately, the true robustness of a model. The efficacy of adversarial training is often gauged by evaluating the resulting model on the test or validation set of a task [98, 181, 247]. However, evaluating the adversarial robustness of a machine learning model solely on those sets may not always be sensible as they often share the same biases as the training data [126]. There are some frameworks, such as CleverHans [325] and FoolBox [353], that provide benchmarks for robustness. These frameworks offer different adversarial attack methods that can be used to evaluate the robustness of a model with a given dataset. While this approach mitigates the issue of evaluating robustness using only test or validation data, it introduces another potential challenge: the adversarial attack methods implemented by the framework may not generate high-quality adversarial examples, potentially skewing the robustness assessment.

13.2.4 Research Gap

XAI-Attack can be classified in the group of word-level data space perturbation methods. Unlike other adversarial example methods, it addresses the weaknesses of the models by extracting indicators from wrong predictions. Thus, it does not require access to the internals and, when combined with adversarial training, can greatly improve robustness. It creates the attacks in a black-box form and only needs the hard labels, i.e. the labels predicted by the model, but not the softmax outputs.

As discussed before, we abstract from any specific importance functions by proposing to use XAI methods that are not only easily replaceable in our framework but also tend to be much more sophisticated. This way, we combine the process of creating adversarial examples with any feature/token attribution XAI method to find the most important words. Importance/influence functions in current research, e.g. [124, 184, 247], simply omit words and calculate how much this changes the prediction score. The XAI method LIME [361], e.g., also omits words but calculates additional weights for the perturbed instances based on similarity to the original instance and then trains a locally interpretable model from which the importance scores are derived. Moreover, LIME does not require prediction results like the others, which makes our method suitable for a hard label environment. In our experiments we show the usage of LIME and SHAP [267].

Furthermore, our method is designed with the objective in mind of making models more adversarially robust so that they are less vulnerable to attacks while maintaining a high performance. Evaluating a model on the test set of the same task on which it was adversarially trained can be problematic due to the presence of biases in all task sets. Frameworks attempting to estimate adversarial robustness by using other adversarial example generators on a given set face the issue that these examples may be of low quality and invalid. To address this, we propose a novel evaluation setting by evaluating adversarial training on the Adversarial GLUE [466] dataset, which contains high quality and valid adversarial examples for GLUE tasks.

13.3 Attack Design

13.3.1 Problem Formulation and Requirements

In formulating the problem, we follow the example of Ye et al. [509]. Suppose we have a victim model f and a text instance x consisting of n words that has the label y and is correctly predicted by f. x' is an adversarial example of x, iff it is semantically similar to x and changes the prediction of the victim model f, i.e., $f(x') \neq f(x)$. It is constructed by inserting adversarial words into the instance x.

While our method for generating adversarial examples aims to train a robust deep learning model, it requires only rudimentary access to the model, making it potentially useful for constructing attacks as well. As previously described, the internals of the model cannot be viewed and the user receives no indication of the confidence in the prediction other than the hard label. In terms of the XAI method, we focus on LIME [361] and SHAP [267], which can explain a model prediction using only the hard labels and have proven to be advantageous in XAI research. It is also possible to use different variants that may require more access to the model. XAI-Attack only requires an feature/token attribution XAI method, i.e. one that provides a distribution over tokens indicating the correlation strength between input tokens and output.

In addition, the adversary needs a hold-out set. Ideally, this data does not come from the data the model was trained on, but from new annotated examples or a development set. However, it is also possible to separate a part of the training data for the generation of adversarial examples. Due to the design of our method, the model is afterwards trained with the data kept out anyway.

13.3.2 XAI-Attack

XAI-Attack uses a XAI method to highlight words which indicate the wrong class. In the following, these words are called adversarial words because they are used to create adversarial examples. However, potential and real adversarial words must be distinguished according to our definition of adversarial examples. The words returned by the XAI method are potential adversarial words because their insertion can change the semantics, whereas real adversarial words do not. Therefore, these potential adversarial words are subsequently cleaned by a targeted filtering method. The resulting adversarial words are then used to create adversarial examples. The procedure is demonstrated in Fig. 13.1 and described in detail below.

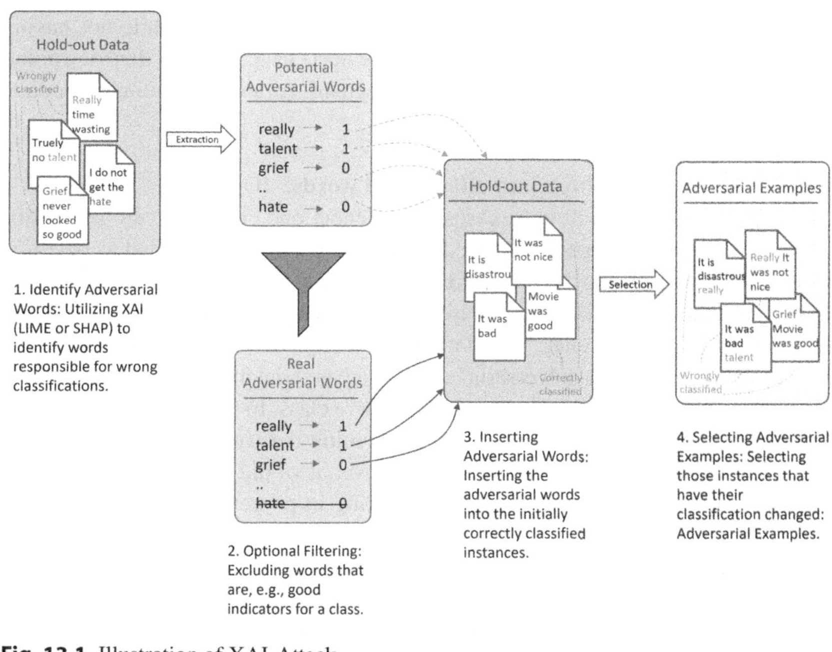

Fig. 13.1 Illustration of XAI-Attack

Step 1: Finding potential adversarial words
The first step of our method is to use a XAI method to find words that are responsible for false predictions. This is achieved by letting the victim model make predictions on the hold-out data. For each instance in this set that is incorrectly predicted, we apply a XAI method, which is able to highlight words responsible for the incorrect prediction. The word highlighted as most indicative of the wrong class is then considered as a potential adversarial word, i.e. a word that can potentially change the label without changing semantics.

This step takes advantage of a common problem with training deep learning models, as they tend to overfit quickly and find spurious correlations in the training data due to sampling and other biases. To illustrate, we anticipate a small example from a sentiment task. Using the XAI method, we can see that an example word responsible for changing the negative sentiment into a positive one is *like*. It stands to reason that the word *like*, in its verb meaning *to find someone or something pleasant or satisfying*, is an indicator of a positive mood. Nevertheless, the word *like*, in its meaning of e.g. the preposition as *similar to*, is not an indicator of positive sentiment. There are many words that are not edge cases like this one, but used as indicators of a class, since they occur frequently in the respective class, while not having semantic significance for the classification (see Sect. 13.4.5). From this viewpoint, it is apparent that our approach leans towards regularisers and eradicating inaccurate bias.

Step 2: Filter for label invariant adversarial words. [Optional]
There are words which in most cases truly denote one class, but can have a different meaning in very specific contexts, and which are also identified as potential adversarial words in step 1. The word *enjoy* in the sentiment task can serve as an illustrative example. While in most cases it is a word used in positive contexts, it can also be used in negative ones: "Hard to say who might enjoy this" (from SST2). If we now see *enjoy* as a real adversarial word, in adversarial training the model would be forced to reject it as an indicator of the positive class. In the best case, the model learns a more robust indicator, e.g., by distributing the weight of the decision of *enjoy* onto the context. In the worst case, however, it would discard a very valuable indicator and even learn further biases, resulting in poorer performance and less robustness.

Therefore, we try to filter out the words that could change the semantics in relation to the label. Besides no filtering of words, we propose two methods, one based on count of label changes and one based on indicator words for the correct class.

Count of label changes. In the count-based method, we analyze how many adversarial examples can be generated with a potential adversarial word. If the count exceeds a certain threshold relative to the class size, we assume that the semantics change in relation to the label and exclude all of those generated adversarial examples.

Indicator words for the correct class. The other method utilizes the additional data that was correctly predicted by the model. This data is explained using the XAI method and the resulting words are matched with the potential adversarial words. Potential adversarial words, which are an indicator of correct prediction, are excluded from further processing as they could semantically alter the instance in terms of the label.

Step 3: Creating adversarial examples
The adversarial words may change the label only for certain instances based on the position or context in the text. Therefore, we now check whether these words also change the labels of instances that were originally predicted correctly. That is, we insert the adversarial words into the correctly predicted instances of the additional data. There are many possibilities for where a word can be inserted, and although we also test random insertion, our main experiments are based on just prefixing them (recall that imperceptibility is not a criterion in this work). For more ideas on inserting an adversarial word into an instance, see Sect. 13.5.1. The instances whose labels have been changed by the insertion of adversarial words are then the resulting adversarial examples.

13.4 Experiments

13.4.1 Experiment Types

Our experimentation perspective lies in the questions (1) how often the method actually generates examples that do not change the label but the prediction, (2) whether a model trained with the examples becomes more robust, and (3) whether the adversarial examples can be transferred to other models.

Human Evaluation: To measure how often the method generates true adversarial examples, we perform a human evaluation in which the generated instances of each dataset are labeled.

Adversarial Testing and Training: The measurement of robustness is based on the benchmark of Wang et al. [466]. The benchmark consists of created and human-rated adversarial examples for various GLUE tasks. For each task, we test whether models trained with XAI-Attack adversarial examples from the normal GLUE tasks are more robust than those trained without or with adversarial examples from other methods. For completeness, we also include adversarial training experiments on standard GLUE tasks.

Adversarial Transfer: In the final setting of our experiments, we predict whether the adversarial examples are transferable to another model. To do this, we create the adversarial examples with distilBERT and then measure the impact on another distilBERT, a BERT and a RoBERTa model.

Unlike some previous work in the field of adversarial examples, we are not investigating the success rate as we believe that the success rate alone is no indication of the quality of the method.

13.4.2 Datasets & Model Settings

The tasks we focus on are the same as in [466] due to the adversarial testing experiment. That is, the experiments are conducted with the datasets SST-2, RTE, QNLI, MNLI and QQP from the GLUE benchmark [464]. For the experiments we use distilBERT [379] as the main model[3]. In terms of robustness measures, we employ only standard methods like weight decay and dropout, without incorporating any additional techniques.

13.4.3 Human Evaluation

For this section, first we go into the details of the quantitative experiment, then some adversarial examples are examined. The adversarial examples used in this section are generated with LIME and the optional filtering step based on the indicator words for the correct class (see Sect. 13.3.2).

For each task, we randomly select 100 instances of the original dataset and 100 instances of the adversarial dataset with uniform class distribution. The resulting

[3] Main experiments: distilBERT-base-uncased—Transfer models: BERT-base and RoBERTa-base—Parameter: Standard Huggingface with 3 epochs, 500 warmup steps, learning rate of 1e-3, and weight decay of 0.01—Implementations: BAE and BERT-Attack ($k = 7$) from TextAttack [295], SMART from the original implementation [181]

200 instances for each task are randomly labeled by two independent annotators. A comparison of human performance can show us whether the adversarial examples are mostly valid if the two metrics for the original and adversarial data of each task are similar (note that we do not test imperceptibility, but only whether an adversarial example changes the gold label). For the rationale and sizes in this experiment, we follow standard practice in the field (see [124, 184, 247]).

Table 13.1 shows the results of the human experiment, measured by the mean accuracy of the two annotators. The results of the original and the adversarial examples are very similar, bearing in mind that the classifier is wrong 100% of the time on the adversarial data. There are even tasks where the adversarial examples match the given labels more than the original data. This is of course due to variance in the selected sample, which was to be expected since not the entire dataset is labeled. Moreover, the agreement between the annotators (Cohen's kappa) is substantial for the non-adversarial data at 0.6681 and for the adversarial data at 0.6465 according to Landis and Koch [232]. Looking at the results as a whole, it is clear that the adversarial examples are of very high quality and only change the semantics in a few cases, but deceive the trained classifier in 100% of the cases.

Table 13.1 Human evaluation: Two annotators assessing 100 original instances and 100 adversarial instances (generated using LIME and indicator filtering) of each task, with the averaged accuracy

Dataset	Original Data	Adversarial Data
SST2	0.8787	0.7534
RTE	0.9150	0.9600
QQP	0.8550	0.7338
QNLI	0.8994	0.7836
MNLI-mm	0.7626	0.8333
MNLI-m	0.7980	0.7828

This can be further illustrated by looking in detail at some adversarial examples. The adversarial examples listed in Table 13.2 are chosen because of the different insights that can be gained from them. The first instance shows a very subtle adversarial example, with only the letter "q" added to the instance. This already makes the classifier predict that the second sentence is a valid answer to the question of the first sentence. The second example also shows that such trained classifiers are very likely to memorise words and phrases from the training data. The word "enjoy" occurs

512 times in the positive training data and only 98 times in the negative training data of the sentiment task. It stands to reason that the model simply predicts almost every instance in which the word "enjoy" occurs as positive. While this is mostly true, there are some exceptions, such as this second example. Likewise, it is very probable that such a word changes the semantics of the instance when it is inserted, as we can see in the next example in the table. In such a case, the count-based filtering method might have excluded the word "enjoy". In sentiment prediction, it is also very odd to see that the word "better" is an adversarial word in the fourth example, turning a positive instance into a negative one, predicted by the classifier. The fifth example is a very interesting case of a wrong adversarial example. Here, the word "Java" is inserted into the first question, causing a semantic shift where the resulting question is a duplicate of the second question. While one might think that "Java" is not a good adversarial word, it is even more interesting to see that the insertion of "Java" in the second question also leads to a change in the prediction, while the semantics are not changed this time.

Table 13.2 Examples of adversarial instances from different datasets (inserted adversarial words shown in blue). Valid adversarial examples are those that do not change the semantics in relation to the label

Task	Instance	Label	Prediction	Valid
QNLI	Question: What issue has been plaguing the civil disobedience movement.	Not Entailment	Entailment	✓
	Sentence: *q* It has been argued that the term "civil disobedience" has always suffered from ambiguity and in modern times, become utterly debased.			
SST2	*enjoy* not only unfunny , but downright repellent .	Negative	Positive	✓
SST2	*enjoy* but it could have been worse .	Negative	Positive	✗
SST2	*better* good old-fashioned slash-and-hack is back !	Positive	Negative	✓
QQP	Question1: *Java* What is abstract class and methods?	Not Duplicate	Duplicate	✗
	Question2: What is abstract class and methods in java?			
QQP	Question1: What is abstract class and methods?	Not Duplicate	Duplicate	✓
	Question2: *Java* What is abstract class and methods in java?			

13.4.4 Adversarial Testing

The rationale behind the adversarial testing experiment is to test if models trained with adversarial examples from XAI-Attack are more robust than those trained with no or other adversarial examples. For this, we first generate adversarial examples for a standard GLUE task. Then, we test a model trained together with the task data and these adversarial examples on a separate dataset, that was deliberately created to measure robustness on the same task. The process is visualized in Fig. 13.2.

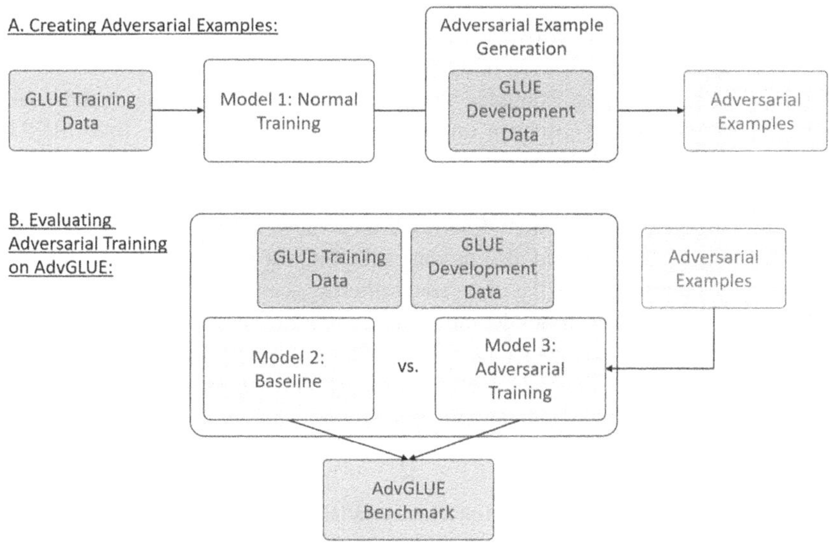

Fig. 13.2 Adversarial testing illustration

In detail, we first train models based on the training data from the original GLUE datasets (non-adversarial). For these models, we then create adversarial examples for the development set of the same GLUE task (non-adversarial). Subsequently, a model based on the training set, the development set and additionally with the adversarial examples is tested on the development sets of the Adversarial GLUE benchmark [466]. For the baseline, we omit the adversarial examples and train only on the GLUE training and development data.

As part of this experiment, we also compare our method with the state-of-the-art word-level adversarial example methods, BAE [124] and BERT-Attack [247], as well as with the adversarial training method SMART [181]. Regarding the ablations, we test five different settings of XAI-Attack. For three of them, the filtering techniques mentioned in Sect. 13.3.2 are used. Based on pre-evaluation experiments, we set the parameters for count-based filtering so that a word is discarded if it changes more than 30% of the class data, and for indicator-based filtering so that a word is an indicator of the correct class if it explains more than 1% of this class. In the fourth ablation scenario, adversarial words are randomly inserted into the instances. Finally, in the fifth scenario, we utilize the SHAP XAI method to identify adversarial words. No filtering is applied in either of the two scenarios.

The accuracy results of the baseline, XAI-Attack with the five different settings, BERT-Attack, BAE, and SMART are shown in Table 13.3. When analysing the results, it is important to bear in mind that the benchmark itself is adversarial, i.e. a trained model gets most examples wrong. The adversarial training method with our XAI-Attack shows significant improvements over the baseline, BAE, BERT-Attack, and SMART. XAI-Attack with indicator and without filtering is able to increase performance on all datasets compared to the baseline. In the case of the QPP dataset, XAI-Attack with no filtering even achieves an improvement of 23.08 accuracy points. This method also produces the highest results overall. While the adversarial example method based on count filtering seems to be the worst filtering type of XAI-Attack, sometimes degrading the results and showing weaker improvements, the indicator filtering method shows the most consistent improvements. The difference between the method without filtering (highest improvements, but somewhat inconsistent) and the method with indicator filtering (more consistent, but not

Table 13.3 Adversarial testing results on the adversarial GLUE tasks (accuracy). Best values are highlighted and arrows represent an increase or a decrease compared to the baseline, respectively

Method	SST2	RTE	QQP	QNLI	MNLI-mm	MNLI-m
Baseline	0.3243	0.5802	0.5513	0.5945	0.2901	0.3636
↳ No Filt.	0.4595 (↑)	**0.8025** (↑)	**0.7821** (↑)	**0.6554** (↑)	0.3827 (↑)	0.5785 (↑)
↳ Count Filt.	0.3851 (↑)	0.5161 (↓)	0.6795 (↑)	0.6351 (↑)	0.3642 (↑)	0.4545 (↑)
↳ Indicator Filt.	0.4527 (↑)	0.6420 (↑)	0.7307 (↑)	0.6283 (↑)	**0.4383** (↑)	**0.5868** (↑)
↳ Rand. Ins.	**0.5067** (↑)	0.6914 (↑)	0.7179 (↑)	0.5608 (↓)	0.3641 (↑)	0.4876 (↑)
↳ SHAP	0.3919 (↑)	0.4691 (↓)	0.5641 (↑)	0.5068 (↓)	0.3086 (↑)	0.3719 (↑)
BAE	0.4459 (↑)	0.5432 (↓)	0.7436 (↑)	0.5676 (↓)	0.3827 (↑)	0.4876 (↑)
BERT-Attack	0.4256 (↑)	0.5556 (↓)	0.6154 (↑)	0.6351 (↑)	0.4136 (↑)	0.4959 (↑)
SMART	0.5000 (↑)	0.5679 (↓)	0.5384 (↓)	0.4527 (↓)	0.3765 (↑)	0.3636 (–)

highest improvements) can also be explained intuitively. Filtering excludes words from the list of potential adversarial words that may be important for predicting the correct class and in fact would change instances semantically, which is why the results consistently improve and the method without filtering may worsen the prediction quality in some cases. On the other hand, words that do not change instances semantically and would have had a high learning effect if they had been included in the adversarial training may also be excluded in this way, resulting in lower improvements than the more open method. In general, if the respective use case does not allow much tuning and validation (or its use in a general framework), we would recommend using the indicator filtering method, as we have noticed the most consistent improvements. However, if the use case allows for tuning and validation, we would also recommend trying to use XAI-Attack without filtering, as it might produce even better results. Besides, the results show that XAI-Attack has only mixed performance with SHAP [267], which is also evident in the adversarial words extracted with the method. We found that these have only little semantics with respect to the label. Finally, the results also highlight that random insertion can further improve already good results, as in the case of SST2, giving rise to further research into more sophisticated insertion methods.

13.4.5 Adversarial Training on Standard GLUE

While we demonstrated that XAI-Attack significantly enhances the resilience of transformers on the Adversarial GLUE benchmark, this subsequent experiment focuses on evaluating the effectiveness of adversarial training on the standard GLUE tasks.

In contrast to the adversarial testing experiment, we split the training data for each task to obtain a hold-out set of 10%. XAI-Attack then creates adversarial examples using LIME and indicator filtering. DistilBERT-base is subsequently trained using both the full training data and additional adversarial examples, and then compared to a model trained only with the full training data.

Table 13.4 displays the results for the validation sets of the tasks. It can be observed that both are quite similar, with an expected smaller decrease as XAI-Attack identifies biases in the models. These biases may originate from the training data and could also be present in the validation data. To further investigate the cause of the decrease in performance, we conduct a qualitative analysis of the adversarial examples in the SST2 sentiment task. Our primary focus is on identifying adversarial words that alter the prediction of most correctly classified instances. These words serve as a clear indicator of a particular class. Words such as "bored," "silly," "charmless," "insightful," and "terrific" are then disregarded due to their clear semantic relevance to sentiment classification. Our investigation highlights words that are overrepresented in both the training and validation datasets, despite their semantic neutrality. For instance, the word "strain" was found to shift 1511 positive instances to negative. This shift occurs because the classifier had erroneously learned to associate "strain" with the negative class, evidenced by its presence in 70 negative class instances compared to just 36 in the positive class within the training data. The benchmark problem then arises because the validation set includes three examples of the negative class and no examples of the positive class. A classifier trained using examples from XAI-Attack or other adversarial methods may, at best, perform equally well in this regard, even if it has overcome bias and learned more complex rules. Additional examples of this phenomenon are detailed in Table 13.5.

Table 13.4 Adversarial training of XAI-Attack (LIME and indicator filtering) on the standard GLUE tasks (accuracy)

Method	SST2	RTE	QQP	QNLI
Baseline	0.9037	0.5884	0.9027	0.8814
XAI-A.	0.9025	0.5704	0.8917	0.8706

Table 13.5 Adversarial words (LIME) from the SST2 dataset that are over-represented in the train and validation set. Train and validation representations are of the form negative (0)—positive (1)

Word	Label Transf.	#Adv. Ex.	Train Repr. 0–1	Val. Repr. 0–1
strain	$1 \Rightarrow 0$	1511	70–36	3–0
pedestrian	$1 \Rightarrow 0$	778	13–3	1–0
slaps	$1 \Rightarrow 0$	849	52–23	2–0
earnest	$0 \Rightarrow 1$	397	42–96	0–3
innocence	$0 \Rightarrow 1$	346	17–28	0–1
provides	$0 \Rightarrow 1$	260	3–63	0–3

13.4.6 Adversarial Transfer

In this section we want to test whether the adversarial examples of one model are transferable to other models, i.e. also valid adversarial examples for other models. To do this, on the original GLUE benchmark, we take the adversarial examples generated by XAI-Attack with LIME and indicator filtering on the distilBERT-base model and test whether they can also fool another distilBERT-base model, a BERT-base model and a RoBERTa-base model trained with the same data.

The results of this experiment are shown in Table 13.6. It is clear that all of the models can be deceived to a large extent and that the adversarial examples are not only valid for a specifically trained model. The distilBERT model remains the one with the lowest results, which was to be expected as it is the same model type. The much more comprehensive BERT model is still very susceptible to the adversarial examples, which is reflected in the low accuracy results. The RoBERTa model, which has been trained for much longer and with much more data, performs slightly better. In the QQP task, it even achieves an acceptable result of 0.7653. In the other tasks, it is still well below the results of the human evaluation. This indicates that XAI-Attack is able to find biases in datasets that are adopted by all transformer models.

Table 13.6 Accuracy results of the adversarial transfer experiment. Adversarial examples (XAI-Attack with LIME and indicator filtering) of a distilBERT model are tested with another distilBERT, a BERT and a RoBERTa model

Dataset	distilBERT	BERT	RoBERTa
SST2	0.1940	0.3421	0.6178
RTE	0.6136	0.6877	0.4293
QQP	0.3299	0.4768	0.7653
QNLI	0.3864	0.6090	0.6962
MNLI-mm	0.2709	0.3670	0.6061
MNLI-m	0.2672	0.3797	0.3797

13.4.7 Summary of the Results

In the human evaluation, the instances of the original datasets and the adversarial examples are labeled and some cases are analyzed in more detail. It shows that the generated adversarial examples are mostly not semantics-changing with regard to the label, i.e. valid adversarial examples. From this, it can also be inferred that the adversarial examples are of high value for adversarial training. To underpin this, we propose an experiment in which we utilize the Adversarial GLUE benchmark [466] to test whether models trained with different adversarial examples are more robust against other adversarial attacks. The results show that adversarial training with XAI-Attack improves the robustness considerably even compared to state-of-the-art word-level adversarial attacks and training methods, such as BERT-Attack, BAE, and SMART. We also test different filtering strategies, of which the absence of filtering achieves the highest improvements and the indicator filtering achieves consistent improvements. Furthermore, the evaluation results of random insertion show that XAI-Attack's standard method of appending the adversarial words at the beginning can be improved (more on this in Sect. 13.5.1). The adversarial training experiments verify that common benchmarks have biases that are present in all task sets and exemplify the need for more out-of-distribution evaluations. In the last experiment we demonstrate that adversarial examples created for one model are transferable to other models, showing that transformer-based models are generally susceptible to XAI-Attack examples.

13.5 Conclusion

Adversarial examples are of great importance in all fields, as they show the flaws of a model and can even be used to attack a system. In this study, a new adversarial example method using XAI is proposed. The method was evaluated in several experiments consisting of a human evaluation, a novel method to assess robustness and a transferability evaluation. These experiments show high quality adversarial examples, significant improvements in the robustness of the models and strong transferability to larger models.

13.5.1 Findings

Besides the apparent contributions of an adversarial example creation/training method based on XAI and a novel way of assessing robustness of adversarial training, by using a specialized adversarial example benchmark, this study also revealed more obscure findings.

Combining XAI and adversarial example research results in two innovations: On the one hand, the use of XAI methods allows for more sophisticated importance scores, a wider function space and fewer constraints on the model (no soft labels required) than the importance functions currently used. On the other hand, focusing on mislearned textual cues by explaining incorrectly predicted instances has a much higher success rate and ultimately the greatest learning effect when combined with adversarial training.

Additionally, this revealed novel insights of **learned biases of transformer models**. One might tend to overestimate the performance of these pre-trained models, as they score very high on common NLP benchmarks. However, inspection of the adversarial examples produced by our method shows that the trained models often make their predictions based on one word only, and do not produce more complex rules, also known as shortcut learning [126]. While showing how fragile a trained model can be, this also shifts the light towards the benchmarks. Due to the high performance of the models on the test data, but the heavy reliance on individual words without semantics in relation to the label, it is evident that the **common test datasets have biases** which are also visible in the training and development data. While Sect. 13.4.5 provides even more evidence for this, we expect further research on this topic. An important implication for practical systems is, that one should always ensure that the validation and test sets should strictly represent the real data (involving constant re-evaluations).

LIMITATIONS

Our study encountered several limitations, in light of which we identify avenues for future research that extend and deepen the understanding of our results. For example, we have not tried XAI-Attack on larger transformer models such as GPT-3 [47]. These models are expected to be much more robust against attacks, but since they can also be brittle in terms of the right prompting, we could imagine XAI-Attack finding adversarial examples for them as well. However, research into smaller language models remains important because, for example, they can be used on one's own hardware, are easier to interpret, can be easily fine-tuned, do not hallucinate and could be just as good or better in certain areas.

With regard to XAI-Attack itself, we would like to emphasise that it requires a hold-out set, which can be the development set or part of the training set. Further research could address the question of how much data is needed for the hold-out set and how the size affects adversarial training and ultimately the robustness of models. This can also be important for the question of how well XAI-Attack generalizes. The adversarial examples in the Adversarial GLUE benchmark are crafted using very diverse methods, resulting in all kinds of adversarial examples. The significant performance increase from adversarial training on these examples is a first indication that the model has indeed become more generally robust. However, we believe that this generalisability depends on how many adversarial examples are found and used for training. With very few adversarial examples, the model might only unlearn the biases for the specific adversarial words used in training. With enough adversarial examples (like in the experiments in the paper), on the other hand, the model tends to become more robust in general.

In its standard implementation, XAI-Attack inserts the adversarial words only at the beginning of the instances, which has already led to good results. However, in our experiments we also looked at inserting the words at a random position, which actually improved the results in one task significantly (see Sect. 13.4.4). Hence, it could be very interesting for further studies to investigate how the words could be inserted in a more informed way (e.g. by using language models to predict the most appropriate and coherent insertion). With this, the instances generated could also be more imperceptible as adversarial examples, which some works, such as Wang et al. [472], consider part of the definition of an adversarial example. Regarding the adversarial training, another direction would be to remove the identified adversarial words of some of the training instances, which could lead to more complex decision rules of the model as long as it is ensured that the label does not change. Taking this this idea even further, it might be interesting to let generative models create instances based on the adversarial words.

In our human evaluation experiments, the annotators' labeling performance on the normal dataset was not perfect. This may be partly due to the fact that our annotators are fluent in English but not native speakers. Furthermore, the results are comparable to the study by Nangia and Bowman [302], who also performed human evaluations on the GLUE benchmark, showing that the tasks are not as easy as they might seem at first glance.

Finally, while we are confident that XAI-Attack will work well with other languages, we have only experimented with English examples. We look forward to adversarial research with other languages.

ETHICS STATEMENT

In our work, we have consistently prioritized ethics, continuously reassessing our approach to ensure responsible conduct. Research in the field of adversarial example generation inherently poses risks of misuse, notably in the form of attacks on machine learning models. This risk is especially pronounced in safety-critical domains, as we have outlined in the beginning. While our method could be repurposed for malicious use, we want to emphasize that the primary goal of this work is to generate adversarial examples that can be used to fortify machine learning models against such and other attacks, as we have demonstrated with the adversarial testing experiment in Sect. 13.4.4.

In addition, we advise reviewing the adversarial examples generated by XAI-Attack. This applies to any adversarial example method, as the examples may introduce new unwanted biases.

Acknowledgements This research work has been funded by the German Federal Ministry of Education and Research and the Hessian Ministry of Higher Education, Research, Science and the Arts within their joint support of the National Research Center for Applied Cybersecurity ATHENE and by the German Federal Ministry of Education and Research in the projects CYWARN (13N15407) and CYLENCE (13N16636). The calculations for this research were conducted on the Lichtenberg high performance computer of the TU Darmstadt.

Bibliography[1]

1. Marah I Abdin, Sam Ade Jacobs, Ammar Ahmad Awan, Jyoti Aneja, Ahmed Awadallah, Hany Awadalla, Nguyen Bach, Amit Bahree, Arash Bakhtiari, Harkirat S. Behl, Alon Benhaim, Misha Bilenko, Johan Bjorck, Sébastien Bubeck, Martin Cai, Caio César Teodoro Mendes, Weizhu Chen, Vishrav Chaudhary, Parul Chopra, Allie Del Giorno, Gustavo de Rosa, Matthew Dixon, Ronen Eldan, Dan Iter, Amit Garg, Abhishek Goswami, Suriya Gunasekar, Emman Haider, Junheng Hao, Russell J. Hewett, Jamie Huynh, Mojan Javaheripi, Xin Jin, Piero Kauffmann, Nikos Karampatziakis, Dongwoo Kim, Mahoud Khademi, Lev Kurilenko, James R. Lee, Yin Tat Lee, Yuanzhi Li, Chen Liang, Weishung Liu, Eric Lin, Zeqi Lin, Piyush Madan, Arindam Mitra, Hardik Modi, Anh Nguyen, Brandon Norick, Barun Patra, Daniel Perez-Becker, Thomas Portet, Reid Pryzant, Heyang Qin, Marko Radmilac, Corby Rosset, Sambudha Roy, Olatunji Ruwase, Olli Saarikivi, Amin Saied, Adil Salim, Michael Santacroce, Shital Shah, Ning Shang, Hiteshi Sharma, Xia Song, Masahiro Tanaka, Xin Wang, Rachel Ward, Guanhua Wang, Philipp Witte, Michael Wyatt, Can Xu, Jiahang Xu, Sonali Yadav, Fan Yang, Ziyi Yang, Donghan Yu, Chengruidong Zhang, Cyril Zhang, Jianwen Zhang, Li Lyna Zhang, Yi Zhang, Yue Zhang, Yunan Zhang, and Xiren Zhou. "Phi-3 Technical Report: A Highly Capable Language Model Locally on Your Phone". In: *CoRR* abs/2404.14219 (2024). DOI: https://doi.org/10.48550/ARXIV.2404.14219. arXiv:2404.14219. URL: https://doi.org/10.48550/arXiv.2404.14219.
2. Fabian Abel, Claudia Hauff, Geert-Jan Houben, Richard Stronkman, and Ke Tao. "Semantics + filtering + search = twitcident. exploring information in social web streams". In: *23rd ACM Conference on Hypertext and Social Media, HT '12, Milwaukee, WI, USA, June 25–28, 2012*. Ed. by Ethan V. Munson and Markus Strohmaier. ACM, 2012, pp. 285–294. DOI: https://doi.org/10.1145/2309996.2310043. URL: https://doi.org/10.1145/2309996.2310043.
3. Fabian Abel, Claudia Hauff, Geert-Jan Houben, Richard Stronkman, and Ke Tao. "Twitcident: fighting fire with information from social web streams". In: *Proceedings of the 21st World Wide Web Conference, WWW 2012, Lyon, France, April 16–20, 2012*

[1] This dissertation, including all associated code, was independently written by the authors. Language and editorial support, as well as code assistance, were provided by OpenAI's ChatGPT.

M. Bayer, *Deep Learning in Textual Low-Data Regimes for Cybersecurity*, Technology, Peace and Security I Technologie, Frieden und Sicherheit, https://doi.org/10.1007/978-3-658-48778-2

(Companion Volume). Ed. by Alain Mille, Fabien Gandon, Jacques Misselis, Michael Rabinovich, and Steffen Staab. ACM, 2012, pp. 305–308. DOI: https://doi.org/10.1145/2187980.2188035. URL: https://doi.org/10.1145/2187980.2188035.

4. Hugo Queiroz Abonizio and Sylvio Barbon Junior. "Pre-trained Data Augmentation for Text Classification". *In: Intelligent Systems—9th Brazilian Conference, BRACIS 2020, Rio Grande, Brazil, October 20–23, 2020, Proceedings, Part I*. Ed. by Ricardo Cerri and Ronaldo C. Prati. Vol. 12319. Lecture Notes in Computer Science. Springer, 2020, pp. 551–565. DOI: https://doi.org/10.1007/978-3-030-61377-8_38. URL: https://doi.org/10.1007/978-3-030-61377-8_38.
5. Md Sahrom Abu, Siti Rahayu Selamat, Aswami Ariffin, and Robiah Yusof. "Cyber threat intelligence–issue and challenges". In: *Indonesian Journal of Electrical Engineering and Computer Science* 10.1 (2018), pp. 371–379.
6. Mistral AI. *Au large*. Feb. 2024. URL: https://www.mistral.ai/news/mistral-large/.
7. Milam Aiken and Mina Park. "The efficacy of round-trip translation for MT evaluation". In: *Translation Journal* 14.1 (2010), pp. 1–10.
8. Firoj Alam, Shafiq R. Joty, and Muhammad Imran. "Domain Adaptation with Adversarial Training and Graph Embeddings". In: *Proceedings of the 56th Annual Meeting of the Association for Computational Linguistics, ACL 2018, Melbourne, Australia, July 15–20, 2018, Volume 1: Long Papers*. Ed. by Iryna Gurevych and Yusuke Miyao. Association for Computational Linguistics, 2018, pp. 1077–1087. DOI: https://doi.org/10.18653/V1/P18-1099. URL: https://www.aclanthology.org/P18-1099/.
9. Firoj Alam, Shafiq R. Joty, and Muhammad Imran. "Graph Based Semi-Supervised Learning with Convolution Neural Networks to Classify Crisis Related Tweets". In: *Proceedings of the Twelfth International Conference on Web and Social Media, ICWSM 2018, Stanford, California, USA, June 25–28, 2018*. AAAI Press, 2018, pp. 556–559. URL: https://www.aaai.org/ocs/index.php/ICWSM/ICWSM18/paper/view/17815.
10. Firoj Alam, Ferda Ofli, and Muhammad Imran. "Descriptive and visual summaries of disaster events using artificial intelligence techniques: case studies of Hurricanes Harvey, Irma, and Maria". In: *Behav. Inf. Technol.* 39.3 (2020), pp. 288–318. DOI: https://doi.org/10.1080/0144929X.2019.1610908. URL: https://doi.org/10.1080/0144929X.2019.1610908.
11. Hunt Allcott and Matthew Gentzkow. "Social Media and Fake News in the 2016 Election". In: *Journal of Economic Perspectives* 31.2 (2017), pp. 211–236. ISSN: 0895–3309.
12. Liz Allen, Alison O'Connell, and Veronique Kiermer. "How can we ensure visibility and diversity in research contributions? How the Contributor Role Taxonomy (CRediT) is helping the shift from authorship to contributorship". In: *Learn. Publ.* 32.1 (2019), pp. 71–74. DOI: https://doi.org/10.1002/LEAP.1210. URL: https://doi.org/10.1002/leap.1210.
13. Noufa Alnajran, Keeley A. Crockett, David McLean, and Annabel Latham. "Cluster Analysis of Twitter Data: A Review of Algorithms". In: *Proceedings of the 9th International Conference on Agents and Artificial Intelligence, ICAART 2017, Volume 2, Porto, Portugal, February 24–26, 2017*. Ed. by H. Jaap van den Herik, Ana Paula Rocha, and Joaquim Filipe. SciTePress, 2017, pp. 239–249. DOI: https://doi.org/10.5220/0006202802390249. URL: https://doi.org/10.5220/0006202802390249.
14. Emily Alsentzer, John R. Murphy, Willie Boag, Wei-Hung Weng, Di Jin, Tristan Naumann, and Matthew B. A. McDermott. "Publicly Available Clinical BERT Embed-

dings". In: *CoRR* abs/1904.03323 (2019). arXiv:1904.03323. URL: http://arxiv.org/abs/1904.03323.

15. Fernando Alves, Ambrose Andongabo, Ilir Gashi, Pedro M. Ferreira, and Alysson Bessani. "Follow the Blue Bird: A Study on Threat Data Published on Twitter". In: *Computer Security—ESORICS 2020—25th European Symposium on Research in Computer Security, ESORICS 2020, Guildford, UK, September 14–18, 2020, Proceedings, Part I*. Ed. by Liqun Chen, Ninghui Li, Kaitai Liang, and Steve A. Schneider. Vol. 12308. Lecture Notes in Computer Science. Springer, 2020, pp. 217–236. DOI: https://doi.org/10.1007/978-3-030-58951-6_11. URL: https://doi.org/10.1007/978-3-030-58951-6_11.
16. Fernando Alves, Aurélien Bettini, Pedro M. Ferreira, and Alysson Bessani. "Processing tweets for cybersecurity threat awareness". In: *Inf. Syst.* 95 (2021), p. 101586. DOI: https://doi.org/10.1016/J.IS.2020.101586. URL: https://doi.org/10.1016/j.is.2020.101586.
17. Moustafa Alzantot, Yash Sharma, Ahmed Elgohary, Bo-Jhang Ho, Mani B. Srivastava, and Kai-Wei Chang. "Generating Natural Language Adversarial Examples". In: *Proceedings of the 2018 Conference on Empirical Methods in Natural Language Processing, Brussels, Belgium, October 31—November 4, 2018*. Ed. by Ellen Riloff, David Chiang, Julia Hocken-maier, and Jun'ichi Tsujii. Association for Computational Linguistics, 2018, pp. 2890–2896. DOI: https://doi.org/10.18653/V1/D18-1316. URL: https://doi.org/10.18653/v1/d18-1316.
18. Ateret Anaby-Tavor, Boaz Carmeli, Esther Goldbraich, Amir Kantor, George Kour, Segev Shlomov, Naama Tepper, and Naama Zwerdling. "Do Not Have Enough Data? Deep Learning to the Rescue!" In: *The Thirty-Fourth AAAI Conference on Artificial Intelligence, AAAI 2020, The Thirty-Second Innovative Applications of Artificial Intelligence Conference, IAAI 2020, The Tenth AAAI Symposium on Educational Advances in Artificial Intelligence, EAAI 2020*, New York, NY, USA, February 7–12, 2020. AAAI Press, 2020, pp. 7383–7390. DOI: https://doi.org/10.1609/AAAI.V34I05.6233. URL: https://doi.org/10.1609/aaai.v34i05.6233.
19. Jacob Andreas. "Good-Enough Compositional Data Augmentation". In: *Proceedings of the 58th Annual Meeting of the Association for Computational Linguistics, ACL 2020, Online, July 5–10, 2020*. Ed. by Dan Jurafsky, Joyce Chai, Natalie Schluter, and Joel R. Tetreault. Association for Computa tional Linguistics, 2020, pp. 7556–7566. DOI: https://doi.org/10.18653/V1/2020.ACL-MAIN.676. URL: https://doi.org/10.18653/v1/2020.acl-main.676.
20. Rohan Anil, Sebastian Borgeaud, Yonghui Wu, Jean-Baptiste Alayrac, Jiahui Yu, Radu Soricut, Johan Schalkwyk, Andrew M. Dai, Anja Hauth, Katie Millican, David Silver, Slav Petrov, Melvin Johnson, Ioannis Antonoglou, Julian Schrittwieser, Amelia Glaese, Jilin Chen, Emily Pitler, Timothy P. Lillicrap, Angeliki Lazaridou, Orhan Firat, James Molloy, Michael Isard, Paul Ronald Barham, Tom Hennigan, Benjamin Lee, Fabio Viola, Malcolm Reynolds, Yuanzhong Xu, Ryan Doherty, Eli Collins, Clemens Meyer, Eliza Rutherford, Erica Moreira, Kareem Ayoub, Megha Goel, George Tucker, Enrique Piqueras, Maxim Krikun, Iain Barr, Nikolay Savinov, Ivo Danihelka, Becca Roelofs, Anaïs White, Anders Andreassen, Tamara von Glehn, Lakshman Yagati, Mehran Kazemi, Lucas Gonzalez, Misha Khalman, Jakub Sygnowski, and et al. "Gemini: A Family of Highly Capable Multimodal Models". In: *CoRR* abs/2312.11805 (2023). DOI: https://doi.org/10.48550/ARXIV.2312.11805. arXiv:2312.11805. URL: https://doi.org/10.48550/arXiv.2312.11805.

21. Fares Antaki, Daniel Milad, Mark A Chia, Charles-douard Gigu re, Samir Touma, Jonathan El-Khoury, Pearse A Keane, and Renaud Duval. "Capabilities of GPT-4 in ophthalmology: an analysis of model entropy and progress towards human-level medical question answering". en. In: British *Journal of Ophthalmology* (Nov. 2023), bjo–2023–324438. ISSN: 0007–1161, 1468–2079. DOI: https://doi.org/10.1136/bjo-2023-324438. URL: https://doi.org/10.1136/bjo-2023-324438 (visited on 02/26/2024).
22. Giovanni Apruzzese, Pavel Laskov, Edgardo Montes de Oca, Wissam Mallouli, Luis Burdalo Rapa, Athanasios Vasileios Grammatopoulos, and Fabio Di Franco. "The Role of Machine Learning in Cybersecurity". In: *CoRR* abs/2206.09707 (2022). DOI: https://doi.org/10.48550/ARXIV.2206.09707. arXiv:2206.09707. URL: https://doi.org/10.48550/arXiv.2206.09707.
23. Segun Taofeek Aroyehun and Alexander F. Gelbukh. "Aggression Detection in Social Media: Using Deep Neural Networks, Data Augmentation, and Pseudo Labeling". In: *Proceedings of the First Workshop on Trolling, Aggression and Cyberbullying, TRAC at COLING 2018, Santa Fe, New Mexico, USA, August 25, 2018*. Ed. by Ritesh Kumar, Atul Kr. Ojha, Marcos Zampieri, and Shervin Malmasi. Association for Computational Linsguistics, 2018, pp. 90–97. URL: https://www.aclanthology.org/W18-4411/.
24. Jason Baldridge and Miles Osborne. "Active Learning and the Total Cost of Annotation". In: *Proceedings of the 2004 Conference on Empirical Methods in Natural Language Processing , EMNLP 2004, A meeting of SIGDAT, a Special Interest Group of the ACL, held in conjunction with ACL 2004, 25–26 July 2004, Barcelona, Spain*. ACL, 2004, pp. 9–16. URL: https://www.aclanthology.org/W04-3202/.
25. Michele Banko and Eric Brill. "Scaling to Very Very Large Corpora for Natural Language Disambiguation". In: *Association for Computational Linguistic, 39th Annual Meeting and 10th Conference of the European Chapter, Proceedings of the Conference, July 9–11, 2001, Toulouse, France*. Morgan Kaufmann Publishers, 2001, pp. 26–33. DOI: https://doi.org/10.3115/1073012.1073017. URL: https://www.aclanthology.org/P01-1005/.
26. Marco Baroni, Georgiana Dinu, and Germán Kruszewski. "Don't count, predict! A systematic comparison of context-counting vs. context-predicting semantic vectors". In: *Proceedings of the 52nd Annual Meeting of the Association for Computational Linguistics, ACL 2014, June 22–27, 2014, Baltimore, MD, USA, Volume 1: Long Papers*. The Association for Computer Linguistics, 2014, pp. 238–247. DOI: https://doi.org/10.3115/V1/P14-1023. URL: https://doi.org/10.3115/v1/p14-1023.
27. Markus Bayer, Tobias Frey, and Christian Reuter. "Multi-level fine-tuning, data augmentation, and few-shot learning for specialized cyber threat intelligence". In: *Comput. Secur.* 134 (2023), p. 103430. DOI: https://doi.org/10.1016/J.COSE.2023.103430. URL: https://doi.org/10.1016/j.cose.2023.103430.
28. Markus Bayer, Marc-André Kaufhold, Bjorn Buchhold, Marcel Keller, Jorg Dallmeyer, and Christian Reuter. "Data augmentation in natural language processing: a novel text generation approach for long and short text classifiers". In: *Int. J. Mach. Learn. Cybern.* 14.1 (2023), pp. 135–150. DOI: https://doi.org/10.1007/S13042-022-01553-3. URL: https://doi.org/10.1007/s13042-022-01553-3.
29. Markus Bayer, Marc-André Kaufhold, and Christian Reuter. "A Survey on Data Augmentation for Text Classification". In: *ACM Comput. Surv.* 55.7 (2023), 146:1–146:39. DOI: https://doi.org/10.1145/3544558. URL: https://doi.org/10.1145/3544558.

30. Markus Bayer, Marc-André Kaufhold, and Christian Reuter. "Information Overload in Crisis Management: Bilingual Evaluation of Embedding Models for Clustering Social Media Posts in Emergencies". In: *29th European Conference on Information Systems—Human Values Crisis in a Digitizing World, ECIS 2021, Marrakech, Morocco, 2020*. Ed. by Frantz Rowe, Redouane El Amrani, Moez Limayem, Sabine Matook, Christoph Rosenkranz, Edgar A. Whitley, and Ali El Quammah. 2021. URL: https://www.aisel.aisnet.org/ecis2021_rp/64.
31. Markus Bayer, Philipp Kuehn, Ramin Shanehsaz, and Christian Reuter. "CySecBERT: A Domain-Adapted Language Model for the Cybersecurity Domain". In: *ACM Trans. Priv. Secur.* 27.2 (Apr. 2024). ISSN: 2471–2566. DOI: https://doi.org/10.1145/3652594. URL: https://doi.org/10.1145/3652594.
32. Markus Bayer, Markus Neiczer, Maximilian Samsinger, Björn Buchhold, and Christian Reuter. "XAI-Attack: Utilizing Explainable AI to Find Incorrectly Learned Patterns for Black-Box Adversarial Example Creation". In: *2024 Joint International Conference on Computational Linguistics, Language Resources and Evaluation (LREC-COLING 2024)*. LREC-COLING '24. 2024.
33. Yonatan Belinkov and Yonatan Bisk. "Synthetic and Natural Noise Both Break Neural Machine Translation". In: *6th International Conference on Learning Representations, ICLR 2018, Vancouver, BC, Canada, April 30—May 3, 2018, Conference Track Proceedings*. OpenReview.net, 2018. URL: https://www.openreview.net/forum?id=BJ8vJebC-.
34. Iz Beltagy, Kyle Lo, and Arman Cohan. "SciBERT: A Pretrained Language Model for Scientific Text". In: *Proceedings of the 2019 Conference on Empirical Methods in Natural Language Processing and the 9th International Joint Conference on Natural Language Processing, EMNLP-IJCNLP 2019, Hong Kong, China, November 3–7, 2019*. Ed. by Kentaro Inui, Jing Jiang, Vincent Ng, and Xiaojun Wan. Association for Computational Linguistics, 2019, pp. 3613–3618. DOI: https://doi.org/10.18653/V1/D19-1371. URL: https://doi.org/10.18653/v1/D19-1371.
35. Emily M. Bender, Timnit Gebru, Angelina McMillan-Major, and Shmargaret Shmitchell. "On the Dangers of Stochastic Parrots: Can Language Models Be Too Big?" In: *FAccT '21: 2021 ACM Conference on Fairness, Accountability, and Transparency, Virtual Event / Toronto, Canada, March 3–10, 2021*. Ed. by Madeleine Clare Elish, William Isaac, and Richard S. Zemel. ACM, 2021, pp. 610–623. DOI: https://doi.org/10.1145/3442188.3445922. URL: https://doi.org/10.1145/3442188.3445922.
36. Battista Biggio and Fabio Roli. "Wild patterns: Ten years after the rise of adversarial machine learning". In: *Pattern Recognit.* 84 (2018), pp. 317–331. DOI: https://doi.org/10.1016/J.PATCOG.2018.07.023. URL: https://doi.org/10.1016/j.patcog.2018.07.023.
37. Sid Black, Stella Biderman, Eric Hallahan, Quentin Anthony, Leo Gao, Laurence Golding, Horace He, Connor Leahy, Kyle McDonell, Jason Phang, Michael Pieler, USVSN Sai Prashanth, Shivanshu Purohit, Laria Reynolds, Jonathan Tow, Ben Wang, and Samuel Weinbach. "GPT-NeoX-20B: An Open-Source Autoregressive Language Model". In: *CoRR* abs/2204.06745 (2022). DOI: https://doi.org/10.48550/ARXIV.2204.06745. arXiv:2204.06745. URL: https://doi.org/10.48550/arXiv.2204.06745.
38. Piotr Bojanowski, Edouard Grave, Armand Joulin, and Tomás Mikolov. "Enriching Word Vectors with Subword Information". In: *Trans. Assoc. Comput. Linguistics* 5 (2017), pp. 135–146. DOI: https://doi.org/10.1162/TACL_A_00051. URL: https://doi.org/10.1162/tacl_a_00051.

39. Sridevi Bonthu, Abhinav Dayal, M Lakshmi, and S Rama Sree. "Effective text augmentation strategy for nlp models". In: *Proceedings of third international conference on sustainable computing*. tex.organization: Springer. 2022, pp. 521–531.
40. Mitchell Bosley, Musashi Jacobs-Harukawa, Hauke Licht, and Alexander Hoyle. "Do We Still Need BERT in the Age of GPT? Comparing the Benefits of Domain-Adaptation and In-Context-Learning Approaches to Using LLMs for Political Science Research." In: *2023 Annual Meeting of the Midwest Political Science Association (MPSA)*. 2023.
41. Amine Boulemtafes, Abdelouahid Derhab, and Yacine Challal. "A review of privacy-preserving techniques for deep learning". In: *Neurocomputing* 384 (2020), pp. 21–45. DOI: https://doi.org/10.1016/J.NEUCOM.2019.11.041. URL: https://doi.org/10.1016/j.neucom.2019.11.041.
42. Samuel R. Bowman, Gabor Angeli, Christopher Potts, and Christopher D. Manning. "A large annotated corpus for learning natural language inference". In: *Proceedings of the 2015 Conference on Empirical Methods in Natural Language Processing, EMNLP 2015, Lisbon, Portugal, September 17–21, 2015*. Ed. by Lluís Màrquez, Chris Callison-Burch, Jian Su, Daniele Pighin, and Yuval Marton. The Association for Computational Linguistics, 2015, pp. 632–642. DOI: https://doi.org/10.18653/V1/D15-1075. URL: https://doi.org/10.18653/v1/d15-1075.
43. Samuel R. Bowman, Luke Vilnis, Oriol Vinyals, Andrew M. Dai, Rafal Józefowicz, and Samy Bengio. "Generating Sentences from a Continuous Space". In: *Proceedings of the 20th SIGNLL Conference on Computational Natural Language Learning, CoNLL 2016, Berlin, Germany, August 11–12, 2016*. Ed. by Yoav Goldberg and Stefan Riezler. ACL, 2016, pp. 10–21. DOI: https://doi.org/10.18653/V1/K16-1002. URL: https://doi.org/10.18653/v1/k16-1002.
44. Jonathan Bragg, Arman Cohan, Kyle Lo, and Iz Beltagy. "FLEX: Unifying Evaluation for Few-Shot NLP". In: *Advances in Neural Information Processing Systems 34: Annual Conference on Neural Information Processing Systems 2021, NeurIPS 2021, December 6–14, 2021, virtual*. Ed. by Marc'Aurelio Ranzato, Alina Beygelzimer, Yann N. Dauphin, Percy Liang, and Jennifer Wortman Vaughan. 2021, pp. 15787–15800. URL: https://www.proceedings.neurips.cc/paper/2021/hash/8493eeaccb772c0878f99d60a0bd2bb3-Abstract.html.
45. Amy Brand, Liz Allen, Micah Altman, Marjorie M. K. Hlava, and Jo Scott. "Beyond authorship: attribution, contribution, collaboration, and credit". In: *Learn. Publ.* 28.2 (2015), pp. 151–155. DOI: https://doi.org/10.1087/20150211. URL: https://doi.org/10.1087/20150211.
46. Tom B. Brown, Dandelion Mané, Aurko Roy, Martín Abadi, and Justin Gilmer. "Adversarial Patch". In: *CoRR* abs/1712.09665 (2017). arXiv:1712.09665. URL: http://arxiv.org/abs/1712.09665.
47. Tom B. Brown, Benjamin Mann, Nick Ryder, Melanie Subbiah, Jared Kaplan, Prafulla Dhariwal, Arvind Neelakantan, Pranav Shyam, Girish Sastry, Amanda Askell, Sandhini Agarwal, Ariel Herbert-Voss, Gretchen Krueger, Tom Henighan, Rewon Child, Aditya Ramesh, Daniel M. Ziegler, Jeffrey Wu, Clemens Winter, Christopher Hesse, Mark Chen, Eric Sigler, Mateusz Litwin, Scott Gray, Benjamin Chess, Jack Clark, Christopher Berner, Sam McCandlish, Alec Radford, Ilya Sutskever, and Dario Amodei. "Language Models are Few-Shot Learners". In: *Advances in Neural Information Processing Systems 33: Annual Conference on Neural Information Processing Systems 2020, NeurIPS 2020, December 6–12, 2020, virtual*. Ed.

by Hugo Larochelle, Marc'Aurelio Ranzato, Raia Hadsell, Maria-Florina Balcan, and Hsuan-Tien Lin. 2020. URL: https://www.proceedings.neurips.cc/paper/2020/hash/1457c0d6bfcb4967418bfb8ac142f64a-Abstract.html.

48. Tom B. Brown, Benjamin Mann, Nick Ryder, Melanie Subbiah, Jared Kaplan, Prafulla Dhariwal, Arvind Neelakantan, Pranav Shyam, Girish Sastry, Amanda Askell, Sandhini Agarwal, Ariel Herbert-Voss, Gretchen Krueger, Tom Henighan, Rewon Child, Aditya Ramesh, Daniel M. Ziegler, Jeffrey Wu, Clemens Winter, Christopher Hesse, Mark Chen, Eric Sigler, Mateusz Litwin, Scott Gray, Benjamin Chess, Jack Clark, Christopher Berner, Sam McCandlish, Alec Radford, Ilya Sutskever, and Dario Amodei. "Language Models are Few-Shot Learners". In: *Advances in Neural Information Processing Systems 33: Annual Conference on Neural Information Processing Systems 2020, NeurIPS 2020, December 6–12, 2020, virtual*. Ed. by Hugo Larochelle, Marc'Aurelio Ranzato, Raia Hadsell, Maria-Florina Balcan, and Hsuan-Tien Lin. 2020. URL: https://www.proceedings.neurips.cc/paper/2020/hash/1457c0d6bfcb4967418bfb8ac142f64a-Abstract.html.
49. BSI. *CERT-Bund*. 2022. URL: https://www.bsi.bund.de/DE/Themen/Cyber-Sicherheit/Aktivitaeten/CERT-Bund/certbund_node.html.
50. Cody Buntain and Jennifer Golbeck. "Automatically Identifying Fake News in Popular Twitter Threads". In: *2017 IEEE International Conference on Smart Cloud, SmartCloud 2017, New York City, NY, USA, November 3–5, 2017*. IEEE Computer Society, 2017, pp. 208–215. DOI: https://doi.org/10.1109/SMARTCLOUD.2017.40. URL: https://doi.org/10.1109/SmartCloud.2017.40.
51. Juan Caballero, Gibran Gomez, Srdjan Matic, Gustavo S nchez, Silvia Sebasti n, and Arturo Villaca as. "The Rise of GoodFATR: A Novel Accuracy Comparison Methodology for Indicator Extraction Tools". en. In: *Future Generation Computer Systems* 144 (July 2023). arXiv:2208.00042 [cs], pp. 74–89. ISSN: 0167739X. DOI: https://doi.org/10.1016/j.future.2023.02.012. URL: http://arxiv.org/abs/2208.00042 (visited on 04/05/2023).
52. T. Cali ski and J. Harabasz. "A Dendrite Method Foe Cluster Analysis". In: *Communications in Statistics* (1974). ISSN: 00903272. DOI: https://doi.org/10.1080/03610927408827101.
53. Yihan Cao, Yanbin Kang, Chi Wang, and Lichao Sun. *Instruction Mining: When Data Mining Meets Large Language Model Finetuning*. en. arXiv:2307.06290 [cs]. Oct. 2023. URL: http://arxiv.org/abs/2307.06290 (visited on 02/26/2024).
54. Cornelia Caragea, Nathan J. McNeese, Anuj R. Jaiswal, Greg Traylor, Hyun-Woo Kim, Prasenjit Mitra, Dinghao Wu, Andrea H. Tapia, C. Lee Giles, Bernard J. Jansen, and John Yen. "Classifying text messages for the haiti earthquake". In: *8th Proceedings of the International Conference on Information Systems for Crisis Response and Management, Lisbon, Portugal, May, 2011*. Ed. by Maria A. Santos, Luísa Sousa, and Eliane Portela. LNEC, Lisbon, Portugal, 2011. URL: http://idl.iscram.org/files/caragea/2011/371_Caragea_etal2011.pdf.
55. Nicholas Carlini, Florian Tramér, Eric Wallace, Matthew Jagielski, Ariel Herbert-Voss, Katherine Lee, Adam Roberts, Tom B. Brown, Dawn Song, Ulfar Erlingsson, Alina Oprea, and Colin Raffel. "Extracting Training Data from Large Language Models". In: *30th USENIX Security Symposium, USENIX Security 2021, August 11–13, 2021*. Ed. by Michael D. Bailey and Rachel Greenstadt. USENIX Association, 2021, pp. 2633–

2650. URL: https://www.usenix.org/conference/usenixsecurity21/presentation/carlini-extracting.

56. Ricardo Carreira and Jaime M. Crato. "Evaluating adaptive user profiles for news classification". In: *Proceedings of the 9th International Conference on Intelligent User Interfaces, IUI 2004, Funchal, Madeira, Portugal, January 13–16, 2004*. Ed. by Jean Vanderdonckt, Nuno Jardim Nunes, and Charles Rich. ACM, 2004, pp. 206–212. DOI: https://doi.org/10.1145/964442.964481. URL: https://doi.org/10.1145/964442.964481.
57. Carlos Castillo, Marcelo Mendoza, and Barbara Poblete. "Information credibility on twitter". In: *Proceedings of the 20th International Conference on World Wide Web, WWW 2011, Hyderabad, India, March 28—April 1, 2011*. Ed. by Sadagopan Srinivasan, Krithi Ramamritham, Arun Kumar, M. P. Ravindra, Elisa Bertino, and Ravi Kumar. ACM, 2011, pp. 675–684. DOI: https://doi.org/10.1145/1963405.1963500. URL: https://doi.org/10.1145/1963405.1963500.
58. Carlos Castillo, Marcelo Mendoza, and Barbara Poblete. "Predicting information credibility in time-sensitive social media". In: *Internet Res.* 23.5 (2013), pp. 560–588. DOI: https://doi.org/10.1108/INTR-05-2012-0095. URL: https://doi.org/10.1108/IntR-05-2012-0095.
59. Ilias Chalkidis, Manos Fergadiotis, Prodromos Malakasiotis, Nikolaos Aletras, and Ion Androutsopoulos. "LEGAL-BERT: The Muppets straight out of Law School". In: *CoRR* abs/2010.02559 (2020). arXiv:2010.02559. URL: https://www.arxiv.org/abs/2010.02559.
60. Ernie Chang, Xiaoyu Shen, Dawei Zhu, Vera Demberg, and Hui Su. "Neural Data-to-Text Generation with LM-based Text Augmentation". In: *Proceedings of the 16th Conference of the European Chapter of the Association for Computational Linguistics: Main Volume, EACL 2021, Online, April 19–23, 2021*. Ed. by Paola Merlo, Jörg Tiedemann, and Reut Tsarfaty. Association for Computational Linguistics, f2021, pp. 758–768. DOI: https://doi.org/10.18653/V1/2021.EACL-MAIN.64. URL: https://doi.org/10.18653/v1/2021.eacl-main.64.
61. Samrat Chatterjee and Shital A. Thekdi. "An iterative learning and inference approach to managing dynamic cyber vulnerabilities of complex systems". In: *Reliab. Eng. Syst. Saf.* 193 (2020), p. 106664. DOI: https://doi.org/10.1016/J.RESS.2019.106664. URL: https://doi.org/10.1016/j.ress.2019.106664.
62. Iti Chaturvedi, Yew-Soon Ong, Ivor W. Tsang, Roy E. Welsch, and Erik Cambria. "Learning word dependencies in text by means of a deep recurrent belief network". In: *Knowl. Based Syst.* 108 (2016), pp. 144–154. DOI: https://doi.org/10.1016/J.KNOSYS.2016.07.019. URL: https://doi.org/10.1016/j.knosys.2016.07.019.
63. Nitesh V. Chawla, Kevin W. Bowyer, Lawrence O. Hall, and W. Philip Kegelmeyer. "SMOTE: Synthetic Minority Over-sampling Technique". In: *J. Artif. Intell. Res.* 16 (2002), pp. 321–357. DOI: https://doi.org/10.1613/JAIR.953. URL: https://doi.org/10.1613/jair.953.
64. Banghao Chen, Zhaofeng Zhang, Nicolas Langrené, and Shengxin Zhu. "Unleashing the potential of prompt engineering in Large Language Models: a comprehensive review". In: *CoRR* abs/2310.14735 (2023). DOI: https://doi.org/10.48550/ARXIV.2310.14735. arXiv:2310.14735. URL: https://doi.org/10.48550/arXiv.2310.14735.
65. Haipeng Chen, Rui Liu, Noseong Park, and V. S. Subrahmanian. "Using Twitter to Predict When Vulnerabilities will be Exploited". In: *Proceedings of the 25th ACM SIGKDD*

International Conference on Knowledge Discovery & Data Mining, KDD 2019, Anchorage, AK, USA, August 4–8, 2019. Ed. by Ankur Teredesai, Vipin Kumar, Ying Li, Romer Rosales, Evimaria Terzi, and George Karypis. ACM, 2019, pp. 3143–3152. DOI: https://doi.org/10.1145/3292500.3330742. URL: https://doi.org/10.1145/3292500.3330742.

66. Jiaao Chen, Zhenghui Wang, Ran Tian, Zichao Yang, and Diyi Yang. "Local Additivity Based Data Augmentation for Semi-supervised NER". In: *Proceedings of the 2020 Conference on Empirical Methods in Natural Language Processing, EMNLP 2020, Online, November 16–20, 2020*. Ed. by Bonnie Webber, Trevor Cohn, Yulan He, and Yang Liu. Association for Computational Linguistics, 2020, pp. 1241–1251. DOI: https://doi.org/10.18653/V1/2020.EMNLP-MAIN.95. URL: https://doi.org/10.18653/v1/2020.emnlp-main.95.
67. Jiaao Chen, Zichao Yang, and Diyi Yang. "MixText: Linguistically-Informed Interpolation of Hidden Space for Semi-Supervised Text Classification". In: *Proceedings of the 58th Annual Meeting of the Association for Computational Linguistics, ACL 2020, Online, July 5–10, 2020*. Ed. by Dan Jurafsky, Joyce Chai, Natalie Schluter, and Joel R. Tetreault. Association for Computational Linguistics, 2020, pp. 2147–2157. DOI: https://doi.org/10.18653/V1/2020.ACL-MAIN.194. URL: https://doi.org/10.18653/v1/2020.acl-main.194.
68. Liangyu Chen, Yutong Bai, Siyu Huang, Yongyi Lu, Bihan Wen, Alan L. Yuille, and Zongwei Zhou. "Making Your First Choice: To Address Cold Start Problem in Medical Active Learning". In: *Imaging with Deep Learning, MIDL 2023, 10–12 July 2023, Nashville, TN, USA*. Ed. by Ipek Oguz, Jack H. Noble, Xiaoxiao Li, Martin Styner, Christian Baumgartner, Mirabela Rusu, Tobias Heimann, Despina Kontos, Bennett A. Landman, and Benoit M. Dawant. Vol. 227. Proceedings of Machine Learning Research. PMLR, 2023, pp. 496–525. URL: https://www.proceedings.mlr.press/v227/chen24a.html.
69. Yizong Cheng. "Mean Shift, Mode Seeking, and Clustering". In: *IEEE Trans. Pattern Anal. Mach. Intell.* 17.8 (1995), pp. 790–799. DOI: https://doi.org/10.1109/34.400568. URL: https://doi.org/10.1109/34.400568.
70. Yong Cheng, Lu Jiang, and Wolfgang Macherey. "Robust Neural Ma chine Translation with Doubly Adversarial Inputs". In: *Proceedings of the 57th Conference of the Association for Computational Linguistics, ACL 2019, Florence, Italy, July 28- August 2, 2019, Volume 1: Long Papers*. Ed. by Anna Korhonen, David R. Traum, and Lluís Màrquez. Association for Computational Linguistics, 2019, pp. 4324–4333. DOI: https://doi.org/10.18653/V1/P19-1425. URL: https://doi.org/10.18653/v1/p19-1425.
71. Seungtaek Choi, Myeongho Jeong, Hojae Han, and Seung-won Hwang. "C2L: Causally Contrastive Learning for Robust Text Classification". In: *Thirty-Sixth AAAI Conference on Artificial Intelligence, AAAI 2022, Thirty-Fourth Conference on Innovative Applications of Artificial Intelligence, IAAI 2022, The Twelveth Symposium on Educational Advances in Artificial Intelligence, EAAI 2022 Virtual Event, February 22—March 1, 2022*. AAAI Press, 2022, pp. 10526–10534. DOI: https://doi.org/10.1609/AAAI.V36I10.21296. URL: https://doi.org/10.1609/aaai.v36i10.21296.
72. François Chollet. "On the Measure of Intelligence". In: *CoRR* (2019). arXiv:1911.01547. URL: http://arxiv.org/abs/1911.01547.
73. François Chollet and Dwarkesh Patel (Host). *François Chollet—LLMs won't lead to AGI—$1,000,000 Prize to find true solution [Video podcast]*. 2024. URL: https://www.youtube.com/watch?v=UakqL6Pj9xo.

74. Aakanksha Chowdhery, Sharan Narang, Jacob Devlin, Maarten Bosma, Gaurav Mishra, Adam Roberts, Paul Barham, Hyung Won Chung, Charles Sutton, Sebastian Gehrmann, Parker Schuh, Kensen Shi, Sasha Tsvyashchenko, Joshua Maynez, Abhishek Rao, Parker Barnes, Yi Tay, Noam Shazeer, Vinodkumar Prabhakaran, Emily Reif, Nan Du, Ben Hutchinson, Reiner Pope, James Bradbury, Jacob Austin, Michael Isard, Guy Gur-Ari, Pengcheng Yin, Toju Duke, Anselm Levskaya, Sanjay Ghemawat, Sunipa Dev, Henryk Michalewski, Xavier Garcia, Vedant Misra, Kevin Robinson, Liam Fedus, Denny Zhou, Daphne Ippolito, David Luan, Hyeontaek Lim, Barret Zoph, Alexander Spiridonov, Ryan Sepassi, David Dohan, Shivani Agrawal, Mark Omernick, Andrew M. Dai, Thanumalayan Sankaranarayana Pillai, Marie Pellat, Aitor Lewkowycz, Erica Moreira, Rewon Child, Oleksandr Polozov, Katherine Lee, Zongwei Zhou, Xuezhi Wang, Brennan Saeta, Mark Diaz, Orhan Firat, Michele Catasta, Jason Wei, Kathy Meier-Hellstern, Douglas Eck, Jeff Dean, Slav Petrov, and Noah Fiedel. "PaLM: Scaling Language Modeling with Pathways". In: *J. Mach. Learn. Res.* 24 (2023), 240:1–240:113. URL: http://jmlr.org/papers/v24/22-1144.html.
75. Vincent Claveau, Antoine Chaffin, and Ewa Kijak. "La generation de textes artificiels en substitution ou en compl ment de donn es d apprentissage". In: *Traitement automatique des langues naturelles*. 2021, pp. 124–136.
76. David A. Cohn, Les E. Atlas, and Richard E. Ladner. "Improving Generalization with Active Learning". In: *Mach. Learn.* 15.2 (1994), pp. 201–221. DOI: https://doi.org/10.1007/BF00993277. URL: https://doi.org/10.1007/BF00993277.
77. Patrick L. Combettes and Jean-Christophe Pesquet. "Proximal Splitting Methods in Signal Processing". In: *Fixed-Point Algorithms for Inverse Problems in Science and Engineering*. Ed. by Heinz H. Bauschke, Regina Sandra Burachik, Patrick L. Combettes, Veit Elser, D. Russell Luke, and Henry Wolkowicz. Vol. 49. Springer Optimization and Its Applications. Springer, 2011, pp. 185–212. DOI: https://doi.org/10.1007/978-1-4419-9569-8_10. URL: https://doi.org/10.1007/978-1-4419-9569-8_10.
78. Carmela Comito, Agostino Forestiero, and Clara Pizzuti. "Word Embedding based Clustering to Detect Topics in Social Media". In: *2019 IEEE/WIC/ACM International Conference on Web Intelligence, WI 2019, Thessaloniki, Greece, October 14–17, 2019*. Ed. by Payam M. Barnaghi, Georg Gottlob, Yannis Manolopoulos, Theodoros Tzouramanis, and Athena Vakali. ACM, 2019, pp. 192–199. DOI: https://doi.org/10.1145/3350546.3352518. URL: https://doi.org/10.1145/3350546.3352518.
79. Alexis Conneau, Douwe Kiela, Holger Schwenk, Loïc Barrault, and Antoine Bordes. "Supervised Learning of Universal Sentence Representations from Natural Language Inference Data". In: *Proceedings of the 2017 Conference on Empirical Methods in Natural Language Processing, EMNLP 2017, Copenhagen, Denmark, September 9-11, 2017*. Ed. by Martha Palmer, Rebecca Hwa, and Sebastian Riedel. Association for Computational Linguistics, 2017, pp. 670–680. DOI: https://doi.org/10.18653/V1/D17-1070. URL: https://doi.org/10.18653/v1/d17-1070.
80. Nadia K. Conroy, Victoria L. Rubin, and Yimin Chen. "Automatic deception detection: Methods for finding fake news". In: *Proceedings of ASIS&T* 52.1 (2015), pp. 1–4. ISSN: 2373–9231, 2373–9231.
81. Claude Coulombe. "Text Data Augmentation Made Simple By Leveraging NLP Cloud APIs". In: *CoRR* abs/1812.04718 (2018). arXiv:1812.04718. URL: http://arxiv.org/abs/1812.04718.

82. Xiaodong Cui, Vaibhava Goel, and Brian Kingsbury. "Data Augmentation for deep neural network acoustic modeling". In: *IEEE International Conference on Acoustics, Speech and Signal Processing, ICASSP 2014, Florence, Italy, May 4–9, 2014*. IEEE, 2014, pp. 5582–5586. DOI: https://doi.org/10.1109/ICASSP.2014.6854671. URL: https://doi.org/10.1109/ICASSP.2014.6854671.
83. Stephan A. Curiskis, Barry Drake, Thomas R. Osborn, and Paul J. Kennedy. "An evaluation of document clustering and topic modelling in two online social networks: Twitter and Reddit". In: *Information Processing and Management* (2020). ISSN: 03064573. DOI: https://doi.org/10.1016/j.ipm.2019.04.002.
84. Xiangfeng Dai, Marwan Bikdash, and Bradley Meyer. "From social media to public health surveillance: Word embedding based clustering method for twitter classification". In: *Conference Proceedings—IEEE SOUTHEASTCON*. 2017. ISBN: 978-1-5386-1539-3. DOI: https://doi.org/10.1109/SECON.2017.7925400.
85. Dipankar Dasgupta, Zahid Akhtar, and Sajib Sen. "Machine learning in cybersecurity: a comprehensive survey". en. In: *The Journal of Defense Modeling and Simulation: Applications, Methodology, Technology* 19.1 (Jan. 2022), pp. 57–106. ISSN: 1548–5129, 1557–380X. DOI: https://doi.org/10.1177/1548512920951275. (Visited on 12/08/2023).
86. David L. Davies and Donald W. Bouldin. "A Cluster Separation Measure". In: *IEEE Trans. Pattern Anal. Mach. Intell.* 1.2 (1979), pp. 224–227. DOI: https://doi.org/10.1109/TPAMI.1979.4766909. URL: https://doi.org/10.1109/TPAMI.1979.4766909.
87. Chunyuan Deng, Yilun Zhao, Xiangru Tang, Mark Gerstein, and Arman Cohan. "Investigating Data Contamination in Modern Benchmarks for Large Language Models". In: *CoRR* abs/2311.09783 (2023). DOI: https://doi.org/10.48550/ARXIV.2311.09783. arXiv:2311.09783. URL: https://doi.org/10.48550/arXiv.2311.09783.
88. Tim Dettmers, Artidoro Pagnoni, Ari Holtzman, and Luke Zettle moyer. "QLoRA: Efficient Finetuning of Quantized LLMs". In: *Advances in Neural Information Processing Systems 36: Annual Conference on Neural Information Processing Systems 2023, NeurIPS 2023, New Orleans, LA, USA, December 10–16, 2023*. Ed. by Alice Oh, Tristan Naumann, Amir Globerson, Kate Saenko, Moritz Hardt, and Sergey Levine. 2023. URL: http://papers.nips.cc/paper_files/paper/2023/hash/1feb87871436031bdc0f2beaa62a049b-Abstract-Conference.html.
89. Jacob Devlin, Ming-Wei Chang, Kenton Lee, and Kristina Toutanova. "BERT: Pre-training of Deep Bidirectional Transformers for Language Understanding". In: *Proceedings of the 2019 Conference of the North American Chapter of the Association for Computational Linguistics: Human Language Technologies, NAACL-HLT 2019, Minneapolis, MN, USA, June 2–7, 2019, Volume 1 (Long and Short Papers)*. Ed. by Jill Burstein, Christy Doran, and Thamar Solorio. Association for Computational Linguistics, 2019, pp. 4171–4186. DOI: https://doi.org/10.18653/V1/N19-1423. URL: https://doi.org/10.18653/v1/n19-1423.
90. Kaustubh D. Dhole, Varun Gangal, Sebastian Gehrmann, Aadesh Gupta, Zhenhao Li, Saad Mahamood, Abinaya Mahendiran, Simon Mille, Ashish Srivastava, Samson Tan, Tongshuang Wu, Jascha Sohl-Dickstein, Jinho D. Choi, Eduard H. Hovy, Ondrej Dusek, Sebastian Ruder, Sajant Anand, Nagender Aneja, Rabin Banjade, Lisa Barthe, Hanna Behnke, Ian Berlot-Attwell, Connor Boyle, Caroline Brun, Marco Antonio Sobrevilla Cabezudo, Samuel Cahyawijaya, Emile Chapuis, Wanxiang Che, Mukund Choudhary, Christian Clauss, Pierre Colombo, Filip Cornell, Gautier Dagan, Mayukh Das, Tanay Dixit, Thomas Dopierre, Paul-Alexis Dray, Suchitra Dubey, Tatiana Ekeinhor, Marco

Di Giovanni, Rishabh Gupta, Rishabh Gupta, Louanes Hamla, Sang Han, Fabrice Harel-Canada, Antoine Honore, Ishan Jindal, Przemyslaw K. Joniak, Denis Kleyko, Venelin Kovatchev, and et al. "NL-Augmenter: A Framework for Task-Sensitive Natural Language Augmentation". In: *CoRR* abs/2112.02721 (2021). arXiv:2112.02721. URL: https://www.arxiv.org/abs/2112.02721.

91. Bosheng Ding, Linlin Liu, Lidong Bing, Canasai Kruengkrai, Thien Hai Nguyen, Shafiq R. Joty, Luo Si, and Chunyan Miao. "DAGA: Data Augmentation with a Generation Approach for Low-resource Tagging Tasks". In: *CoRR* abs/2011.01549 (2020). arXiv:2011.01549. URL: https://www.arxiv.org/abs/2011.01549.
92. Bosheng Ding, Chengwei Qin, Ruochen Zhao, Tianze Luo, Xinze Li, Guizhen Chen, Wenhan Xia, Junjie Hu, Anh Tuan Luu, and Shafiq Joty. "Data Augmentation using LLMs: Data Perspectives, Learning Paradigms and Challenges". In: *Findings of the Association for Computational Linguistics: ACL 2024*. Ed. by Lun-Wei Ku, Andre Martins, and Vivek Srikumar. Bangkok, Thailand: Association for Computational Linguistics, Aug. 2024, pp. 1679–1705. DOI: https://doi.org/10.18653/v1/2024.findings-acl.97. URL: https://www.aclanthology.org/2024.findings-acl.97/.
93. Nuno Dionísio, Fernando Alves, Pedro M. Ferreira, and Alysson Bessani. "Towards end-to-end Cyberthreat Detection from Twitter using Multi Task Learning". In: *2020 International Joint Conference on Neural Networks, IJCNN 2020, Glasgow, United Kingdom, July 19–24, 2020*. IEEE, 2020, pp. 1–8. DOI: https://doi.org/10.1109/IJCNN48605.2020.9207159. URL: https://doi.org/10.1109/IJCNN48605.2020.9207159.
94. Shi Dong, Ping Wang, and Khushnood Abbas. "A survey on deep learning and its applications". In: *Comput. Sci. Rev.* 40 (2021), p. 100379. DOI: https://doi.org/10.1016/J.COSREV.2021.100379. URL: https://doi.org/10.1016/j.cosrev.2021.100379.
95. Ying Dong, Wenbo Guo, Yueqi Chen, Xinyu Xing, Yuqing Zhang, and Gang Wang. "Towards the Detection of Inconsistencies in Public Security Vulnerability Reports". In: *28th USENIX Security Symposium, USENIX Security 2019, Santa Clara, CA, USA, August 14–16, 2019*. Ed. by Nadia Heninger and Patrick Traynor. USENIX Association, 2019, pp. 869–885. URL: https://www.usenix.org/conference/usenixsecurity19/presentation/dong.
96. Ruixue Duan, Dan Li, Qiang Tong, Tao Yang, Xiaotong Liu, and Xiulei Liu. "A Survey of Few-Shot Learning: An Effective Method for Intrusion Detection". en. In: *Security and Communication Networks* 2021 (Oct. 2021). Ed. by Yunchuan Guo, pp. 1–10. ISSN: 1939–0122, 1939–0114. DOI: https://doi.org/10.1155/2021/4259629. URL: https://www.hindawi.com/journals/scn/2021/4259629/ (visited on 12/19/2023).
97. Xinhuan Duan, Elham Naghizade, Damiano Spina, and Xiuzhen Zhang. "RMIT at PAN-CLEF 2020: Pro?ling Fake News Spreaders on Twitter". In: *CLEF 2020* (2020).
98. Javid Ebrahimi, Daniel Lowd, and Dejing Dou. "On Adversarial Examples for Character-Level Neural Machine Translation". In: *Proceedings of the 27th International Conference on Computational Linguistics, COLING 2018, Santa Fe, New Mexico, USA, August 20–26, 2018*. Ed. by Emily M. Bender, Leon Derczynski, and Pierre Isabelle. Association for Computational Linguistics, 2018, pp. 653–663. URL: https://www.aclanthology.org/C18-1055/.
99. Javid Ebrahimi, Anyi Rao, Daniel Lowd, and Dejing Dou. "HotFlip: White-Box Adversarial Examples for Text Classification". In: *Proceedings of the 56th Annual Meeting of the Association for Computational Linguistics, ACL 2018, Melbourne, Australia, July*

15–20, 2018, Volume 2: Short Papers. Ed. by Iryna Gurevych and Yusuke Miyao. Association for Computational Linguistics, 2018, pp. 31–36. DOI: https://doi.org/10.18653/V1/P18-2006. URL: https://www.aclanthology.org/P18-2006/.

100. Liat Ein-Dor, Alon Halfon, Ariel Gera, Eyal Shnarch, Lena Dankin, Leshem Choshen, Marina Danilevsky, Ranit Aharonov, Yoav Katz, and Noam Slonim. "Active Learning for BERT: An Empirical Study". In: *Proceedings of the 2020 Conference on Empirical Methods in Natural Language Processing, EMNLP 2020, Online, November 16–20, 2020*. Ed. by Bonnie Webber, Trevor Cohn, Yulan He, and Yang Liu. Association for Compu tational Linguistics, 2020, pp. 7949–7962. DOI: https://doi.org/10.18653/V1/2020.EMNLP-MAIN.638. URL: https://doi.org/10.18653/v1/2020.emnlp-main.638.
101. Kathrin Eismann, Oliver Posegga, and Kai Fischbach. "Decision Making in Emergency Management: The Role of Social Media". In: *26th European Conference on Information Systems: Beyond Digitization—Facets of Socio Technical Change, ECIS 2018, Portsmouth, UK, June 23–28, 2018*. Ed. by Peter M. Bednar, Ulrich Frank, and Karlheinz Kautz. 2018, p. 152. URL: https://www.aisel.aisnet.org/ecis2018_rp/152.
102. Martin J. Eppler and Jeanne Mengis. "The Concept of Information Overload: A Review of Literature from Organization Science, Accounting, Marketing, MIS, and Related Disciplines". In: *Inf. Soc.* 20.5 (2004), pp. 325–344. DOI: https://doi.org/10.1080/01972240490507974. URL: https://doi.org/10.1080/01972240490507974.
103. SONG Tae Eun. "Cyber Warfare in the Russo-Ukrainian War: Assessment and Implications". en. In: *Institute of Foreign Affairs and National Security* (2022), p. 2.
104. R. Evtimov, M. Falli, and A. Maiwald. *Anti Social Online Behaviour Detection with BERT*. Feb. 2020. URL: https://www.humboldt-wi.github.io/blog/research/information_systems_1920/bert_blog_post/.
105. Alexander R. Fabbri, Simeng Han, Haoyuan Li, Haoran Li, Marjan Ghazvininejad, Shafiq R. Joty, Dragomir R. Radev, and Yashar Mehdad. "Improving Zero and Few-Shot Abstractive Summarization with Intermediate Fine-tuning and Data Augmentation". In: *Proceedings of the 2021 Conference of the North American Chapter of the Association for Computational Linguistics: Human Language Technologies, NAACL-HLT 2021, Online, June 6–11, 2021*. Ed. by Kristina Toutanova, Anna Rumshisky, Luke Zettlemoyer, Dilek Hakkani-Tür, Iz Beltagy, Steven Bethard, Ryan Cotterell, Tanmoy Chakraborty, and Yichao Zhou. Association for Computational Linguistics, 2021, pp. 704–717. DOI: https://doi.org/10.18653/V1/2021.NAACL-MAIN.57. URL: https://doi.org/10.18653/v1/2021.naacl-main.57.
106. Marzieh Fadaee, Arianna Bisazza, and Christof Monz. "Data Augmentation for Low-Resource Neural Machine Translation". In: *Proceedings of the 55th Annual Meeting of the Association for Computational Linguistics, ACL 2017, Vancouver, Canada, July 30—August 4, Volume 2: Short Papers*. Ed. by Regina Barzilay and Min-Yen Kan. Association for Computational Linguistics, 2017, pp. 567–573. DOI: https://doi.org/10.18653/V1/P17-2090. URL: https://doi.org/10.18653/v1/P17-2090.
107. Adil Fahad, Najlaa Alshatri, Zahir Tari, Abdullah Alamri, Ibrahim Khalil, Albert Y. Zomaya, Sebti Foufou, and Abdelaziz Bouras. "A Survey of Clustering Algorithms for Big Data: Taxonomy and Empirical Analysis". In: *IEEE Trans. Emerg. Top. Comput.* 2.3 (2014), pp. 267–279. DOI: https://doi.org/10.1109/TETC.2014.2330519. URL: https://doi.org/10.1109/TETC.2014.2330519.
108. Weiguo Fan and Michael D Gordon. "Unveiling the Power of Social Media Analytic". In: Communications of the ACM 12.JUNE 2014 (2013), pp. 1–26. ISSN: 00010782.

DOI: https://doi.org/10.1145/2602574. URL: http://www.researchgate.net/publication/259148570_Unveiling_the_Power_of_Social_Media_Analytics.

109. Hongchao Fang and Pengtao Xie. "CERT: Contrastive Self-supervised Learning for Language Understanding". In: *CoRR* abs/2005.12766 (2020). arXiv:2005.12766. URL: https://www.arxiv.org/abs/2005.12766.
110. Yong Fang, Jian Gao, Zhonglin Liu, and Cheng Huang. "Detecting Cyber Threat Event from Twitter Using IDCNN and BiLSTM". In: *Applied Sciences* 10.17 (Jan. 2020), p. 5922. DOI: https://doi.org/10.3390/app10175922. URL: https://www.mdpi.com/2076-3417/10/17/5922.
111. Manaal Faruqui, Yulia Tsvetkov, Pushpendre Rastogi, and Chris Dyer. "Problems With Evaluation of Word Embeddings Using Word Similarity Tasks". In: *Proceedings of the 1st Workshop on Evaluating Vector-Space Representations for NLP, RepEval at ACL 2016, Berlin, Germany, August 2016*. Association for Computational Linguistics, 2016, pp. 30–35. DOI: https://doi.org/10.18653/V1/W16-2506. URL: https://doi.org/10.18653/v1/W16-2506.
112. Steven Y. Feng, Varun Gangal, Dongyeop Kang, Teruko Mitamura, and Eduard H. Hovy. "GenAug: Data Augmentation for Finetuning Text Generators". In: *CoRR* abs/2010.01794 (2020). arXiv:2010.01794. URL: https://www.arxiv.org/abs/2010.01794.
113. Steven Y. Feng, Varun Gangal, Jason Wei, Sarath Chandar, Soroush Vosoughi, Teruko Mitamura, and Eduard H. Hovy. "A Survey of Data Augmentation Approaches for NLP". In: *Findings of the Association for Computational Linguistics: ACL/IJCNLP 2021, Online Event, August 1–6, 2021*. Ed. by Chengqing Zong, Fei Xia, Wenjie Li, and Roberto Navigli. Vol. ACL/IJCNLP 2021. Findings of ACL. Association for Computational Linguistics, 2021, pp. 968–988. DOI: https://doi.org/10.18653/V1/2021.FINDINGS-ACL.84. URL: https://doi.org/10.18653/v1/2021.findings-acl.84.
114. Steven Y. Feng, Aaron W. Li, and Jesse Hoey. "Keep Calm and Switch On! Preserving Sentiment and Fluency in Semantic Text Exchange". In: *Proceedings of the 2019 Conference on Empirical Methods in Natural Language Processing and the 9th International Joint Conference on Natural Language Processing, EMNLP-IJCNLP 2019, Hong Kong, China, November 3–7, 2019*. Ed. by Kentaro Inui, Jing Jiang, Vincent Ng, and Xiaojun Wan. Association for Computational Linguistics, 2019, pp. 2701–2711. DOI: https://doi.org/10.18653/V1/D19-1272. URL: https://doi.org/10.18653/v1/D19-1272.
115. John R Firth. "A synopsis of linguistic theory, 1930–1955". In: *Studies in Linguistic Analysis*. 1957. ISBN: 90-04-10265-5.
116. Diana Fischer, Oliver Posegga, and Kai Fischbach. "Communication Barriers in Crisis Management: a literature Review". In: *24th European Conference on Information Systems, ECIS 2016, Istanbul, Turkey, June 12–15, 2016*. 2016, Research Paper 168. URL: http://aisel.aisnet.org/ecis2016_rp/168.
117. A. J. Flanagin and Miriam J. Metzger. "Digital media and youth: Unparalled opportunity and unprecedented responsibility". In: *Digital media, youth, and credibility*. Ed. by A. J. Flanagin and Miriam J. Metzger. 2008, pp. 5–28. ISBN: 978-0-262-56232-4.
118. Sabina-Adriana Floria, Florin Leon, and Doina Logofatu. "A Credibility-Based Analysis of Information Diffusion in Social Networks". In: *Artificial Neural Networks and Machine Learning—ICANN 2018—27th International Conference on Artificial Neural Networks, Rhodes, Greece, October 4–7, 2018, Proceedings, Part III*. Ed. by Vera Kurková, Yannis Manolopoulos, Barbara Hammer, Lazaros S. Iliadis, and Ilias

Maglogiannis. Vol. 11141. Lecture Notes in Computer Science. Springer, 2018, pp. 828–838. DOI: https://doi.org/10.1007/978-3-030-01424-7_80. URL: https://doi.org/10.1007/978-3-030-01424-7_80.

119. B. J. Fogg and Hsiang Tseng. "The Elements of Computer Credibility". In: *Proceeding of the CHI '99 Conference on Human Factors in Computing Systems: The CHI is the Limit, Pittsburgh, PA, USA, May 15–20, 1999*. Ed. by Marian G. Williams and Mark W. Altom. ACM, 1999, pp. 80–87. DOI: https://doi.org/10.1145/302979.303001. URL: https://doi.org/10.1145/302979.303001.
120. Maayan Frid-Adar, Idit Diamant, Eyal Klang, Michal Amitai, Jacob Gold-berger, and Hayit Greenspan. "GAN-based Synthetic Medical Image Augmentation for increased CNN Performance in Liver Lesion Classification". In: *CoRR* abs/1803.01229 (2018). arXiv:1803.01229. URL: http://arxiv.org/abs/1803.01229.
121. Merna Gamal, Hala M. Abbas, Nour Moustafa, Elena Sitnikova, and Rowayda A. Sadek. "Few-Shot Learning for Discovering Anomalous Behaviors in Edge Networks". en. In: *Computers, Materials & Continua* 69.2 (2021), pp. 1823–1837. ISSN: 1546-2226. DOI: https://doi.org/10.32604/cmc.2021.012877. URL: https://www.techscience.com/cmc/v69n2/43839 (visited on 12/19/2023).
122. Fei Gao, Jinhua Zhu, Lijun Wu, Yingce Xia, Tao Qin, Xueqi Cheng, Wengang Zhou, and Tie-Yan Liu. "Soft Contextual Data Augmentation for Neural Machine Translation". In: *Proceedings of the 57th Conference of the Association for Computational Linguistics, ACL 2019, Florence, Italy, July 28- August 2, 2019, Volume 1: Long Papers*. Ed. by Anna Korhonen, David R. Traum, and Lluís Màrquez. Association for Computational ' Linguistics, 2019, pp. 5539–5544. DOI: https://doi.org/10.18653/V1/P19-1555. URL: https://doi.org/10.18653/v1/p19-1555.
123. Tianyu Gao, Adam Fisch, and Danqi Chen. "Making Pre-trained Language Models Better Few-shot Learners". In: *Proceedings of the 59th Annual Meeting of the Association for Computational Linguistics and the 11th International Joint Conference on Natural Language Processing, ACL/IJCNLP 2021, (Volume 1: Long Papers), Virtual Event, August 1–6, 2021*. Ed. by Chengqing Zong, Fei Xia, Wenjie Li, and Roberto Navigli. Association for Computational Linguistics, 2021, pp. 3816–3830. DOI: https://doi.org/10.18653/V1/2021.ACL-LONG.295. URL: https://doi.org/10.18653/v1/2021.acl-long.295.
124. Siddhant Garg and Goutham Ramakrishnan. "BAE: BERT-based Adversarial Examples for Text Classification". In: *Proceedings of the 2020 Conference on Empirical Methods in Natural Language Processing, EMNLP 2020, Online, November 16–20, 2020*. Ed. by Bonnie Webber, Trevor Cohn, Yulan He, and Yang Liu. Association for Computational Linguistics, 2020, pp. 6174–6181. DOI: https://doi.org/10.18653/V1/2020.EMNLP-MAIN.498. URL: https://doi.org/10.18653/v1/2020.emnlp-main.498.
125. Sebastian Gehrmann, Tosin P. Adewumi, Karmanya Aggarwal, Pawan Sasanka Ammanamanchi, Aremu Anuoluwapo, Antoine Bosselut, Khyathi Raghavi Chandu, Miruna-Adriana Clinciu, Dipanjan Das, Kaustubh D. Dhole, Wanyu Du, Esin Durmus, Ondrej Dusek, Chris Emezue, Varun Gangal, Cristina Garbacea, Tatsunori Hashimoto, Yufang Hou, Yacine Jernite, Harsh Jhamtani, Yangfeng Ji, Shailza Jolly, Dhruv Kumar, Faisal Ladhak, Aman Madaan, Mounica Maddela, Khyati Mahajan, Saad Mahamood, Bodhisattwa Prasad Majumder, Pedro Henrique Martins, Angelina McMillan-Major, Simon Mille, Emiel van Miltenburg, Moin Nadeem, Shashi Narayan, Vitaly Nikolaev, Rubungo Andre Niyongabo, Salomey Osei, Ankur P. Parikh, Laura Perez-Beltrachini,

Niranjan Ramesh Rao, Vikas Raunak, Juan Diego Rodriguez, Sashank Santhanam, João Sedoc, Thibault Sellam, Samira Shaikh, Anastasia Shi-morina, Marco Antonio Sobrevilla Cabezudo, Hendrik Strobelt, Nishant Subramani, Wei Xu, Diyi Yang, Akhila Yerukola, and Jiawei Zhou. "The GEM Benchmark: Natural Language Generation, its Evaluation and Metrics". In: *CoRR* abs/2102.01672 (2021). arXiv:2102.01672. URL: https://www.arxiv.org/abs/2102.01672.

126. Robert Geirhos, Jörn-Henrik Jacobsen, Claudio Michaelis, Richard Zemel, Wieland Brendel, Matthias Bethge, and Felix A. Wichmann. "Shortcut Learning in Deep Neural Networks". en. In: *Nature Machine Intelligence* 2.11 (Nov. 2020). arXiv:2004.07780 [cs, q-bio], pp. 665–673. ISSN: 2522–5839. DOI: https://doi.org/10.1038/s42256-020-00257-z. URL: http://arxiv.org/abs/2004.07780 (visited on 03/19/2024).
127. Stuart Geman, Elie Bienenstock, and René Doursat. "Neural Networks and the Bias/Variance Dilemma". In: *Neural Comput.* 4.1 (1992), pp. 1–58. DOI: https://doi.org/10.1162/NECO.1992.4.1.1. URL: https://doi.org/10.1162/neco.1992.4.1.1.
128. Fr deric Godin, Baptist Vandersmissen, Wesley De Neve, and Rik Van de Walle. "Multimedia Lab *at* ACL WNUT NER Shared Task: Named Entity Recognition for Twitter Microposts using Distributed Word Representations". In: 2015. DOI: https://doi.org/10.18653/v1/w15-4322.
129. Yoav Goldberg. "Assessing BERT's Syntactic Abilities". In: *CoRR* abs/1901.05287 (2019). arXiv:1901.05287. URL: http://arxiv.org/abs/1901.05287.
130. Micah Goldblum, Liam Fowl, and Tom Goldstein. "Adversarially Robust Few-Shot Learning: A Meta-Learning Approach". In: *Advances in Neural Information Processing Systems 33: Annual Conference on Neural Information Processing Systems 2020, NeurIPS 2020, Decem ber 6–12, 2020, virtual*. Ed. by Hugo Larochelle, Marc'Aurelio Ranzato, Raia Hadsell, Maria-Florina Balcan, and Hsuan-Tien Lin. 2020. URL: https://www.proceedings.neurips.cc/paper/2020/hash/cfee398643cbc3dc5eefc89334cacdc1-Abstract.html.
131. Tom Goldstein, Christoph Studer, and Richard G. Baraniuk. "A Field Guide to Forward-Backward Splitting with a FASTA Implementation". In: *CoRR* abs/1411.3406 (2014). arXiv:1411.3406. URL: http://arxiv.org/abs/1411.3406.
132. Ian J. Goodfellow, Jonathon Shlens, and Christian Szegedy. "Explaining and Harnessing Adversarial Examples". In: *3rd International Conference on Learning Representations, ICLR 2015, San Diego, CA, USA, May 7–9, 2015, Conference Track Proceedings*. Ed. by Yoshua Bengio and Yann LeCun. 2015. URL: http://arxiv.org/abs/1412.6572.
133. Shreya Goyal, Sumanth Doddapaneni, Mitesh M. Khapra, and Balaraman Ravindran. "A Survey of Adversarial Defenses and Robustness in NLP". In: *ACM Comput. Surv.* 55.14s (2023), 332:1–332:39. DOI: https://doi.org/10.1145/3593042. URL: https://doi.org/10.1145/3593042.
134. Daniel Grießhaber, Johannes Maucher, and Ngoc Thang Vu. "Fine-tuning BERT for Low-Resource Natural Language Understanding via Active Learning". In: *Proceedings of the 28th International Conference on Computational Linguistics, COLING 2020, Barcelona, Spain (Online), December 8–13, 2020*. Ed. by Donia Scott, Núria Bel, and Chengqing Zong. International Committee on Computational Linguistics, 2020, pp. 1158–1171. DOI: https://doi.org/10.18653/V1/2020.COLING-MAIN.100. URL: https://doi.org/10.18653/v1/2020.coling-main.100.
135. Sabine Gründer-Fahrer, Antje Schlaf, Gregor Wiedemann, and Gerhard Heyer. "Topics and topical phases in German social media communication during a disas-

ter". In: *Nat. Lang. Eng.* 24.2 (2018), pp. 221–264. DOI: https://doi.org/10.1017/S1351324918000025. URL: https://doi.org/10.1017/S1351324918000025.

136. Chuan Guo, Geoff Pleiss, Yu Sun, and Kilian Q. Weinberger. "On Calibration of Modern Neural Networks". In: *Proceedings of the 34th International Conference on Machine Learning, ICML 2017, Sydney, NSW, Australia, 6–11 August 2017*. Ed. by Doina Precup and Yee Whye Teh. Vol. 70. Proceedings of Machine Learning Research. PMLR, 2017, pp. 1321–1330. URL: http://proceedings.mlr.press/v70/guo17a.html.
137. Demi Guo, Yoon Kim, and Alexander M. Rush. "Sequence-Level Mixed Sample Data Augmentation". In: *Proceedings of the 2020 Conference on Empirical Methods in Natural Language Processing, EMNLP 2020, Online, November 16–20, 2020*. Ed. by Bonnie Webber, Trevor Cohn, Yulan He, and Yang Liu. Association for Computational Linguistics, 2020, pp. 5547–5552. DOI: https://doi.org/10.18653/V1/2020.EMNLP-MAIN.447. URL: https://doi.org/10.18653/v1/2020.emnlp-main.447.
138. Hongyu Guo, Yongyi Mao, and Richong Zhang. "Augmenting Data with Mixup for Sentence Classification: An Empirical Study". In: *CoRR* abs/1905.08941 (2019). arXiv:1905.08941. URL: http://arxiv.org/abs/1905.08941.
139. Aditi Gupta, Ponnurangam Kumaraguru, Carlos Castillo, and Patrick Meier. "TweetCred: Real-Time Credibility Assessment of Content on Twitter". In: *Social Informatics - 6th International Conference, SocInfo 2014, Barcelona, Spain, November 11–13, 2014. Proceedings*. Ed. by Luca Maria Aiello and Daniel A. McFarland. Vol. 8851. Lecture Notes in Computer Science. Springer, 2014, pp. 228–243. DOI: https://doi.org/10.1007/978-3-319-13734-6_16. URL: https://doi.org/10.1007/978-3-319-13734-6_16.
140. Suchin Gururangan, Ana Marasovic, Swabha Swayamdipta, Kyle Lo, Iz Beltagy, Doug Downey, and Noah A. Smith. "Don't Stop Pretraining: Adapt Language Models to Domains and Tasks". In: *Proceedings of the 58th Annual Meeting of the Association for Computational Linguistics, ACL 2020, Online, July 5–10, 2020*. Ed. by Dan Jurafsky, Joyce Chai, Natalie Schluter, and Joel R. Tetreault. Association for Computational Linguistics, 2020, pp. 8342–8360. DOI: https://doi.org/10.18653/V1/2020.ACL-MAIN.740. URL: https://doi.org/10.18653/v1/2020.acl-main.740.
141. Kelvin Guu, Tatsunori B. Hashimoto, Yonatan Oren, and Percy Liang. "Generating Sentences by Editing Prototypes". In: *Trans. Assoc. Comput. Linguistics* 6 (2018), pp. 437–450. DOI: https://doi.org/10.1162/TACL_A_00030. URL: https://doi.org/10.1162/tacl_a_00030.
142. Matthias Habdank, Nikolai Rodehutskors, and Rainer Koch. "Relevancy assessment of tweets using supervised learning techniques: Mining emergency related tweets for automated relevancy classification". In: *4th International Conference on Information and Communication Technologies for Disaster Management, ICT-DM 2017, Müfvnster, Germany, December 11–13, 2017*. Ed. by Bernd Hellingrath and Ivan Gojmerac. IEEE, 2017, pp. 1–8. DOI: https://doi.org/10.1109/ICT-DM.2017.8275670. URL: https://doi.org/10.1109/ICT-DM.2017.8275670.
143. Christine Hagar. "Crisis Informatics: Introduction". In: *Bulletin of the American Society for Information Science and Technology* 36.5 (2010), pp. 10–12. DOIhttps://doi.org/10.1002/bult.2010.1720360504.
144. Awni Y. Hannun, Carl Case, Jared Casper, Bryan Catanzaro, Greg Diamos, Erich Elsen, Ryan Prenger, Sanjeev Satheesh, Shubho Sengupta, Adam Coates, and Andrew Y. Ng. "Deep Speech: Scaling up end-to-end speech recognition". In: *CoRR* abs/1412.5567 (2014). arXiv:1412.5567. URL: http://arxiv.org/abs/1412.5567.

145. Zellig S. Harris. "Distributional Structure". In: *Papers on Syntax*. 1954. DOI: https://doi.org/10.1080/00437956.1954.11659520.
146. A. Hartigan and M. A. Wong. "A K-Means Clustering Algorithm". In: *Journal of the Royal Statistical Society* (1979). ISSN: 00359254. DOI: https://doi.org/10.2307/2346830.
147. Katrin Hartwig, Stefka Schmid, Tom Biselli, Helene Pleil, and Christian Reuter. "Misleading information in crises: exploring content-specific indicators on Twitter from a user perspective". In: *Behaviour & Information Technology* 0.0 (2024), pp. 1–34. DOI: https://doi.org/10.1080/0144929X.2024.2373166. eprint: https://doi.org/10.1080/0144929X.2024.2373166. URL: https://doi.org/10.1080/0144929X.2024.2373166.
148. Doaa Hassan. "A Text Mining Approach for Evaluating Event Credibility on Twitter". In: *27th IEEE International Conference on Enabling Technologies: Infrastructure for Collaborative Enterprises, WETICE 2018, Paris, France, June 27–29, 2018*. Ed. by Layth Sliman, Ismael Bouassida Rodriguez, and Kaori Yoshida. IEEE Computer Society, 2018, pp. 171–174. DOI: https://doi.org/10.1109/WETICE.2018.00039. URL: https://doi.org/10.1109/WETICE.2018.00039.
149. Stefan Helmstetter and Heiko Paulheim. "Weakly Supervised Learning for Fake News Detection on Twitter". In: *IEEE/ACM 2018 International Conference on Advances in Social Networks Analysis and Mining, ASONAM 2018, Barcelona, Spain, August 28–31, 2018*. Ed. by Ulrik Brandes, Chandan Reddy, and Andrea Tagarelli. IEEE Computer Society, 2018, pp. 274–277. DOI: https://doi.org/10.1109/ASONAM.2018.8508520. URL: https://doi.org/10.1109/ASONAM.2018.8508520.
150. Dan Hendrycks, Norman Mu, Ekin Dogus Cubuk, Barret Zoph, Justin Gilmer, and Balaji Lakshminarayanan. "AugMix: A Simple Data Processing Method to Improve Robustness and Uncertainty". In: *8th International Conference on Learning Representations, ICLR 2020, Addis Ababa, Ethiopia, April 26–30, 2020*. OpenReview.net, 2020. URL: https://www.openreview.net/forum?id=S1gmrxHFvB.
151. Alex Hernández-García and Peter König. "Data augmentation instead of explicit regularization". In: *CoRR* abs/1806.03852 (2018). arXiv:1806.03852. URL: http://arxiv.org/abs/1806.03852.
152. Starr Roxanne Hiltz and Linda Plotnick. "Dealing with information overload when using social media for emergency management: Emerging solutions". In: *10th Proceedings of the International Conference on Information Systems for Crisis Response and Management, Baden-Baden, Germany, May 12–15, 2013*. Ed. by Tina Comes, Frank Fiedrich, Simon Fortier, Jutta Geldermann, and Tim Müller. ISCRAM Association, 2013. URL: http://idl.iscram.org/files/hiltz/2013/583_Hiltz+Plotnick2013.pdf.
153. Eoin Hinchy. *Voice of the SOC Analyst*. en-US. Tech. rep. Tines, 2022, p. 39. URL: https://www.tines.com/reports/voice-of-the-soc-analyst/ (visited on 03/04/2022).
154. Neil Houlsby, Ferenc Huszar, Zoubin Ghahramani, and Máté Lengyel. "Bayesian Active Learning for Classification and Preference Learning". In: *CoRR* abs/1112.5745 (2011). arXiv:1112.5745. URL: http://arxiv.org/abs/1112.5745.
155. Jeremy Howard and Sylvain Gugger. "Fastai: A Layered API for Deep Learning". In: *Inf.* 11.2 (2020), p. 108. DOI: https://doi.org/10.3390/INFO11020108. URL: https://doi.org/10.3390/info11020108.
156. Jeremy Howard and Sebastian Ruder. "Universal Language Model Fine-tuning for Text Classification". In: *Proceedings of the 56th Annual Meeting of the Association for Computational Linguistics, ACL 2018, Melbourne, Australia, July 15–20, 2018, Volume*

1: Long Papers. Ed. by Iryna Gurevych and Yusuke Miyao. Association for Computational Linguistics, 2018, pp. 328–339. DOI: https://doi.org/10.18653/V1/P18-1031. URL: https://www.aclanthology.org/P18-1031/.

157. Cheng-Ping Hsieh, Simeng Sun, Samuel Kriman, Shantanu Acharya, Dima Rekesh, Fei Jia, Yang Zhang, and Boris Ginsburg. RULER: *What's the Real Context Size of Your Long-Context Language Models?* 2024. arXiv:2404.06654 [cs.CL].
158. Edward J. Hu, Yelong Shen, Phillip Wallis, Zeyuan Allen-Zhu, Yuanzhi Li, Shean Wang, Lu Wang, and Weizhu Chen. "LoRA: Low-Rank Adaptation of Large Language Models". In: *The Tenth International Conference on Learning Representations, ICLR 2022, Virtual Event, April 25–29*, 2022. OpenReview.net, 2022. URL: https://www.openreview.net/forum?id=nZeVKeeFYf9.
159. Yi-Qi Hu and Yang Yu. "A technical view on neural architecture search". In: *Int. J. Mach. Learn. Cybern.* 11.4 (2020), pp. 795–811. DOI: https://doi.org/10.1007/S13042-020-01062-1. URL: https://doi.org/10.1007/s13042-020-01062-1.
160. Zhiting Hu, Bowen Tan, Ruslan Salakhutdinov, Tom M. Mitchell, and Eric P. Xing. "Learning Data Manipulation for Augmentation and Weighting". In: *Advances in Neural Information Processing Systems 32: Annual Conference on Neural Information Processing Systems 2019, NeurIPS 2019, December 8-14, 2019, Vancouver, BC, Canada*. Ed. by Hanna M. Wallach, Hugo Larochelle, Alina Beygelzimer, Florence d'Alché-Buc, Emily B. Fox, and Roman Garnett. 2019, pp. 15738–15749. URL: https://www.proceedings.neurips.cc/paper/2019/hash/671f0311e2754fcdd37f70a8550379bc-Abstract.html.
161. Amanda Lee Hughes and Leysia Palen. "Twitter adoption and use in mass convergence and emergency events". In: *Proceedings of the Information Systems for Crisis Response and Management (ISCRAM)*. Ed. by J. Landgren and S. Jul. Vol. 6. Gothenburg, 2009. DOI: https://doi.org/10.1504/IJEM.2009.031564.
162. Thien Ho Huong and Vinh Truong Hoang. "A data augmentation techsnique based on text for Vietnamese sentiment analysis". In: *IAIT 2020: The 11th International Conference on Advances in Information Technology, Bangkok, Thailand, July 1–3, 2020*. Ed. by Kriengkrai Porkaew, Mark H. Chignell, S. Fong, and Bunthit Watanapa. ACM, 2020, 13:1–13:5. DOI: https://doi.org/10.1145/3406601.3406618. URL: https://doi.org/10.1145/3406601.3406618.
163. Ghaith Husari, Ehab Al-Shaer, Mohiuddin Ahmed, Bill Chu, and Xi Niu. "TTPDrill: Automatic and Accurate Extraction of Threat Actions from Unstructured Text of CTI Sources". In: *Proceedings of the 33rd Annual Computer Security Applications Conference, Orlando, FL, USA, December 4–8, 2017*. ACM, 2017, pp. 103–115. DOI: https://doi.org/10.1145/3134600.3134646. URL: https://doi.org/10.1145/3134600.3134646.
164. Clayton J. Hutto and Eric Gilbert. "VADER: A Parsimonious Rule-Based Model for Sentiment Analysis of Social Media Text". In: *Proceedings of the Eighth International Conference on Weblogs and Social Media, ICWSM 2014, Ann Arbor, Michigan, USA, June 1–4, 2014*. Ed. by Eytan Adar, Paul Resnick, Munmun De Choudhury, Bernie Hogan, and Alice Oh. The AAAI Press, 2014. URL: http://www.aaai.org/ocs/index.php/ICWSM/ICWSM14/paper/view/8109.
165. Ayei E. Ibor, Olusoji B. Okunoye, Florence A. Oladeji, and Khadeejah A. Abdulsalam. "Novel Hybrid Model for Intrusion Prediction on Cyber Physical Systems' Communication Networks based on Bio-inspired Deep Neural Network Structure". In: *J. Inf. Secur. Appl.* 65 (2022), p. 103107. DOI: https://doi.org/10.1016/J.JISA.2021.103107. URL: https://doi.org/10.1016/j.jisa.2021.103107.

166. Adrian Iftene, Daniela Gîfu, Andrei-Remus Miron, and Mihai-Stefan Dudu. "A Real-Time System for Credibility on Twitter". In: *Proceedings of The 12th Language Resources and Evaluation Conference, LREC 2020, Marfseille, France, May 11–16, 2020*. Ed. by Nicoletta Calzolari, Frédéric Béchet, Philippe Blache, Khalid Choukri, Christopher Cieri, Thierry Declerck, Sara Goggi, Hitoshi Isahara, Bente Maegaard, Joseph Mariani, Heléné Mazo, Asunción Moreno, Jan Odijk, and Stelios Piperidis. European Language Resources Association, 2020, pp. 6166–6173. URL: https://www.aclanthology.org/2020.lrec-1.757/.
167. Muhammad Imran, Carlos Castillo, Fernando Diaz, and Sarah Vieweg. "Processing Social Media Messages in Mass Emergency: A Survey". In: *ACM Comput. Surv.* 47.4 (2015), 67:1–67:38. DOI: https://doi.org/10.1145/2771588. URL: https://doi.org/10.1145/2771588.
168. Muhammad Imran, Shady Elbassuoni, Carlos Castillo, Fernando Diaz, and Patrick Meier. "Practical extraction of disaster-relevant information from social media". In: *22nd International World Wide Web Conference, WWW'3, Rio de Janeiro, Brazil, May 13–17, 2013, Companion Volume*. Ed. by Leslie Carr, Alberto H. F. Laender, Bernadette Farias Loscio, Irwin King, Marcus Fontoura, Denny Vrandecic, Lora Aroyo, José Palazzo M. de Oliveira, Fernanda Lima, and Erik Wilde. International World Wide Web Conferences Steering Committee / ACM, 2013, pp. 1021–1024. DOI: https://doi.org/10.1145/2487788.2488109. URL: https://doi.org/10.1145/2487788.2488109.
169. Muhammad Imran, Prasenjit Mitra, and Carlos Castillo. "Twitter as a Lifeline: Human-annotated Twitter Corpora for NLP of Crisis-related Messages". In: *Proceedings of the Tenth International Conference on Language Resources and Evaluation LREC 2016, Portorož, Slovenia, May 23–28, 2016*. Ed. by Nicoletta Calzolari, Khalid Choukri, Thierry Declerck, Sara Goggi, Marko Grobelnik, Bente Maegaard, Joseph Mariani, Hélène Mazo, Asunción Moreno, Jan Odijk, and Stelios Piperidis. European Language Resources Association (ELRA), 2016. URL: http://www.lrec-conf.org/proceedings/lrec2016/summaries/842.html.
170. Muhammad Imran, Prasenjit Mitra, and Jaideep Srivastava. "Cross Language Domain Adaptation for Classifying Crisis-Related Short Mes sages". In: *13th Proceedings of the International Conference on Information Systems for Crisis Response and Management, Rio de Janeiro, Brasil, May 22–25, 2016*. Ed. by Andrea H. Tapia, Pedro Antunes, Victor A. Bañuls, Kathleen A. Moore, and João Porto de Albuquerque. ISCRAM Association, 2016. URL: http://idl.iscram.org/files/muhammadimran/2016/1396_MuhammadImran_etal2016.pdf.
171. Muhammad Imran, Prasenjit Mitra, and Jaideep Srivastava. "Enabling Rapid Classification of Social Media Communications During Crises". In: *Int. J. Inf. Syst. Crisis Response Manag.* 8.3 (2016), pp. 1–17. DOI: https://doi.org/10.4018/IJISCRAM.2016070101. URL: https://doi.org/10.4018/IJISCRAM.2016070101.
172. Hiroshi Inoue. "Data Augmentation by Pairing Samples for Images Classification". In: *CoRR* abs/1801.02929 (2018). arXiv:1801.02929. URL: http://arxiv.org/abs/1801.02929.
173. Mohit Iyyer, John Wieting, Kevin Gimpel, and Luke Zettlemoyer. "Adversarial Example Generation with Syntactically Controlled Paraphrase Networks". In: *Proceedings of the 2018 Conference of the North American Chapter of the Association for Computational Linguistics: Human Language Technologies, NAACL-HLT 2018, New Orleans, Louisiana, USA, June 1-6, 2018, Volume 1 (Long Papers)*. Ed. by Marilyn A. Walker,

Heng Ji, and Amanda Stent. Association for Computational Linguistics, 2018, pp. 1875–1885. DOI: https://doi.org/10.18653/V1/N18-1170. URL: https://doi.org/10.18653/v1/n18-1170.

174. Pieter Floris Jacobs, Gideon Maillette de Buy Wenniger, Marco A. Wiering, and Lambert Schomaker. "Active Learning for Reducing Labeling Effort in Text Classification Tasks". In: *Artificial Intelligence and Machine Learning—33rd Benelux Conference on Artificial Intelligence, BNAIC/Bene-learn 2021, Esch-sur-Alzette, Luxembourg, November 10–12, 2021, Revised Selected Papers*. Ed. by Luis A. Leiva, Cédric Pruski, Réka Markovich, Amro Najjar, and Christoph Schommer. Vol. 1530. Communications in Computer and Information Science. Springer, 2021, pp. 3–29. DOI: https://doi.org/10.1007/978-3-030-93842-0_1. URL: https://doi.org/10.1007/978-3-030-93842-0_1.

175. Anil K. Jain. "Data clustering: 50 years beyond K-means". In: *Pattern Recognit. Lett.* 31.8 (2010), pp. 651–666. DOI: https://doi.org/10.1016/J.PATREC.2009.09.011. URL: https://doi.org/10.1016/j.patrec.2009.09.011.

176. Praphula Kumar Jain, Waris Quamer, Vijayalakshmi Saravanan, and Rajendra Pamula. "Employing BERT-DCNN with sentic knowledge base for social media sentiment analysis". In: *J. Ambient Intell. Humaniz. Comput.* 14.8 (2023), pp. 10417–10429. DOI: https://doi.org/10.1007/S12652-022-03698-Z. URL: https://doi.org/10.1007/s12652-022-03698-z.

177. Navdeep Jaitly and Geoffrey E Hinton. "Vocal tract length perturbation (VTLP) improves speech recognition". In: *Proc. ICML Workshop on Deep Learning for Audio, Speech and Language*. Vol. 117. 2013, p. 21.

178. Ganesh Jawahar, Benoîst Sagot, and Djam'e Seddah. "What Does BERT Learn about the Structure of Language?" In: *Proceedings of the 57th Conference of the Association for Computational Linguistics, ACL 2019, Florence, Italy, July 28- August 2, 2019, Volume 1: Long Papers*. Ed. by Anna Korho nen, David R. Traum, and Lluís Màrquez. Association for Computational ' Linguistics, 2019, pp. 3651–3657. DOI: https://doi.org/10.18653/V1/P19-1356. URL: https://doi.org/10.18653/v1/p19-1356.

179. Robin Jia and Percy Liang. "Adversarial Examples for Evaluating Reading Comprehension Systems". In: *Proceedings of the 2017 Conference on Empirical Methods in Natural Language Processing, EMNLP 2017, Copenhagen, Denmark, September 9–11, 2017*. Ed. by Martha Palmer, Rebecca Hwa, and Sebastian Riedel. Association for Computational Linguistics, 2017, pp. 2021–2031. DOI: https://doi.org/10.18653/V1/D17-1215. URL: https://doi.org/10.18653/v1/d17-1215.

180. Albert Q. Jiang, Alexandre Sablayrolles, Antoine Roux, Arthur Mensch, Blanche Savary, Chris Bamford, Devendra Singh Chaplot, Diego de Las Casas, Emma Bou Hanna, Florian Bressand, Gianna Lengyel, Guillaume Bour, Guillaume Lample, Lélio Renard Lavaud, Lucile Saulnier, Marie-Anne Lachaux, Pierre Stock, Sandeep Subramanian, Sophia Yang, Szymon Antoniak, Teven Le Scao, Théophile Gervet, Thibaut Lavril, Thomas Wang, Timothée Lacroix, and William El Sayed. "Mixtral of Experts". In: *CoRR* abs/2401.04088 (2024). DOI: https://doi.org/10.48550/ARXIV.2401.04088. arXiv:2401.04088. URL: https://doi.org/10.48550/arXiv.2401.04088.

181. Haoming Jiang, Pengcheng He, Weizhu Chen, Xiaodong Liu, Jianfeng Gao, and Tuo Zhao. "SMART: Robust and Efficient Fine-Tuning for Pre-trained Natural Language Models through Principled Regularized Optimization". In: *Proceedings of the 58th Annual Meeting of the Association for Computational Linguistics, ACL 2020, Online, July 5–10, 2020*. Ed. by Dan Jurafsky, Joyce Chai, Natalie Schluter, and Joel R. Tetreault.

Association for Computational Linguistics, 2020, pp. 2177–2190. DOI: https://doi.org/10.18653/V1/2020.ACL-MAIN.197. URL: https://doi.org/10.18653/v1/2020.acl-main.197.

182. Shaohua Jiang, Shan Zhao, Kai Hou, Yang Liu, Li Zhang, et al. "A BERT-BiLSTM-CRF model for Chinese electronic medical records named entity recognition". In: *2019 12th International Conference on Intelligent Computa tion Technology and Automation (ICICTA)*. IEEE. 2019, pp. 166–169.
183. Xiaoqi Jiao, Yichun Yin, Lifeng Shang, Xin Jiang, Xiao Chen, Linlin Li, Fang Wang, and Qun Liu. "TinyBERT: Distilling BERT for Natural Language Understanding". In: *Findings of the Association for Computational Linguistics: EMNLP 2020, Online Event, 16–20 November 2020*. Ed. by Trevor Cohn, Yulan He, and Yang Liu. Vol. EMNLP 2020. Findings of ACL. Association for Computational Linguistics, 2020, pp. 4163–4174. DOI: https://doi.org/10.18653/V1/2020.FINDINGS-EMNLP.372. URL: https://doi.org/10.18653/v1/2020.findings-emnlp.372.
184. Di Jin, Zhijing Jin, Joey Tianyi Zhou, and Peter Szolovits. "Is BERT Really Robust? A Strong Baseline for Natural Language Attack on Text Classification and Entailment". In: *The Thirty-Fourth AAAI Conference on Artificial Intelligence, AAAI 2020, The Thirty-Second Innovative Applications of Artificial Intelligence Conference, IAAI 2020, The Tenth AAAI Symposium on Educational Advances in Artificial Intelligence, EAAI 2020, New York, NY, USA, February 7–12, 2020*. AAAI Press, 2020, pp. 8018–8025. DOI: https://doi.org/10.1609/AAAI.V34I05.6311. URL: https://doi.org/10.1609/aaai.v34i05.6311.
185. Armand Joulin, Edouard Grave, Piotr Bojanowski, Matthijs Douze, Hervé Jégou, and Tomás Mikolov. "FastText.zip: Compressing text classification models". In: *CoRR* abs/1612.03651 (2016). arXiv:1612.03651. URL: http://arxiv.org/abs/1612.03651.
186. Armand Joulin, Edouard Grave, Piotr Bojanowski, and Tomás Mikolov. "Bag of Tricks for Efficient Text Classification". In: *Proceedings of the 15th Conference of the European Chapter of the Association for Computational Linguistics, EACL 2017, Valencia, Spain, April 3–7, 2017, Volume 2: Short Papers*. Ed. by Mirella Lapata, Phil Blunsom, and Alexander Koller. Association for Computational Linguistics, 2017, pp. 427–431. DOI: https://doi.org/10.18653/V1/E17-2068. URL: https://doi.org/10.18653/v1/e17-2068.
187. Michal Jungiewicz and Aleksander Smywinski-Pohl. "Towards textual data augmentation for neural networks: Synonyms and maximum loss". In: *Computer Science* (2019). ISSN: 23007036. DOI: https://doi.org/10.7494/csci.2019.20.1.3023.
188. Kushal Kafle, Mohammed A. Yousefhussien, and Christopher Kanan. "Data Augmentation for Visual Question Answering". In: *Proceedings of the 10th International Conference on Natural Language Generation, INLG 2017, Santiago de Compostela, Spain, September 4–7, 2017*. Ed. by José Maria Alonso, Alberto Bugarín, and Ehud Reiter. Association for Computational Linguistics, 2017, pp. 198–202. DOI: https://doi.org/10.18653/V1/W17-3529. URL: https://doi.org/10.18653/v1/w17-3529.
189. Hacer Karacan and Mehmet Sevri. "A Novel Data Augmentation Technique and Deep Learning Model for Web Application Security". In: *IEEE Access* 9 (2021), pp. 150781–150797. DOI: https://doi.org/10.1109/ACCESS.2021.3125785. URL: https://doi.org/10.1109/ACCESS.2021.3125785.
190. Omid Kashefi and Rebecca Hwa. "Quantifying the Evaluation of Heuristic Methods for Textual Data Augmentation". In: *Proceedings of the Sixth Workshop on Noisy User-generated Text, W-NUT at EMNLP 2020 Online, November 19, 2020*. Ed. by Wei Xu,

Alan Ritter, Tim Baldwin, and Afshin Rahimi. Association for Computational Linguistics, 2020, pp. 200–208. DOI: https://doi.org/10.18653/V1/2020.WNUT-1.26. URL: https://doi.org/10.18653/v1/2020.wnut-1.26.

191. Marc-Andr Kaufhold and Christian Reuter. "Cultural violence and peace in social media". In: *Information technology for peace and security—IT—Applications and infrastructures in conflicts, crises, war, and peace*. Ed. by Christian Reuter. 2019, pp. 361–381. ISBN: 978-3-658-25652-4.
192. Marc-André Kaufhold. "Information refinement technologies for crisis informatics: user expectations and design principles for social media and mobile apps". PhD thesis. Technical University of Darmstadt, Germany, 2020. ISBN: 978-3-658-33343-0. URL: https://www.d-nb.info/1230698515.
193. Marc-André Kaufhold, Ali Sercan Basyurt, Kaan Eyilmez, Marc Stöttinger, and Christian Reuter. "Cyber Threat Observatory: Design and Evaluation of an Interactive Dashboard for Computer Emergency Response Teams". In: *30th European Conference on Information Systems—New Horizons in Digitally United Societies, ECIS 2022, Timisoara, Romania, June 18–24, 2022*. Ed. by Roman Beck, Dana Petcu, Marin Fotache, Sabine Matook, Remko Helms, Martin Wiener, Lazar Rusu, and Tuure Tuunanen. 2022. URL: https://www.aisel.aisnet.org/ecis2022_rp/99.
194. Marc-André Kaufhold, Markus Bayer, Daniel Hartung, and Christian Reuter. "Design and Evaluation of Deep Learning Models for Real-Time Credibility Assessment in Twitter". In: *Artificial Neural Networks and Machine Learning—ICANN 2021—30th International Conference on Artificial Neural Networks, Bratislava, Slovakia, September 14–17, 2021, Proceedings, Part V.* Ed. by Igor Farkas, Paolo Masulli, Sebastian Otte, and Stefan Wermter. Vol. 12895. Lecture Notes in Computer Science. Springer, 2021, pp. 396–408. DOI: https://doi.org/10.1007/978-3-030-86383-8_32. URL: https://doi.org/10.1007/978-3-030-86383-8_32.
195. Marc-André Kaufhold, Jennifer Fromm, Thea Riebe, Milad Mirbabaie, Philipp Kühn, Ali Sercan Basyurt, Markus Bayer, Marc Stöttinger, Kaan Eyilmez, Reinhard Möller, Christoph Fuchß, Stefan Stieglitz, and Christian Reuter. "CYWARN: Strategy and Technology Development for Cross-Platform Cyber Situational Awareness and Actor-Specific Cyber Threat Communication". In: *Mensch und Computer 2021—Workshopband, Ingolstadt, Germany, September 5–8, 2021*. Ed. by Carolin Wienrich, Philipp Wintersberger, and Benjamin Weyers. Gesellschaft für Informatik e.V., 2021. DOI: https://doi.org/10.18420/MUC2021-MCI-WS08-263. URL: https://doi.org/10.18420/muc2021-mci-ws08-263.
196. Marc-André Kaufhold, Thea Riebe, Markus Bayer, and Christian Reuter. "'We Do Not Have the Capacity to Monitor All Media': A Design Case Study on Cyber Situational Awareness in Computer Emergency Response Teams". In: *Proceedings of the CHI Conference on Human Factors in Computing Systems, CHI 2024, Honolulu, HI, USA, May 11–16, 2024*. Ed. by Florian 'Floyd' Mueller, Penny Kyburz, Julie R. Williamson, Corina Sas, Max L. Wilson, Phoebe O. Toups Dugas, and Irina Shklovski. ACM, 2024, 580:1–580:16. DOI: https://doi.org/10.1145/3613904.3642368. URL: https://doi.org/10.1145/3613904.3642368.
197. Marc-André Kaufhold, Nicola Rupp, Christian Reuter, and Matthias Habdank. "Mitigating information overload in social media during conflicts and crises: design and evaluation of a cross-platform alerting system". In: *Behav. Inf. Technol.* 39.3 (2020),

pp. 319–342. DOI: https://doi.org/10.1080/0144929X.2019.1620334. URL: https://doi.org/10.1080/0144929X.2019.1620334.

198. Marc-André Kaufhold, Markus Bayer, and Christian Reuter. "Rapid relevance classification of social media posts in disasters and emergencies: A system and evaluation featuring active, incremental and online learning". In: *Inf. Process. Manag.* 57.1 (2020). DOI: https://doi.org/10.1016/J.IPM.2019.102132. URL: https://doi.org/10.1016/j.ipm.2019.102132.
199. Marc André Kaufhold and Christian Reuter. "The Self-Organization of Digital Volunteers across Social Media: The Case of the 2013 European Floods in Germany". In: *Journal of Homeland Security and Emergency Management 13.1* (2016), pp. 137–166. ISSN: 15477355. DOI: https://doi.org/10.1515/jhsem-2015-0063.
200. Daniel Keim, Gennady Andrienko, Jean-daniel Fekete, G Carsten, and Guy Melan. "Visual Analytics: Definition, Process and Challenges". In: *Information Visualization—Human-Centered Issues and Perspectives* (2008), pp. 154–175. ISSN: 0302–9743. DOI: https://doi.org/10.1007/978-3-540-70956-5_7.
201. Ronald Kemker, Marc McClure, Angelina Abitino, Tyler L. Hayes, and Christopher Kanan. "Measuring Catastrophic Forgetting in Neural Networks". In: *Proceedings of the Thirty-Second AAAI Conference on Artificial Intelligence, (AAAI-18), the 30th innovative Applications of Artificial Intelligence (IAAI-18), and the 8th AAAI Symposium on Educational Advances in Artificial Intelligence (EAAI-18), New Orleans, Louisiana, USA, February 2–7, 2018*. Ed. by Sheila A. McIlraith and Kilian Q. Weinberger. AAAI Press, 2018, pp. 3390–3398. DOI: https://doi.org/10.1609/AAAI.V32I1.11651. URL: https://doi.org/10.1609/aaai.v32i1.11651.
202. Bilal Khan. *Generate your own text with OpenAI's GPT-2*. 2019. URL: https://www.kaggle.com/bkkaggle/generate-your-own-text-with-openai-s-gpt-2-117m.
203. Omar Khattab, Arnav Singhvi, Paridhi Maheshwari, Zhiyuan Zhang, Keshav Santhanam, Sri Vardhamanan, Saiful Haq, Ashutosh Sharma, Thomas T. Joshi, Hanna Moazam, Heather Miller, Matei Zaharia, and Christopher Potts. "DSPy: Compiling Declarative Language Model Calls into Self-Improving Pipelines". In: *CoRR* abs/2310.03714 (2023). DOI: https://doi.org/10.48550/ARXIV.2310.03714. arXiv:2310.03714. URL: https://doi.org/10.48550/arXiv.2310.03714.
204. Aparup Khatua, Apalak Khatua, and Erik Cambria. "A tale of two epidemics: Contextual Word2Vec for classifying twitter streams during outbreaks". In: *Inf. Process. Manag.* 56.1 (2019), pp. 247–257. DOI: https://doi.org/10.1016/J.IPM.2018.10.010. URL: https://doi.org/10.1016/j.ipm.2018.10.010.
205. Hazel H. Kim, Daecheol Woo, Seong Joon Oh, Jeong-Won Cha, and Yo-Sub Han. "ALP: Data Augmentation Using Lexicalized PCFGs for Few-Shot Text Classification". In: *Thirty-Sixth AAAI Conference on Artificial Intelligence, AAAI 2022, Thirty-Fourth Conference on Innovative Applications of Artificial Intelligence, IAAI 2022, The Twelveth Symposium on Educational Advances in Artificial Intelligence, EAAI 2022 Virtual Event, February 22—March 1, 2022*. AAAI Press, 2022, pp. 10894–10902. DOI: https://doi.org/10.1609/AAAI.V36I10.21336. URL: https://doi.org/10.1609/aaai.v36i10.21336.
206. Jiyeon Kim, Jiwon Kim, Hyunjung Kim, Minsun Shim, and Eunjung Choi. "CNN-Based Network Intrusion Detection against Denial-of-Service Attacks". en. In: *Electronics* 9.6 (June 2020), p. 916. ISSN: 2079–9292. DOI: https://doi.org/10.3390/electronics9060916. URL: https://www.mdpi.com/2079-9292/9/6/916 (visited on 11/22/2023).

207. Diederik P. Kingma and Jimmy Ba. "Adam: A Method for Stochastic Optimization". In: *3rd International Conference on Learning Representations, ICLR 2015, San Diego, CA, USA, May 7–9, 2015, Conference Track Proceedings*. Ed. by Yoshua Bengio and Yann LeCun. 2015. URL: http://arxiv.org/abs/1412.6980.
208. Tom Ko, Vijayaditya Peddinti, Daniel Povey, and Sanjeev Khudanpur. "Audio augmentation for speech recognition". In: *INTERSPEECH 2015, 16th Annual Conference of the International Speech Communication Association, Dresden, Germany, September 6–10, 2015*. ISCA, 2015, pp. 3586–3589. DOI: https://doi.org/10.21437/INTERSPEECH.2015-711. URL: https://doi.org/10.21437/Interspeech.2015-711.
209. Sosuke Kobayashi. "Contextual Augmentation: Data Augmentation by Words with Paradigmatic Relations". In: *Proceedings of the 2018 Conference of the North American Chapter of the Association for Computational Linguis tics: Human Language Technologies, NAACL-HLT, New Orleans, Louisiana, USA, June 1–6, 2018, Volume 2 (Short Papers)*. Ed. by Marilyn A. Walker, Heng Ji, and Amanda Stent. Association for Computational Linguis tics, 2018, pp. 452–457. DOI: https://doi.org/10.18653/V1/N18-2072. URL: https://doi.org/10.18653/v1/n18-2072.
210. Pang Wei Koh and Percy Liang. "Understanding Black-box Predictions via Influence Functions". In: *Proceedings of the 34th International Conference on Machine Learning, ICML 2017, Sydney, NSW, Australia, 6–11 August 2017*. Ed. by Doina Precup and Yee Whye Teh. Vol. 70. Proceedings of Machine Learning Research. PMLR, 2017, pp. 1885–1894. URL: http://proceedings.mlr.press/v70/koh17a.html.
211. Harsh Kohli. "Transfer Learning and Augmentation for Word Sense Disambiguation". In: *Advances in Information Retrieval—43rd European Conference on IR Research, ECIR 2021, Virtual Event, March 28—April 1, 2021, Proceedings, Part II*. Ed. by Djoerd Hiemstra, Marie-Francine Moens, Josiane Mothe, Raffaele Perego, Martin Potthast, and Fabrizio Sebastiani. Vol. 12657. Lecture Notes in Computer Science. Springer, 2021, pp. 303–311. DOI: https://doi.org/10.1007/978-3-030-72240-1_29. URL: https://doi.org/10.1007/978-3-030-72240-1_29.
212. Takeshi Kojima, Shixiang Shane Gu, Machel Reid, Yutaka Matsuo, and Yusuke Iwasawa. "Large Language Models are Zero-Shot Reasoners". In: *Advances in Neural Information Processing Systems 35: Annual Conference on Neural Information Processing Systems 2022, NeurIPS 2022, New Orleans, LA, USA, November 28 - December 9, 2022*. Ed. by Sanmi Koyejo, S. Mohamed, A. Agarwal, Danielle Belgrave, K. Cho, and A. Oh. 2022. URL: http://papers.nips.cc/paper_files/paper/2022/hash/8bb0d291acd4acf06ef112099c16f326-Abstract-Conference.html.
213. Takeshi Kojima, Shixiang Shane Gu, Machel Reid, Yutaka Matsuo, and Yusuke Iwasawa. "Large Language Models are Zero-Shot Reasoners". In: *Advances in Neural Information Processing Systems 35: Annual Conference on Neural Information Processing Systems 2022, NeurIPS 2022, New Orleans, LA, USA, November 28—December 9, 2022*. Ed. by Sanmi Koyejo, S. Mohamed, A. Agarwal, Danielle Belgrave, K. Cho, and A. Oh. 2022. URL: http://papers.nips.cc/paper_files/paper/2022/hash/8bb0d291acd4acf06ef112099c16f326-Abstract-Conference.html.
214. Oleksandr Kolomiyets, Steven Bethard, and Marie-Francine Moens. "Model-Portability Experiments for Textual Temporal Analysis". In: *The 49th Annual Meeting of the Association for Computational Linguistics: Human Language Technologies, Proceedings of the Conference, 19–24 June, 2011, Portland, Oregon, USA—Short Papers*. The Asso-

ciation for Computer Linguistics, 2011, pp. 271–276. URL: https://www.aclanthology.org/P11-2047/.

215. Paris Koloveas, Thanasis Chantzios, Sofia Alevizopoulou, Spiros Ski-adopoulos, and Christos Tryfonopoulos. "inTIME: A Machine Learning-Based Framework for Gathering and Leveraging Web Data to Cyber-Threat Intelligence". en. In: *Electronics* 10.7 (Mar. 2021), p. 818. ISSN: 2079–9292. DOI: https://doi.org/10.3390/electronics10070818. URL: https://www.mdpi.com/2079-9292/10/7/818 (visited on 12/20/2023).
216. Jan-David Krieger, Timo Spinde, Terry Ruas, Juhi Kulshrestha, and Bela Gipp. "A domain-adaptive pre-training approach for language bias detection in news". In: *JCDL '22: The ACM/IEEE Joint Conference on Digital Libraries in 2022, Cologne, Germany, June 20–24, 2022*. Ed. by Akiko Aizawa, Thomas Mandl, Zeljko Carevic, Annika Hinze, Philipp Mayr, and Philipp Schaer. ACM, 2022, p. 3. DOI: https://doi.org/10.1145/3529372.3530932. URL: https://doi.org/10.1145/3529372.3530932.
217. G Krishnalal, S Babu Rengarajan, and K G Srinivasagan. "A New Text Mining Approach Based on HMM-SVM for Web News Classification". In: *International Journal of Computer Applications* (2010). DOI: https://doi.org/10.5120/395-589.
218. Anna Kruspe, Jens Kersten, Matti Wiegmann, Benno Stein, and Friederike Klan. "Classification of Incident-related Tweets : Tackling Imbalanced Training Data using Hybrid CNNs and Translation-based Data Augmentation". In: *Notebook papers of the TREC 2018 conference*. 2018.
219. Philipp Kuehn, Markus Bayer, Marc Wendelborn, and Christian Reuter. "OVANA: An Approach to Analyze and Improve the Information Qual ity of Vulnerability Databases". In: *ARES 2021: The 16th International Conference on Availability, Reliability and Security, Vienna, Austria, August 17–20, 2021*. Ed. by Delphine Reinhardt and Tilo Müller. ACM, 2021, 22:1–22:11. DOI: https://doi.org/10.1145/3465481.3465744. URL: https://doi.org/10.1145/3465481.3465744.
220. Philipp Kuehn, Moritz Kerk, Marc Wendelborn, and Christian Reuter. "Clustering of Threat Information to Mitigate Information Overload for Computer Emergency Response Teams". In: *CoRR* abs/2210.14067 (2022). DOI: https://doi.org/10.48550/ARXIV.2210.14067. arXiv:2210.14067. URL: https://doi.org/10.48550/arXiv.2210.14067.
221. Philipp Kuehn, Thea Riebe, Lynn Apelt, Max Jansen, and Christian Reuter. "Sharing of Cyber Threat Intelligence between States". en. In: *Sicherheit & Frieden* 38.1 (2020), pp. 22–28. ISSN: 0175–274X. DOI: https://doi.org/10.5771/0175-274X-2020-1-22. URL: https://www.nomos-elibrary.de/index.php?doi=10.5771/0175-274X-2020-1-22 (visited on 04/05/2023).
222. Prabhat Kumar, Govind P. Gupta, Rakesh Tripathi, Sahil Garg, and Mohammad Mehedi Hassan. "DLTIF: Deep Learning-Driven Cyber Threat Intelligence Modeling and Identification Framework in IoT-Enabled Maritime Transportation Systems". In: *IEEE Trans. Intell. Transp. Syst.* 24.2 (2023), pp. 2472–2481. DOI: https://doi.org/10.1109/TITS.2021.3122368. URL: https://doi.org/10.1109/TITS.2021.3122368.
223. Varun Kumar, Ashutosh Choudhary, and Eunah Cho. "Data Augmentation using Pre-trained Transformer Models". In: *CoRR* abs/2003.02245 (2020). arXiv:2003.02245. URL: https://www.arxiv.org/abs/2003.02245.
224. Varun Kumar, Hadrien Glaude, Cyprien de Lichy, and William Campbell. "A Closer Look At Feature Space Data Augmentation For Few-Shot Intent Classification". In:

Proceedings of the 2nd Workshop on Deep Learning Approaches for Low-Resource NLP, DeepLo at EMNLP-IJCNLP 2019, Hong Kong, China, November 3, 2019. Ed. by Colin Cherry, Greg Durrett, George F. Foster, Reza Haffari, Shahram Khadivi, Nanyun Peng, Xiang Ren, and Swabha Swayamdipta. Association for Computational Linguistics, 2019, pp. 1–10. DOI: https://doi.org/10.18653/V1/D19-6101. URL: https://doi.org/10.18653/v1/D19-6101.

225. Alexey Kurakin, Ian J. Goodfellow, and Samy Bengio. "Adversarial examples in the physical world". In: *5th International Conference on Learning Representations, ICLR 2017, Toulon, France, April 24–26, 2017, Workshop Track Proceedings*. OpenReview.net, 2017. URL: https://www.openreview.net/forum?id=HJGU3Rodl.
226. Gakuto Kurata, Bing Xiang, and Bowen Zhou. "Labeled Data Generation with Encoder-Decoder LSTM for Semantic Slot Filling". In: *Interspeech 2016, 17th Annual Conference of the International Speech Communication Association, San Francisco, CA, USA, September 8–12, 2016*. Ed. by Nelson Morgan. ISCA, 2016, pp. 725–729. DOI: https://doi.org/10.21437/INTERSPEECH.2016-727. URL: https://doi.org/10.21437/Interspeech.2016-727.
227. Dilek Küçük Matci and Ugur Avdan. "Comparative analysis of unsupervised classification methods for mapping burned forest areas". In: *Arabian Journal of Geosciences* 13.15 (Aug. 2020), p. 711. ISSN: 1866–7511, 1866–7538. DOI: https://doi.org/10.1007/s12517-020-05670-7. URL: https://www.link.springer.com/10.1007/s12517-020-05670-7 (visited on 09/09/2022).
228. Lab42. *About ARC*. URL: https://www.lab42.global/arc/ (visited on 02/26/2024).
229. Rabindra Lamsal, Maria Rodriguez Read, and Shanika Karunasekera. "CrisisTransformers: Pre-trained language models and sentence encoders for crisis-related social media texts". In: *CoRR* abs/2309.05494 (2023). DOI: https://doi.org/10.48550/ARXIV.2309.05494. arXiv:2309.05494. URL: https://doi.org/10.48550/arXiv.2309.05494.
230. Zhenzhong Lan, Mingda Chen, Sebastian Goodman, Kevin Gimpel, Piyush Sharma, and Radu Soricut. "ALBERT: A Lite BERT for Self-supervised Learning of Language Representations". In: *8th International Conference on Learning Representations, ICLR 2020, Addis Ababa, Ethiopia, April 26–30, 2020*. OpenReview.net, 2020. URL: https://www.openreview.net/forum?id=H1eA7AEtvS.
231. Max Landauer, Florian Skopik, Markus Wurzenberger, and Andreas Rauber. "Dealing with Security Alert Flooding: Using Machine Learning for Domain-independent Alert Aggregation". In: *ACM Trans. Priv. Secur.* 25.3 (2022), 18:1–18:36. DOI: https://doi.org/10.1145/3510581. URL: https://doi.org/10.1145/3510581.
232. J. Richard Landis and Gary G. Koch. "The Measurement of Observer Agreement for Categorical Data". en. In: *Biometrics* 33.1 (Mar. 1977), p. 159. ISSN: 0006341X. DOI: https://doi.org/10.2307/2529310. URL: https://www.jstor.org/stable/2529310?origin=crossref (visited on 09/14/2023).
233. Simon Lansmann and Stefan Klein. "How much Collaboration? Balancing the Needs for Collaborative and Uninterrupted Work". In: *26th European Conference on Information Systems: Beyond Digitization—Facets of Socio-Technical Change, ECIS 2018, Portsmouth, UK, June 23–28, 2018*. Ed. by Peter M. Bednar, Ulrich Frank, and Karlheinz Kautz. 2018, p. 118. URL: https://www.aisel.aisnet.org/ecis2018_rp/118.
234. Pedro Lara-Benítez, Manuel Carranza-García, Jorge García-Gutiérrez, and José C. Riquelme. "Asynchronous dual-pcne data stream classification". In: *Integr. Comput.*

Aided Eng. 27.2 (2020), pp. 101–119. DOI: https://doi.org/10.3233/ICA-200617. URL: https://doi.org/10.3233/ICA-200617.

235. David M. J. Lazer, Matthew A. Baum, Yochai Benkler, Adam J. Berinsky, Kelly M. Greenhill, Filippo Menczer, Miriam J. Metzger, Brendan Nyhan, Gordon Pennycook, David M. Rothschild, Michael Schudson, Steven A. Sloman, Cass R. Sunstein, Emily A. Thorson, Duncan J. Watts, and Jonathan L. Zittrain. "The science of fake news". In: *CoRR* abs/2307.07903 (2023). DOI: https://doi.org/10.48550/ARXIV.2307.07903. arXiv:2307.07903. URL: https://doi.org/10.48550/arXiv.2307.07903.
236. Rémi Leblond, Florent Altché, and Felix Gimeno. *AlphaCode2_Tech_Report.pdf*. 2023. URL: https://www.storage.googleapis.com/deepmind-media/AlphaCode2/AlphaCode2_Tech_Report.pdf.
237. Yann LeCun, Léon Bottou, Yoshua Bengio, and Patrick Haffner. "Gradient-based learning applied to document recognition". In: *Proc. IEEE* 86.11 (1998), pp. 2278–2324. DOI: https://doi.org/10.1109/5.726791. URL: https://doi.org/10.1109/5.726791.
238. Jinhyuk Lee, Wonjin Yoon, Sungdong Kim, Donghyeon Kim, Sunkyu Kim, Chan Ho So, and Jaewoo Kang. "BioBERT: a pre-trained biomedical language representation model for biomedical text mining". In: *Bioinform.* 36.4 (2020), pp. 1234–1240. DOI: https://doi.org/10.1093/BIOINFORMATICS/BTZ682. URL: https://doi.org/10.1093/bioinformatics/btz682.
239. Kenton Lee, Kelvin Guu, Luheng He, Tim Dozat, and Hyung Won Chung. "Neural Data Augmentation via Example Extrapolation". In: *CoRR* abs/2102.01335 (2021). arXiv:2102.01335. URL: https://www.arxiv.org/abs/2102.01335.
240. Younghoo Lee, Joshua Saxe, and Richard E. Harang. "CATBERT: Context-Aware Tiny BERT for Detecting Social Engineering Emails". In: *CoRR* abs/2010.03484 (2020). arXiv:2010.03484. URL: https://www.arxiv.org/abs/2010.03484.
241. V. I. Levenshtein. "Binary codes capable of correcting deletions, insertions and reversals". In: *Soviet Physics Doklady* 10 (Feb. 1966), p. 707.
242. David D. Lewis and Jason Catlett. "Heterogeneous Uncertainty Sampling for Supervised Learning". In: *Machine Learning, Proceedings of the Eleventh International Conference, Rutgers University, New Brunswick, NJ, USA, July 10–13, 1994*. Ed. by William W. Cohen and Haym Hirsh. Morgan Kaufmann, 1994, pp. 148–156. DOI: https://doi.org/10.1016/B978-1-55860-335-6.50026-X. URL: https://doi.org/10.1016/b978-1-55860-335-6.50026-x.
243. Ang Li, Fangyuan Zhang, Shuangjiao Li, Tianhua Chen, Pan Su, and Hongtao Wang. "Efficiently generating sentence-level textual adversarial examples with Seq2seq Stacked Auto-Encoder". In: *Expert Syst. Appl.* 213.Part (2023), p. 119170. DOI: https://doi.org/10.1016/J.ESWA.2022.119170. URL: https://doi.org/10.1016/j.eswa.2022.119170.
244. Hongmin Li, Doina Caragea, Cornelia Caragea, and Nic Herndon. "Disaster response aided by tweet classification with a domain adaptation approach". In: *Journal of Contingencies and Crisis Management* (2018). ISSN: 14685973. DOI: https://doi.org/10.1111/1468-5973.12194.
245. Hongmin Li, Xukun Li, Doina Caragea, and Cornelia Caragea. "Comparison of Word Embeddings and Sentence Encodings as Generalized Representations for Crisis Tweet Classification Tasks". In: *Proceedings of the ISCRAM Asian Pacific 2018 Conference* (2018). URL: https://www.google.com/url?client=internal-element-cse&cx=009511351313755808885:dkx-7mlm6ni&q=https://www.cs.uic.edu/~cornelia/

papers/iscram_asian18.pdf&sa=U&ved=2ahUKEwipgLHA0YmFAxW3bmwGHca_DoIQFnoECAEQAQ&usg=AOvVaw35YBZYn5ZiXaD0I9PzWEre.

246. Ke Li, Hui Wen, Hong Li, Hongsong Zhu, and Limin Sun. "Security OSIF: Toward Automatic Discovery and Analysis of Event Based Cyber Threat Intelligence". In: *2018 IEEE SmartWorld, Ubiquitous Intelligence & Computing, Advanced & Trusted Computing, Scalable Computing & Communications, Cloud & Big Data Computing, Internet of People and Smart City Innovation, SmartWorld/SCALCOM/UIC/ATC/CBDCom/IOP/SCI 2018, Guangzhou, China, October 8–12, 2018*. Ed. by Guojun Wang, Qi Han, Md. Zakirul Alam Bhuiyan, Xiaoxing Ma, Frédéric Loulergue, Peng Li, Manuel Roveri, and Lei Chen. IEEE, 2018, pp. 741–747. DOI: https://doi.org/10.1109/SMARTWORLD.2018.00142. URL: https://doi.org/10.1109/SmartWorld.2018.00142.

247. Linyang Li, Ruotian Ma, Qipeng Guo, Xiangyang Xue, and Xipeng Qiu. "BERT-ATTACK: Adversarial Attack Against BERT Using BERT". In: *Proceedings of the 2020 Conference on Empirical Methods in Natural Language Processing, EMNLP 2020*, Online, November 16-20, 2020. Ed. by Bonnie Webber, Trevor Cohn, Yulan He, and Yang Liu. Association for Computational Linguistics, 2020, pp. 6193–6202. DOI: https://doi.org/10.18653/V1/2020.EMNLP-MAIN.500. URL: https://doi.org/10.18653/v1/2020.emnlp-main.500.

248. Tao Li, Yongjin Hu, Ankang Ju, and Zhuoran Hu. "Adversarial Active Learning for Named Entity Recognition in Cybersecurity". en. In: *Computers, Materials & Continua* 66.1 (2020), pp. 407–420. ISSN: 1546–2226. DOI: https://doi.org/10.32604/cmc.2020.012023. URL: https://www.techscience.com/cmc/v66n1/40455 (visited on 12/20/2023).

249. Yang Li, Quan Pan, Suhang Wang, Tao Yang, and Erik Cambria. "A Generative Model for category text generation". In: *Inf. Sci.* 450 (2018), pp. 301–315. DOI: https://doi.org/10.1016/J.INS.2018.03.050. URL: https://doi.org/10.1016/j.ins.2018.03.050.

250. Yitong Li, Trevor Cohn, and Timothy Baldwin. "Robust Training under Linguistic Adversity". In: *Proceedings of the 15th Conference of the European Chapter of the Association for Computational Linguistics, EACL 2017, Valencia, Spain, April 3–7, 2017, Volume 2: Short Papers*. Ed. by Mirella Lapata, Phil Blunsom, and Alexander Koller. Association for Computational Linguistics, 2017, pp. 21–27. DOI: https://doi.org/10.18653/V1/E17-2004. URL: https://doi.org/10.18653/v1/e17-2004.

251. Znaonui Liang, Gang Zhang, Jimmy Xiangji Huang, and Qmming Vivian Hu. "Deep learning for healthcare decision making with EMRs". In: *Proceedings—2014 IEEE International Conference on Bioinformatics and Biomedicine, IEEE BIBM 2014*. 2014. ISBN: 978-1-4799-5669-2. DOI: https://doi.org/10.1109/BIBM.2014.6999219.

252. Xiaojing Liao, Kan Yuan, XiaoFeng Wang, Zhou Li, Luyi Xing, and Raheem A. Beyah. "Acing the IOC Game: Toward Automatic Discovery and Analysis of Open-Source Cyber Threat Intelligence". In: *Proceedings of the 2016 ACM SIGSAC Conference on Computer and Communications Security, Vienna, Austria, October 24–28, 2016*. Ed. by Edgar R. Weippl, Stefan Katzenbeisser, Christopher Kruegel, Andrew C. Myers, and Shai Halevi. ACM, 2016, pp. 755–766. DOI: https://doi.org/10.1145/2976749.2978315. URL: https://doi.org/10.1145/2976749.2978315.

253. Bing Liu and Lei Zhang. "A Survey of Opinion Mining and Sentiment Analysis". In: *Mining Text Data*. Ed. by Charu C. Aggarwal and ChengXiang Zhai. Springer, 2012, pp. 415–463. DOI: https://doi.org/10.1007/978-1-4614-3223-4_13. URL: https://doi.org/10.1007/978-1-4614-3223-4_13.

254. Dayiheng Liu, Yeyun Gong, Jie Fu, Yu Yan, Jiusheng Chen, Jiancheng Lv, Nan Duan, and Ming Zhou. "Tell Me How to Ask Again: Question Data Augmentation with Controllable Rewriting in Continuous Space". In: *Proceedings of the 2020 Conference on Empirical Methods in Natural Language Processing, EMNLP 2020, Online, November 16–20, 2020*. Ed. by Bonnie Webber, Trevor Cohn, Yulan He, and Yang Liu. Association for Computational Linguistics, 2020, pp. 5798–5810. DOI: https://doi.org/10.18653/V1/2020.EMNLP-MAIN.467. URL: https://doi.org/10.18653/v1/2020.emnlp-main.467.
255. Nelson F. Liu, Kevin Lin, John Hewitt, Ashwin Paranjape, Michele Bevilacqua, Fabio Petroni, and Percy Liang. "Lost in the Middle: How Language Models Use Long Contexts". In: *Trans. Assoc. Comput.* Lin guistics 12 (2024), pp. 157–173. DOI: https://doi.org/10.1162/TACL_A_00638. URL: https://doi.org/10.1162/tacl_a_00638.
256. Ruibo Liu, Guangxuan Xu, Chenyan Jia, Weicheng Ma, Lili Wang, and Soroush Vosoughi. "Data Boost: Text Data Augmentation Through Re inforcement Learning Guided Conditional Generation". In: *Proceedings of the 2020 Conference on Empirical Methods in Natural Language Process ing, EMNLP 2020, Online, November 16–20, 2020*. Ed. by Bonnie Webber, Trevor Cohn, Yulan He, and Yang Liu. Association for Computational Linguistics, 2020, pp. 9031–9041. DOI: https://doi.org/10.18653/V1/2020.EMNLPMAIN.726. URL: https://doi.org/10.18653/v1/2020.emnlpmain.726.
257. Xiao Liu, Fanjin Zhang, Zhenyu Hou, Li Mian, Zhaoyu Wang, Jing Zhang, and Jie Tang. "Self-Supervised Learning: Generative or Contrastive". In: *IEEE Trans. Knowl. Data Eng.* 35.1 (2023), pp. 857–876. DOI: https://doi.org/10.1109/TKDE.2021.3090866. URL: https://doi.org/10.1109/TKDE.2021.3090866.
258. Xiaodong Liu, Hao Cheng, Pengcheng He, Weizhu Chen, Yu Wang, Hoifung Poon, and Jianfeng Gao. "Adversarial Training for Large Neural Language Models". In: *CoRR* abs/2004.08994 (2020). arXiv:2004.08994. URL: https://www.arxiv.org/abs/2004.08994.
259. Xiaomo Liu, Armineh Nourbakhsh, Quanzhi Li, Rui Fang, and Sameena Shah. "Real-time Rumor Debunking on Twitter". In: *Proceedings of the 24th ACM International Conference on Information and Knowledge Management, CIKM 2015, Melbourne, VIC, Australia, October 19–23, 2015*. Ed. by James Bailey, Alistair Moffat, Charu C. Aggarwal, Maarten de Rijke, Ravi Kumar, Vanessa Murdock, Timos K. Sellis, and Jeffrey Xu Yu. ACM, 2015, pp. 1867–1870. DOI: https://doi.org/10.1145/2806416.2806651. URL: https://doi.org/10.1145/2806416.2806651.
260. Yang Liu and Yi-fang Brook Wu. "Early Detection of Fake News on Social Media Through Propagation Path Classification with Recurrent and Con volutional Networks". In: *Proceedings of the Thirty-Second AAAI Conference on Artificial Intelligence, (AAAI-18), the 30th innovative Applications of ArLi, Sishuo Chen, Hao Zhoutificial Intelligence (IAAI-18), and the 8th AAAI Symposium on Educational Advances in Artificial Intelligence (EAAI-18), New Orleans, Louisiana, USA, February 2–7, 2018*. Ed. by Sheila A. McIlraith and Kilian Q. Weinberger. AAAI Press, 2018, pp. 354–361. DOI: https://doi.org/10.1609/AAAI.V32I1.11268. URL: https://doi.org/10.1609/aaai.v32i1.11268.
261. Yi Liu, Lianzhe Huang, Shicheng Li, Sishuo Chen, Hao Zhou, Fan-dong Meng, Jie Zhou, and Xu Sun. "RECALL: A Benchmark for LLMs Robustness against External Counterfactual Knowledge". In: *CoRR* abs/2311.08147 (2023). DOI: https://doi.org/10.48550/ARXIV.2311.08147. arXiv:2311.08147. URL: https://doi.org/10.48550/arXiv.2311.08147.

262. Yinhan Liu, Myle Ott, Naman Goyal, Jingfei Du, Mandar Joshi, Danqi Chen, Omer Levy, Mike Lewis, Luke Zettlemoyer, and Veselin Stoyanov. "RoBERTa: A Robustly Optimized BERT Pretraining Approach". In: *CoRR* abs/1907.11692 (2019). arXiv:1907.11692. URL: http://arxiv.org/abs/1907.11692.
263. Yinhan Liu, Myle Ott, Naman Goyal, Jingfei Du, Mandar Joshi, Danqi Chen, Omer Levy, Mike Lewis, Luke Zettlemoyer, and Veselin Stoyanov. "RoBERTa: A Robustly Optimized BERT Pretraining Approach". In: *CoRR* abs/1907.11692 (2019). arXiv:1907.11692. URL: http://arxiv.org/abs/1907.11692.
264. Llama Team, AI @ Meta. *The Llama 3 Herd of Models*. July 2024. URL: https://www.ai.meta.com/research/publications/the-llama3-herd-of-models/.
265. Shayne Longpre, Yu Wang, and Chris DuBois. "How Effective is Task-Agnostic Data Augmentation for Pretrained Transformers?" In: *Findings of the Association for Computational Linguistics: EMNLP 2020, Online Event, 16–20 November 2020*. Ed. by Trevor Cohn, Yulan He, and Yang Liu. Vol. EMNLP 2020. Findings of ACL. Association for Computa tional Linguistics, 2020, pp. 4401–4411. DOI: https://doi.org/10.18653/V1/2020.FINDINGS-EMNLP.394. URL: https://doi.org/10.18653/v1/2020.findings-emnlp.394.
266. Xiaoyu Sean Lu and MengChu Zhou. "Analyzing the evolution of rare events via social media data and k-means clustering algorithm". In: *13th IEEE International Conference on Networking, Sensing, and Control, ICNSC 2016, Mexico City, Mexico, April 28–30, 2016*. IEEE, 2016, pp. 1–6. DOI: https://doi.org/10.1109/ICNSC.2016.7479041. URL: https://doi.org/10.1109/ICNSC.2016.7479041.
267. Scott M. Lundberg and Su-In Lee. "A Unified Approach to Interpreting Model Predictions". In: *Advances in Neural Information Processing Systems 30: Annual Conference on Neural Information Processing Systems 2017, December 4–9, 2017, Long Beach, CA, USA*. Ed. by Isabelle Guyon, Ulrike von Luxburg, Samy Bengio, Hanna M. Wallach, Rob Fergus, S. V. N. Vishwanathan, and Roman Garnett. 2017, pp. 4765–4774. URL: https://www.proceedings.neurips.cc/paper/2017/hash/8a20a8621978632d76c43dfd28b67767-Abstract.html.
268. Son T. Luu, Kiet Van Nguyen, and Ngan Luu-Thuy Nguyen. "Empirical Study of Text Augmentation on Social Media Text in Vietnamese". In: *Proceedings of the 34th Pacific Asia Conference on Language, Information and Computation, PACLIC 2020, Hanoi, Vietnam, October 24–26, 2020*. Ed. by Minh Le Nguyen, Mai Chi Luong, and Sanghoun Song. Association for Computational Linguistics, 2020, pp. 462–470. URL: https://www.aclanthology.org/2020.paclic-1.53/.
269. Jing Ma, Wei Gao, Prasenjit Mitra, Sejeong Kwon, Bernard J. Jansen, Kam-Fai Wong, and Meeyoung Cha. "Detecting Rumors from Microblogs with Recurrent Neural Networks". In: *Proceedings of the Twenty-Fifth Interna tional Joint Conference on Artificial Intelligence, IJCAI 2016*, New York, NY, USA, 9–15 July 2016. Ed. by Subbarao Kambhampati. IJCAI/AAAI Press, 2016, pp. 3818–3824. URL: http://www.ijcai.org/Abstract/16/537.
270. Yukun Ma, Haiyun Peng, Tahir Khan, Erik Cambria, and Amir Hussain. "Sentic LSTM: a Hybrid Network for Targeted Aspect-Based Sentiment Analysis". In: *Cogn. Comput.* 10.4 (2018), pp. 639–650. DOI: https://doi.org/10.1007/S12559-018-9549-X. URL: https://doi.org/10.1007/s12559-018-9549-x.
271. Rabeeh Karimi Mahabadi, Luke Zettlemoyer, James Henderson, Marzieh Saeidi, Lambert Mathias, Veselin Stoyanov, and Majid Yazdani. "PER FECT: Prompt-free and Effi-

cient Few-shot Learning with Language Models". In: CoRR abs/2204.01172 (2022). DOI: https://doi.org/10.48550/ARXIV.2204.01172. arXiv:2204.01172. URL: https://doi.org/10.48550/arXiv.2204.01172.

272. Rabeeh Karimi Mahabadi, Luke Zettlemoyer, James Henderson, Marzieh Saeidi, Lambert Mathias, Veselin Stoyanov, and Majid Yazdani. "PER FECT: Prompt-free and Efficient Few-shot Learning with Language Models". In: *CoRR* abs/2204.01172 (2022). DOI: https://doi.org/10.48550/ARXIV.2204.01172. arXiv:2204.01172. URL: https://doi.org/10.48550/arXiv.2204.01172.
273. Samaneh Mahdavifar and Ali A. Ghorbani. "Application of deep learn ing to cybersecurity: A survey". In: *Neurocomputing* 347 (2019), pp. 149–176. DOI: https://doi.org/10.1016/J.NEUCOM.2019.02.056. URL: https://doi.org/10.1016/j.neucom.2019.02.056.
274. Nikolaos Malandrakis, Minmin Shen, Anuj Kumar Goyal, Shuyang Gao, Abhishek Sethi, and Angeliki Metallinou. "Controlled Text Generation for Data Augmentation in Intelligent Artificial Agents". In: *of the 3rd Workshop on Neural Generation and Translation at EMNLP-IJCNLP 2019, Hong Kong, November 4, 2019*. Ed. by Alexandra Birch, Andrew M. Finch, Hiroaki Hayashi, Ioannis Konstas, Thang Luong, Graham Neubig, Yusuke Oda, and Katsuhito Sudoh. Association for Computational Linguistics, 2019, pp. 90–98. DOI: https://doi.org/10.18653/V1/D19-5609. URL: https://doi.org/10.18653/v1/D19-5609.
275. Vukosi Marivate and Tshephisho Sefara. "Improving Short Text Classification Through Global Augmentation Methods". In: *Machine Learning and Knowledge Extraction—4th IFIP TC 5, TC 12, WG 8.4, WG 8.9, WG 12.9 International Cross-Domain Conference, CD-MAKE 2020, Dublin, Ireland, August 25–28, 2020, Proceedings*. Ed. by Andreas Holzinger, Peter Kieseberg, A Min Tjoa, and Edgar R. Weippl. Vol. 12279. Lecture Notes in Computer Science. Springer, 2020, pp. 385–399. DOI: https://doi.org/10.1007/978-3-030-57321-8_21. URL: https://doi.org/10.1007/978-3-030-57321-8_21.
276. Louis Martin, Benjamin Muller, Pedro Javier Ortiz Suárez, Yoann Dupont, Laurent Romary, Éric de la Clergerie, Djamé Seddah, and Benoît Sagot. "CamemBERT: a Tasty French Language Model". In: *Proceedings of the 58th Annual Meeting of the Association for Computational Linguistics*. Online: Association for Computational Linguistics, July 2020, pp. 7203–7219. URL: https://www.aclweb.org/anthology/2020.aclmain.645.
277. Michael McCloskey and Neal J Cohen. "Catastrophic interference in connectionist networks: The sequential learning problem". In: *Psychology of learning and motivation*. Vol. 24. Elsevier, 1989, pp. 109–165.
278. Tom McCoy, Ellie Pavlick, and Tal Linzen. "Right for the Wrong Reasons: Diagnosing Syntactic Heuristics in Natural Language Inference". In: *Proceedings of the 57th Conference of the Association for Computational Linguistics, ACL 2019, Florence, Italy, July 28- August 2, 2019, Volume 1: Long Papers*. Ed. by Anna Korhonen, David R. Traum, and Lluís Màrquez. Association for Computational Linguistics, 2019, pp. 3428–3448. DOI: https://doi.org/10.18653/V1/P19-1334. URL: https://doi.org/10.18653/v1/p19-1334.
279. Rob McMillan. *Definition: Threat Intelligence*. 2013. DOI: https://www.gartner.com/en/documents/2487216.
280. Walaa Medhat, Ahmed Hassan, and Hoda Korashy. "Sentiment analysis algorithms and applications: A survey". In: *Ain Shams Engineering Journal* 5.4 (2014), pp. 1093–1113. ISSN: 2090–4479. URL: https://www.doi.org/10.1016/j.asej.2014.04.011.
281. Otgonpurev Mendsaikhan, Hirokazu Hasegawa, Yukiko Yamaguchi, Hajime Shimada, and Enkhbold Bataa. "Identification of Cybersecurity Specific Content Using Different

Language Models". In: *J. Inf. Process.* 28 (2020), pp. 623–632. DOI: https://doi.org/10.2197/IPSJJIP.28.623. URL: https://doi.org/10.2197/ipsjjip.28.623.

282. Stephen Merity, Nitish Shirish Keskar, and Richard Socher. "Regularizing and Optimizing LSTM Language Models". In: *6th International Conference on Learning Representations, ICLR 2018, Vancouver, BC, Canada, April 30—May 3, 2018, Conference Track Proceedings*. OpenReview.net, 2018. URL: https://www.openreview.net/forum?id=SyyGPP0TZ.
283. Tomas Mikolov, Kai Chen, Greg Corrado, and Jeffrey Dean. "Efficient Estimation of Word Representations in Vector Space". In: *1st International Conference on Learning Representations, ICLR 2013, Scottsdale, Arizona, USA, May 2–4, 2013, Workshop Track Proceedings*. Ed. by Yoshua Bengio and Yann LeCun. 2013. URL: http://arxiv.org/abs/1301.3781.
284. George A Miller. *The magical number seven, plus or minus two: some limits on our capacity for processing information*. Vol. 63. 2. US: American Psychological Association, 1956. ISBN: 1939–1471(Electronic),0033–295X(Print). DOI: https://doi.org/10.1037/h0043158.
285. George A. Miller, Richard Beckwith, Christiane Fellbaum, Derek Gross, and Katherine J. Miller. "Introduction to wordnet: An on-line lexical database". In: *International Journal of Lexicography* (1990). ISSN: 09503846. DOI: https://doi.org/10.1093/ijl/3.4.235.
286. Erxue Min, Xifeng Guo, Qiang Liu, Gen Zhang, Jianjing Cui, and Jun Long. "A Survey of Clustering With Deep Learning: From the Perspec tive of Network Architecture". In: *IEEE Access* 6 (2018), pp. 39501–39514. DOI: https://doi.org/10.1109/ACCESS.2018.2855437. URL: https://doi.org/10.1109/ACCESS.2018.2855437.
287. Junghyun Min, R. Thomas McCoy, Dipanjan Das, Emily Pitler, and Tal Linzen. "Syntactic Data Augmentation Increases Robustness to Inference Heuristics". In: *Proceedings of the 58th Annual Meeting of the Association for Computational Linguistics, ACL 2020, Online, July 5–10, 2020*. Ed. by Dan Jurafsky, Joyce Chai, Natalie Schluter, and Joel R. Tetreault. Association for Computational Linguistics, 2020, pp. 2339–2352. DOI: https://doi.org/10.18653/V1/2020.ACL-MAIN.212. URL: https://doi.org/10.18653/v1/2020.acl-main.212.
288. Shervin Minaee, Nal Kalchbrenner, Erik Cambria, Narjes Nikzad, Meysam Chenaghlu, and Jianfeng Gao. "Deep Learning-based Text Classification: A Comprehensive Review". In: *ACM Comput. Surv.* 54.3 (2022), 62:1–62:40. DOI: https://doi.org/10.1145/3439726. URL: https://doi.org/10.1145/3439726.
289. Milad Mirbabaie and Elisa Zapatka. "Sensemaking in Social Media Crisis Communication—a Case Study on the Brussels Bombings in 2016". In: *25th European Conference on Information Systems, ECIS 2017, Guimarães, Portugal, June 5–10, 2017*. Ed. by Isabel Ramos, Virpi Tuunainen, and Helmut Krcmar. 2017, p. 138. URL: https://www.aisel.aisnet.org/ecis2017_rp/138.
290. Tanushree Mitra and Eric Gilbert. "CREDBANK: A Large-Scale Social Media Corpus With Associated Credibility Annotations". In: *Proceedings of the Ninth International Conference on Web and Social Media, ICWSM 2015, University of Oxford, Oxford, UK, May 26–29, 2015*. Ed. by Meeyoung Cha, Cecilia Mascolo, and Christian Sandvig. AAAI Press, 2015, pp. 258–267. URL: http://www.aaai.org/ocs/index.php/ICWSM/ICWSM15/paper/view/10582.

291. Sudip Mittal, Prajit Kumar Das, Varish Mulwad, Anupam Joshi, and Tim Finin. "Cyber-Twitter: Using Twitter to generate alerts for cyber-security threats and vulnerabilities". In: *2016 IEEE/ACM International Conference on Advances in Social Networks Analysis and Mining, ASONAM 2016, San Francisco, CA, USA, August 18–21, 2016*. Ed. by Ravi Kumar, James Caverlee, and Hanghang Tong. IEEE Computer Society, 2016, pp. 860–867. DOI: https://doi.org/10.1109/ASONAM.2016.7752338. URL: https://doi.org/10.1109/ASONAM.2016.7752338.
292. Takeru Miyato, Andrew M. Dai, and Ian J. Goodfellow. "Adversarial Training Methods for Semi-Supervised Text Classification". In: *5th International Conference on Learning Representations, ICLR 2017, Toulon, France, April 24–26, 2017, Conference Track Proceedings*. OpenReview.net, 2017. URL: https://www.openreview.net/forum?id=r1X3g2_xl.
293. Takeru Miyato, Shin-ichi Maeda, Masanori Koyama, Ken Nakae, and Shin Ishii. *Distributional Smoothing with Virtual Adversarial Training*. en. arXiv:1507.00677 [cs, stat]. June 2016. URL: http://arxiv.org/abs/1507.00677 (visited on 03/12/2024).
294. Christoph Molnar. *Interpretable Machine Learning. A Guide for Making Black Box Models Explainable*. 2019.
295. John X. Morris, Eli Lifland, Jin Yong Yoo, Jake Grigsby, Di Jin, and Yanjun Qi. "TextAttack: A Framework for Adversarial Attacks, Data Augmentation, and Adversarial Training in NLP". In: *Proceedings of the 2020 Conference on Empirical Methods in Natural Language Processing: System Demonstrations, EMNLP 2020—Demos, Online, November 16–20*, 2020. Ed. by Qun Liu and David Schlangen. Association for Computational Linguistics, 2020, pp. 119–126. DOI: https://doi.org/10.18653/V1/2020.EMNLP-DEMOS.16. URL: https://doi.org/10.18653/v1/2020.emnlp-demos.16.
296. Marius Mosbach, Maksym Andriushchenko, and Dietrich Klakow. "On the Stability of Fine-tuning BERT: Misconceptions, Explanations, and Strong Baselines". In: *9th International Conference on Learning Representations, ICLR 2021, Virtual Event, Austria, May 3–7, 2021*. OpenReview.net, 2021. URL: https://www.openreview.net/forum?id=nzpLWnVAyah.
297. Anna V. Mosolova, Vadim V. Fomin, and Ivan Yu Bondarenko. "Text augmentation for neural networks". In: *CEUR Workshop Proceedings* 2268 (2018), pp. 104–109. ISSN: 16130073.
298. Nikola Mrksic, Diarmuid Ó Séaghdha, Blaise Thomson, Milica Gasic, Lina Maria Rojas-Barahona, Pei-Hao Su, David Vandyke, Tsung-Hsien Wen, and Steve J. Young. "Counter-fitting Word Vectors to Linguistic Constraints". In: *NAACL HLT 2016, The 2016 Conference of the North American Chapter of the Association for Computational Linguistics: Human Language Technologies, San Diego California, USA, June 12–17, 2016*. Ed. by Kevin Knight, Ani Nenkova, and Owen Rambow. The Association for Computational Linguistics, 2016, pp. 142–148. DOI: https://doi.org/10.18653/V1/N16-1018. URL: https://doi.org/10.18653/v1/n16-1018.
299. Subhabrata Mukherjee, Arindam Mitra, Ganesh Jawahar, Sahaj Agarwal, Hamid Palangi, and Ahmed Awadallah. "Orca: Progressive Learning from Complex Explanation Traces of GPT-4". In: *CoRR* abs/2306.02707 (2023). DOI: https://doi.org/10.48550/ARXIV.2306.02707. arXiv:2306.02707. URL: https://doi.org/10.48550/arXiv.2306.02707.
300. Rafael Müller, Simon Kornblith, and Geoffrey E. Hinton. "When does label smoothing help?" In: *Advances in Neural Information Processing Systems 32: Annual Con-*

ference on Neural Information Processing Systems 2019, NeurIPS 2019, December 8–14, 2019, Vancouver, BC, Canada. Ed. by Hanna M. Wallach, Hugo Larochelle, Alina Beygelzimer, Florence d'Alché-Buc, Emily B. Fox, and Roman Garnett. 2019, pp. 4696–4705. URL: https://www.proceedings.neurips.cc/paper/2019/hash/f1748d6b0fd9d439f71450117eba2725-Abstract.html.

301. Sara Mumtaz, Carlos Rodríguez, Boualem Benatallah, Mortada Al-Banna, and Shayan Zamanirad. "Learning Word Representation for the Cyber Security Vulnerability Domain". In: *2020 International Joint Conference on Neural Networks, IJCNN 2020, Glasgow, United Kingdom, July 19–24, 2020*. IEEE, 2020, pp. 1–8. DOI: https://doi.org/10.1109/IJCNN48605.2020.9207140. URL: https://doi.org/10.1109/IJCNN48605.2020.9207140.
302. Nikita Nangia and Samuel R. Bowman. "Human vs. Muppet: A Con servative Estimate of Human Performance on the GLUE Benchmark". In: *Proceedings of the 57th Conference of the Association for Computational Linguistics, ACL 2019, Florence, Italy, July 28- August 2, 2019, Volume 1: Long Papers*. Ed. by Anna Korhonen, David R. Traum, and Lluís Màrquez. Association for Computational Linguistics, 2019, pp. 4566–4575. DOI: https://doi.org/10.18653/V1/P19-1449. URL: https://doi.org/10.18653/v1/p19-1449.
303. Neha Nayak, Gabor Angeli, and Christopher D. Manning. "Evaluating Word Embeddings Using a Representative Suite of Practical Tasks". In: *Proceedings of the 1st Workshop on Evaluating Vector-Space Representations for NLP, RepEval at ACL 2016, Berlin, Germany, August 2016*. Association for Computational Linguistics, 2016, pp. 19–23. DOI: https://doi.org/10.18653/V1/W16-2504. URL: https://doi.org/10.18653/v1/W16-2504.
304. Dat Tien Nguyen, Kamla Al-Mannai, Shafiq R. Joty, Hassan Sajjad, Muhammad Imran, and Prasenjit Mitra. "Rapid Classification of Crisis-Related Data on Social Networks using Convolutional Neural Networks". In: *CoRR* abs/1608.03902 (2016). arXiv:1608.03902. URL: http://arxiv.org/abs/1608.03902.
305. Dat Tien Nguyen, Ferda Ofli, Muhammad Imran, and Prasenjit Mitra. "Damage Assessment from Social Media Imagery Data During Disasters". In: *Proceedings of the 2017 IEEE/ACM International Conference on Advances in Social Networks Analysis and Mining 2017, Sydney, Australia, July 31—August 03, 2017*. Ed. by Jana Diesner, Elena Ferrari, and Guandong Xu. ACM, 2017, pp. 569–576. DOI: https://doi.org/10.1145/3110025.3110109. URL: https://doi.org/10.1145/3110025.3110109.
306. Minh Van Nguyen, Nghia Trung Ngo, Bonan Min, and Thien Huu Nguyen. "FAMIE: A Fast Active Learning Framework for Multilin-gual Information Extraction". In: *CoRR* abs/2202.08316 (2022). arXiv:2202.08316. URL: https://www.arxiv.org/abs/2202.08316.
307. Amirreza Niakanlahiji, Lida Safarnejad, Reginald Harper, and Bei-Tseng Chu. "IoCMiner: Automatic Extraction of Indicators of Compromise from Twitter". In: *2019 IEEE International Conference on Big Data (IEEE BigData), Los Angeles, CA, USA, December 9–12, 2019*. Ed. by Chaitanya K. Baru, Jun Huan, Latifur Khan, Xiaohua Hu, Ronay Ak, Yuanyuan Tian, Roger S. Barga, Carlo Zaniolo, Kisung Lee, and Yanfang (Fanny) Ye. IEEE, 2019, pp. 4747–4754. DOI: https://doi.org/10.1109/BIGDATA47090.2019.9006562. URL: https://doi.org/10.1109/BigData47090.2019.9006562.
308. Harsha Nori, Nicholas King, Scott Mayer McKinney, Dean Carignan, and Eric Horvitz. "Capabilities of GPT-4 on Medical Challenge Problems". In: *CoRR* abs/2303.13375

(2023). DOI: https://doi.org/10.48550/ARXIV.2303.13375. arXiv:2303.13375. URL: https://doi.org/10.48550/arXiv.2303.13375.

309. Stavros Ntalampiras and Ilyas Potamitis. "Few-shot learning for modeling cyber physical systems in non-stationary environments". In: *Neural Comput. Appl.* 35.5 (2023), pp. 3853–3863. DOI: https://doi.org/10.1007/S00521-022-07903-0. URL: https://doi.org/10.1007/s00521-022-07903-0.
310. Olusoji B. Okunoye and Ayei E. Ibor. "Hybrid fake news detection technique with genetic search and deep learning". In: *Comput. Electr.* Eng. 103 (2022), p. 108344. DOI: https://doi.org/10.1016/J.COMPELECENG.2022.108344. URL: https://doi.org/10.1016/j.compeleceng.2022.108344.
311. Anaïs Ollagnier and Hywel T. P. Williams. "Text Augmentation Techniques for Clinical Case Classification". In: *Working Notes of CLEF 2020—Conference and Labs of the Evaluation Forum, Thessaloniki, Greece, September 22–25, 2020.* Ed. by Linda Cappellato, Carsten Eickhoff, Nicola Ferro, and Aurélie Névéol. Vol. 2696. CEUR Workshop Proceedings. CEUR-WS.org, 2020. URL: https://www.ceur-ws.org/Vol-2696/paper_166.pdf.
312. Ekaterina Olshannikova, Thomas Olsson, Jukka Huhtamäki, and Hannu Kärkkäinen. "Conceptualizing Big Social Data". In: *J. Big Data* 4 (2017), p. 3. DOI: https://doi.org/10.1186/S40537-017-0063-X. URL: https://doi.org/10.1186/s40537-017-0063-x.
313. Alexandra Olteanu, Sarah Vieweg, and Carlos Castillo. "What to Expect When the Unexpected Happens: Social Media Communications Across Crises". *In: Proceedings of the 18th ACM Conference on Computer Supported Cooperative Work & Social Computing, CSCW 2015, Vancouver, BC, Canada, March 14–18, 2015.* Ed. by Dan Cosley, Andrea Forte, Luigina Ciolfi, and David McDonald. ACM, 2015, pp. 994–1009. DOI: https://doi.org/10.1145/2675133.2675242. URL: https://doi.org/10.1145/2675133.2675242.
314. Teresa Onorati, Paloma Díaz, and Belen Carrion Recio. "From social networks to emergency operation centers: A semantic visualization approach". *In: Future Gener. Comput. Syst.* 95 (2019), pp. 829–840. DOI: https://doi.org/10.1016/J.FUTURE.2018.01.052. URL: https://doi.org/10.1016/j.future.2018.01.052.
315. OpenAI. "GPT-4 Technical Report". In: *CoRR* abs/2303.08774 (2023). DOI: https://doi.org/10.48550/ARXIV.2303.08774. arXiv:2303.08774. URL: https://doi.org/10.48550/arXiv.2303.08774.
316. OpenAI. *Learning to reason with LLMS*. 2024. URL: https://www.openai.com/index/learning-to-reason-with-llms.
317. FY Osisanwo, JET Akinsola, O Awodele, JO Hinmikaiye, O Olakanmi, J Akinjobi, et al. "Supervised machine learning algorithms: classification and comparison". In: *International Journal of Computer Trends and Technology (IJCTT)* 48.3 (2017), pp. 128–138.
318. Daniel W. Otter, Julian R. Medina, and Jugal K. Kalita. "A Survey of the Usages of Deep Learning for Natural Language Processing". In: *IEEE Trans. Neural Networks Learn. Syst.* 32.2 (2021), pp. 604–624. DOI: https://doi.org/10.1109/TNNLS.2020.2979670. URL: https://doi.org/10.1109/TNNLS.2020.2979670.
319. Long Ouyang, Jeffrey Wu, Xu Jiang, Diogo Almeida, Carroll L. Wain-wright, Pamela Mishkin, Chong Zhang, Sandhini Agarwal, Katarina Slama, Alex Ray, John Schulman, Jacob Hilton, Fraser Kelton, Luke Miller, Maddie Simens, Amanda Askell, Peter Welinder, Paul F. Christiano, Jan Leike, and Ryan Lowe. "Training language models to follow instructions with human feedback". In: *Advances in Neu-*

ral Information Processing Systems 35: Annual Conference on Neural Information Processing Systems 2022, NeurIPS 2022, New Orleans, LA, USA, November 28—December 9, 2022. Ed. by Sanmi Koyejo, S. Mohamed, A. Agarwal, Danielle Belgrave, K. Cho, and A. Oh. 2022. URL: http://papers.nips.cc/paper_files/paper/2022/hash/b1efde53be364a73914f58805a001731-AbstractConference.html.

320. Matteo Pagliardini, Prakhar Gupta, and Martin Jaggi. "Unsupervised Learning of Sentence Embeddings Using Compositional n-Gram Fea tures". In: *Proceedings of the 2018 Conference of the North American Chapter of the Association for Computational Linguistics: Human Language Technologies, NAACL-HLT 2018, New Orleans, Louisiana, USA, June 1–6, 2018, Volume 1 (Long Papers)*. Ed. by Marilyn A. Walker, Heng Ji, and Amanda Stent. Association for Computational Linguistics, 2018, pp. 528–540. DOI: https://doi.org/10.18653/V1/N18-1049. URL: https://doi.org/10.18653/v1/n18-1049.
321. Leysia Palen and Kenneth M Anderson. "Crisis informatics: New data for extraordinary times". In: *Science (New York, N.Y.)* 353.6296 (2016), pp. 224–225. ISSN: 0036-8075. DOI: https://doi.org/10.1126/science.aag2579.
322. Leysia Palen and Amanda L Hughes. "Social Media in Disaster Communication". In: *Handbook of Disaster Research*. Ed. by Havid n Rodr guez, William Donner, and Joseph E Trainor. Cham: Springer International Pub lishing, 2018, pp. 497–518. ISBN: 978-3-319-63254-4. DOI: https://doi.org/10.1007/978-3-319-63254-4_24.
323. Sinno Jialin Pan. "Transfer Learning". In: *Data Classification: Algorithms and Applications*. Ed. by Charu C. Aggarwal. CRC Press, 2014, pp. 537–570. DOIhttps://doi.org/10.1201/B17320-22. URL: http://www.crcnetbase.com/doi/abs/10.1201/b17320-22.
324. Zoe Papakipos and Joanna Bitton. "AugLy: Data Augmentations for Robustness". In: *CoRR* abs/2201.06494 (2022). arXiv:2201.06494. URL: https://www.arxiv.org/abs/2201.06494.
325. Nicolas Papernot, Fartash Faghri, Nicholas Carlini, Ian Goodfellow, Reuben Feinman, Alexey Kurakin, Cihang Xie, Yash Sharma, Tom Brown, Aurko Roy, Alexander Matyasko, Vahid Behzadan, Karen Hambardzumyan, Zhishuai Zhang, Yi-Lin Juang, Zhi Li, Ryan Sheatsley, Abhibhav Garg, Jonathan Uesato, Willi Gierke, Yinpeng Dong, David Berthelot, Paul Hendricks, Jonas Rauber, Rujun Long, and Patrick McDaniel. *Technical Report on the CleverHans v2.1.0 Adversarial Examples Library*. arXiv:1610.00768 [cs, stat]. June 2018. URL: http://arxiv.org/abs/1610.00768 (visited on 03/08/2024).
326. Nicolas Papernot, Patrick D. McDaniel, Ian J. Goodfellow, Somesh Jha, Z. Berkay Celik, and Ananthram Swami. "Practical Black-Box Attacks against Machine Learning". In: *Proceedings of the 2017 ACM on Asia Conference on Computer and Communications Security, AsiaCCS 2017, Abu Dhabi, United Arab Emirates, April 2–6, 2017*. Ed. by Ramesh Karri, Ozgur Sinanoglu, Ahmad-Reza Sadeghi, and Xun Yi. ACM, 2017, pp. 506–519. DOI: https://doi.org/10.1145/3052973.3053009. URL: https://doi.org/10.1145/3052973.3053009.
327. Ankur P. Parikh, Oscar Täckström, Dipanjan Das, and Jakob Uszkoreit. "A Decomposable Attention Model for Natural Language Inference". In: *Proceedings of the 2016 Conference on Empirical Methods in Natural Language Processing, EMNLP 2016, Austin, Texas, USA, November 1–4, 2016*. Ed. by Jian Su, Xavier Carreras, and Kevin Duh. The Association for Computational Linguistics, 2016, pp. 2249–2255. DOI: https://doi.org/10.18653/V1/D16-1244. URL: https://doi.org/10.18653/v1/d16-1244.

328. David A. Patterson, Joseph Gonzalez, Quoc V. Le, Chen Liang, Lluis-Miquel Munguia, Daniel Rothchild, David R. So, Maud Texier, and Jeff Dean. "Carbon Emissions and Large Neural Network Training". In: *CoRR* abs/2104.10350 (2021). arXiv:2104.10350. URL: https://www.arxiv.org/abs/2104.10350.
329. Kellin Pelrine, Anne Imouza, Camille Thibault, Meilina Reksoprodjo, Caleb Gupta, Joel Christoph, Jean-François Godbout, and Reihaneh Rabbany. "Towards Reliable Misinformation Mitigation: Generalization, Uncertainty, and GPT-4". In: *Proceedings of the 2023 Conference on Empirical Methods in Natural Language Processing, EMNLP 2023, Singapore, December 6–10, 2023*. Ed. by Houda Bouamor, Juan Pino, and Kalika Bali. Association for Computational Linguistics, 2023, pp. 6399–6429. DOI: https://doi.org/10.18653/V1/2023.EMNLP-MAIN.395. URL: https://doi.org/10.18653/v1/2023.emnlp-main.395.
330. Jeffrey Pennington, Richard Socher, and Christopher D. Manning. "Glove: Global Vectors for Word Representation". In: *Proceedings of the 2014 Conference on Empirical Methods in Natural Language Processing, EMNLP 2014, October 25–29, 2014, Doha, Qatar, A meeting of SIGDAT, a Special Interest Group of the ACL*. Ed. by Alessandro Moschitti, Bo Pang, and Walter Daelemans. ACL, 2014, pp. 1532–1543. DOI: https://doi.org/10.3115/V1/D14-1162. URL: https://doi.org/10.3115/v1/d14-1162.
331. Thananya Phreeraphattanakarn and Boonserm Kijsirikul. "Text data-augmentation using Text Similarity with Manhattan Siamese long short-term memory for Thai language". In: *Journal of Physics: Conference Series*. Vol. 1780. 1. IOP Publishing. 2021, p. 012018.
332. Francesco Pierri and Stefano Ceri. "False News On Social Media: A Data-Driven Survey". In: SIGMOD Rec. 48.2 (2019), pp. 18–27. DOI: https://doi.org/10.1145/3377330.3377334. URL: https://doi.org/10.1145/3377330.3377334.
333. Daniela Pohl, Abdelhamid Bouchachia, and Hermann Hellwagner. "Social media for crisis management: clustering approaches for sub-event detection". In: *Multim. Tools Appl.* 74.11 (2015), pp. 3901–3932. DOI: https://doi.org/10.1007/S11042-013-1804-2. URL: https://doi.org/10.1007/s11042-013-1804-2.
334. Mohammadreza Qaraei and Rohit Babbar. "Adversarial examples for extreme multilabel text classification". In: *Mach. Learn.* 111.12 (2022), pp. 4539–4563. DOI: https://doi.org/10.1007/S10994-022-06263-Z. URL: https://doi.org/10.1007/s10994-022-06263-z.
335. Vahed Qazvinian, Emily Rosengren, Dragomir R. Radev, and Qiaozhu Mei. "Rumor has it: Identifying Misinformation in Microblogs". In: *Proceedings of the 2011 Conference on Empirical Methods in Natural Language Processing, EMNLP 2011, 27–31 July 2011, John McIntyre Conference Centre, Edinburgh, UK, A meeting of SIGDAT, a Special Interest Group of the ACL*. ACL, 2011, pp. 1589–1599. URL: https://www.aclanthology.org/D11-1147/.
336. Siyuan Qiu, Binxia Xu, Jie Zhang, Yafang Wang, Xiaoyu Shen, Gerard de Melo, Chong Long, and Xiaolong Li. "EasyAug: An Automatic Textual Data Augmentation Platform for Classification Tasks". In: *Companion of The 2020 Web Conference 2020, Taipei, Taiwan, April 20–24, 2020*. Ed. by Amal El Fallah Seghrouchni, Gita Sukthankar, Tie-Yan Liu, and Maarten van Steen. ACM / IW3C2, 2020, pp. 249–252. DOI: https://doi.org/10.1145/3366424.3383552. URL: https://doi.org/10.1145/3366424.3383552.
337. Yanru Qu, Dinghan Shen, Yelong Shen, Sandra Sajeev, Weizhu Chen, and Jiawei Han. "CoDA: Contrast-enhanced and Diversity-promoting Data Augmentation for Natural

Language Understanding". In: *9th Interna tional Conference on Learning Representations, ICLR 2021, Virtual Event, Austria, May 3–7, 2021*. OpenReview.net, 2021. URL: https://www.openreview.net/forum?id=Ozk9MrX1hvA.

338. Ella Rabinovich, Raj Nath Patel, Shachar Mirkin, Lucia Specia, and Shuly Wintner. "Personalized Machine Translation: Preserving Original Author Traits". In: *Proceedings of the 15th Conference of the European Chapter of the Association for Computational Linguistics, EACL 2017, Valencia, Spain, April 3–7, 2017, Volume 1: Long Papers*. Ed. by Mirella Lapata, Phil Blunsom, and Alexander Koller. Association for Computational Linguistics, 2017, pp. 1074–1084. DOI: https://doi.org/10.18653/V1/E17-1101. URL: https://doi.org/10.18653/v1/e17-1101.
339. Alec Radford, Karthik Narasimhan, Tim Salimans, and Ilya Sutskever. *Improving Language Understanding by Generative Pre-Training*. en. 2018.
340. Alec Radford, Jeffrey Wu, Rewon Child, David Luan, Dario Amodei, and Ilya Sutskever. "Language Models are Unsupervised Multitask Learners". In: OpenAI Blog (2018). URL: https://www.paperswithcode.com/paper/language-models-are-unsupervised-multitask.
341. Drasko Radovanovic and Slobodan Dukanovic. "Image-Based Plant Disease Detection: A Comparison of Deep Learning and Classical Machine Learning Algorithms". In: *2020 24th International Conference on Information Technology, IT 2020*. 2020. ISBN: 978-1-72815-136-6. DOI: https://doi.org/10.1109/IT48810.2020.9070664.
342. Bhagat Singh Raghuwanshi and Sanyam Shukla. "Classifying imbalanced data using SMOTE based class-specific kernelized ELM". In: *Int. J. Mach. Learn. Cybern.* 12.5 (2021), pp. 1255–1280. DOI: https://doi.org/10.1007/S13042-020-01232-1. URL: https://doi.org/10.1007/s13042-020-01232-1.
343. Anton Ragni, Kate M. Knill, Shakti P. Rath, and Mark J. F. Gales. "Data augmentation for low resource languages". In: *INTERSPEECH 2014, 15th Annual Conference of the International Speech Communication Association, Singapore, September 14–18, 2014*. Ed. by Haizhou Li, Helen M. Meng, Bin Ma, Engsiong Chng, and Lei Xie. ISCA, 2014, pp. 810–814. DOI: https://doi.org/10.21437/INTERSPEECH.2014-207. URL: https://doi.org/10.21437/Interspeech.2014-207.
344. Abir Rahali and Moulay A. Akhloufi. "MalBERT: Using Transformers for Cybersecurity and Malicious Software Detection". In: *CoRR* abs/2103.03806 (2021). arXiv:2103.03806. URL: https://www.arxiv.org/abs/2103.03806.
345. Guillaume Raille, Sandra Djambazovska, and Claudiu Musat. "Fast Cross-domain Data Augmentation through Neural Sentence Editing". In: *CoRR* abs/2003.10254 (2020). arXiv:2003.10254. URL: https://www.arxiv.org/abs/2003.10254.
346. Rajat Raina, Alexis J. Battle, Honglak Lee, Benjamin Packer, and Andrew Y. Ng. "Self-taught learning: transfer learning from unlabeled data". In: *Machine Learning, Proceedings of the Twenty-Fourth International Conference (ICML 2007), Corvallis, Oregon, USA, June 20–24, 2007*. Ed. by Zoubin Ghahramani. Vol. 227. ACM International Conference Proceeding Series. ACM, 2007, pp. 759–766. DOI: https://doi.org/10.1145/1273496.1273592. URL: https://doi.org/10.1145/1273496.1273592.
347. Priyanka Ranade, Aritran Piplai, Anupam Joshi, and Tim Finin. "CyBERT: Contextualized Embeddings for the Cybersecurity Domain". In: *2021 IEEE International Conference on Big Data (Big Data), Orlando, FL, USA, December 15–18, 2021*. Ed. by Yixin Chen, Heiko Ludwig, Yicheng Tu, Usama M. Fayyad, Xingquan Zhu, Xiaohua Hu, Suren Byna, Xiong Liu, Jianping Zhang, Shirui Pan, Vagelis Papalexakis,

Jianwu Wang, Alfredo Cuzzocrea, and Carlos Ordonez. IEEE, 2021, pp. 3334–3342. DOI: https://doi.org/10.1109/BIGDATA52589 . 2021 . 9671824. URL: https://doi.org/10.1109/BigData52589.2021.9671824.

348. Francisco Rangel, Anastasia Giachanou, Bilal Ghanem, and Paolo Rosso. "Overview of the 8th Author Pro?ling Task at PAN 2020: Pro?ling Fake News Spreaders on Twitter". In: (2020).

349. Runqing Rao, Linda Plotnick, and Starr Roxanne Hiltz. "Supporting the Use of Social Media by Emergency Managers: Software Tools to Overcome Information Overload". In: *50th Hawaii International Conference on System Sciences, HICSS 2017, Hilton Waikoloa Village, Hawaii, USA, January 4–7, 2017*. Ed. by Tung Bui. ScholarSpace / AIS Electronic Library (AISeL), 2017, pp. 1–9. URL: https://www.hdl.handle.net/10125/41185.

350. Carl Edward Rasmussen and Zoubin Ghahramani. "Occam's Razor". In: *Advances in Neural Information Processing Systems 13, Papers from Neural Information Processing Systems (NIPS) 2000, Denver, CO, USA*. Ed. by Todd K. Leen, Thomas G. Dietterich, and Volker Tresp. MIT Press, 2000, pp. 294–300. URL: https://www.proceedings.neurips.cc/paper/2000/hash/0950ca92a4dcf426067cfd2246bb5ff3-Abstract.html.

351. Jacob Ratkiewicz, Michael D. Conover, Mark R. Meiss, Bruno Gonçalves, Snehal Patil, Alessandro Flammini, and Filippo Menczer. "Truthy: mapping the spread of astroturf in microblog streams". In: *Proceedings of the 20th International Conference on World Wide Web, WWW 2011, Hyderabad, India, March 28—April 1, 2011 (Companion Volume)*. Ed. by Sadagopan Srinivasan, Krithi Ramamritham, Arun Kumar, M. P. Ravindra, Elisa Bertino, and Ravi Kumar. ACM, 2011, pp. 249–252. DOI: https://doi.org/10.1145/1963192.1963301. URL: https://doi.org/10.1145/1963192.1963301.

352. Alexander J. Ratner, Henry R. Ehrenberg, Zeshan Hussain, Jared Dunnmon, and Christopher Ré. "Learning to Compose Domain-Specific Transformations for Data Augmentation". In: *Advances in Neural Information Processing Systems 30: Annual Conference on Neural Information Processing Systems 2017, December 4–9, 2017, Long Beach, CA, USA*. Ed. by Isabelle Guyon, Ulrike von Luxburg, Samy Bengio, Hanna M. Wallach, Rob Fergus, S. V. N. Vishwanathan, and Roman Garnett. 2017, pp. 3236–3246. URL: https://www.proceedings.neurips.cc/paper/2017/hash/f26dab9bf6a137c3b6782e562794c2f2-Abstract.html.

353. Jonas Rauber, Wieland Brendel, and Matthias Bethge. *Foolbox: A Python toolbox to benchmark the robustness of machine learning models*. arXiv:1707.04131 [cs, stat]. Mar. 2018. URL: http://arxiv.org/abs/1707.04131 (visited on 03/08/2024).

354. Mehdi Regina, Maxime Meyer, and Sebastien Goutal. "Text Data Augmentation: Towards better detection of spear-phishing emails". In: *CoRR* abs/2007.02033 (2020). arXiv:2007.02033. URL: https://www.arxiv.org/abs/2007.02033.

355. Nils Reimers and Iryna Gurevych. "Reporting Score Distributions Makes a Difference: Performance Study of LSTM-networks for Sequence Tagging". In: *Proceedings of the 2017 Conference on Empirical Methods in Natural Language Processing, EMNLP 2017, Copenhagen, Denmark, September 9–11, 2017*. Ed. by Martha Palmer, Rebecca Hwa, and Sebastian Riedel. Association for Computational Linguistics, 2017, pp. 338–348. DOI: https://doi.org/10.18653/V1/D17-1035. URL: https://doi.org/10.18653/v1/d17-1035.

356. Nils Reimers and Iryna Gurevych. "Sentence-BERT: Sentence Embeddings using Siamese BERT-Networks". In: *Proceedings of the 2019 Conference on Empirical Meth-*

ods in Natural Language Processing and the 9th International Joint Conference on Natural Language Processing, EMNLPIJCNLP 2019, Hong Kong, China, November 3–7, 2019. Ed. by Kentaro Inui, Jing Jiang, Vincent Ng, and Xiaojun Wan. Association for Computational Linguistics, 2019, pp. 3980–3990. DOI: https://doi.org/10.18653/V1/D19-1410. URL: https://doi.org/10.18653/v1/D19-1410.

357. Christian Reuter and Marc Andr Kaufhold. "Fifteen years of social media in emergencies: A retrospective review and future directions for crisis Informatics". In: *Journal of Contingencies and Crisis Management* 26.1 (2018), pp. 41–57. ISSN: 14685973. DOI: https://doi.org/10.1111/1468-5973.12196.
358. Christian Reuter, Marc-Andr Kaufhold, Stefka Schmid, Thomas Spielhofer, Anna Sophie Hahne, and others. "The impact of risk cultures: Citizens perception of social media use in emergencies across Europe". In: *Technological Forecasting and Social Change* 148.1 (2019). Publisher: Elsevier, pp. 1–17.
359. Christian Reuter, Thomas Ludwig, Marc-André Kaufhold, and Thomas Spielhofer. "Emergency services' attitudes towards social media: A quantitative and qualitative survey across Europe". In: *Int. J. Hum. Comput. Stud.* 95 (2016), pp. 96–111. DOI: https://doi.org/10.1016/J.IJHCS.2016.03.005. URL: https://doi.org/10.1016/j.ijhcs.2016.03.005.
360. Christian Reuter, Alexandra Marx, and Volkmar Pipek. "Crisis Management 2.0: Towards a Systematization of Social Software Use in Crisis Situations". In: *Int. J. Inf. Syst. Crisis Response Manag.* 4.1 (2012), pp. 1–16. DOI: https://doi.org/10.4018/JISCRM.2012010101. URL: https://doi.org/10.4018/jiscrm.2012010101.
361. Marco Túlio Ribeiro, Sameer Singh, and Carlos Guestrin. ""Why Should I Trust You?": Explaining the Predictions of Any Classifier". In: *Proceedings of the 22nd ACM SIGKDD International Conference on Knowledge Discovery and Data Mining, San Francisco, CA, USA, August 13–17, 2016*. Ed. by Balaji Krishnapuram, Mohak Shah, Alexander J. Smola, Charu C. Aggarwal, Dou Shen, and Rajeev Rastogi. ACM, 2016, pp. 1135–1144. DOI: https://doi.org/10.1145/2939672.2939778. URL: https://doi.org/10.1145/2939672.2939778.
362. Thea Riebe, Marc-André Kaufhold, and Christian Reuter. "The Impact of Organizational Structure and Technology Use on Collaborative Practices in Computer Emergency Response Teams: An Empirical Study". In: *Proc. ACM Hum. Comput. Interact.* 5.CSCW2 (2021), 478:1–478:30. DOI: https://doi.org/10.1145/3479865. URL: https://doi.org/10.1145/3479865.
363. Thea Riebe, Tristan Wirth, Markus Bayer, Philipp Kuehn, Marc-AndreKaufhold, Volker Knauthe, Stefan Guthe, and Christian Reuter. "CySecAlert: An Alert Generation System for Cyber Security Events Using Open Source Intelligence Data". In: *Information and Communications Security—23rd International Conference, ICICS 2021, Chongqing, China, Novem ber 19–21, 2021, Proceedings, Part I*. Ed. by Debin Gao, Qi Li, Xiaohong Guan, and Xiaofeng Liao. Vol. 12918. Lecture Notes in Computer Science. Springer, 2021, pp. 429–446. DOI: https://doi.org/10.1007/978-3-030-86890-1_24. URL: https://doi.org/10.1007/978-3-030-86890-1_24.
364. Slamet Riyanto, Imas Sukaesih Sitanggang, Taufik Djatna, and Tika Dewi Atikah. "Comparative Analysis using Various Performance Metrics in Imbalanced Data for Multi-class Text Classification". In: *International Journal of Advanced Computer Science and Applications* 14.6 (2023). DOI: https://doi.org/10.14569/IJACSA.2023.01406116. URL: https://doi.org/10.14569/IJACSA.2023.01406116.

365. Georgios Rizos, Konstantin Hemker, and Björn W. Schuller. "Augment to Prevent: Short-Text Data Augmentation in Deep Learning for Hate-Speech Classification". In: *Proceedings of the 28th ACM International Conference on Information and Knowledge Management, CIKM 2019, Beijing, China, November 3–7, 2019*. Ed. by Wenwu Zhu, Dacheng Tao, Xueqi Cheng, Peng Cui, Elke A. Rundensteiner, David Carmel, Qi He, and Jeffrey Xu Yu. ACM, 2019, pp. 991–1000. DOI: https://doi.org/10.1145/3357384.3358040. URL: https://doi.org/10.1145/3357384.3358040.
366. Anthony Robinson, R. Dennis Cook, and Sanford Weisberg. "Residuals and Influence in Regression." In: *Journal of the Royal Statistical Society. Series A (General)* 147.1 (1984). ISSN: 00359238. DOI: https://doi.org/10.2307/2981746.
367. Ariel Rodríguez and Koji Okamura. "Generating Real Time Cyber Situational Awareness Information Through Social Media Data Mining". In: *43rd IEEE Annual Computer Software and Applications Conference, COMPSAC 2019, Milwaukee, WI, USA, July 15–19, 2019, Volume 2*. Ed. by Vladimir Getov, Jean-Luc Gaudiot, Nariyoshi Yamai, Stelvio Cimato, J. Morris Chang, Yuuichi Teranishi, Ji-Jiang Yang, Hong Va Leong, Hossain Shahriar, Michiharu Takemoto, Dave Towey, Hiroki Takakura, Atilla Elçi, Susumu Takeuchi, and Satish Puri. IEEE, 2019, pp. 502–507. DOI: https://doi.org/10.1109/COMPSAC.2019.10256. URL: https://doi.org/10.1109/COMPSAC.2019.10256.
368. Luis Gustavo Araujo Rodriguez, Julia Selvatici Trazzi, Victor Fossaluza, Rodrigo Campiolo, and Daniel Macedo Batista. *Analysis of Vulnerability Disclosure Delays from the National Vulnerability Database*. 2018.
369. Subendhu Rongali, Abhyuday Jagannatha, Bhanu Pratap Singh Rawat, and Hong Yu. *Continual Domain-Tuning for Pretrained Language Models*. en. 2021.
370. Peter J. Rousseeuw. "Silhouettes: A graphical aid to the interpretation and validation of cluster analysis". In: *Journal of Computational and Applied Mathematics* (1987). ISSN: 03770427. DOI: https://doi.org/10.1016/0377-0427(87)90125-7.
371. Natali Ruchansky, Sungyong Seo, and Yan Liu. "CSI: A Hybrid Deep Model for Fake News Detection". In: *Proceedings of the 2017 ACM on Conference on Information and Knowledge Management, CIKM 2017, Singapore, November 06—10, 2017*. Ed. by Ee-Peng Lim, Marianne Winslett, Mark Sanderson, Ada Wai-Chee Fu, Jimeng Sun, J. Shane Culpepper, Eric Lo, Joyce C. Ho, Debora Donato, Rakesh Agrawal, Yu Zheng, Carlos Castillo, Aixin Sun, Vincent S. Tseng, and Chenliang Li. ACM, 2017, pp. 797–806. DOI: https://doi.org/10.1145/3132847.3132877. URL: https://doi.org/10.1145/3132847.3132877.
372. Sebastian Ruder, Matthew E. Peters, Swabha Swayamdipta, and Thomas Wolf. "Transfer Learning in Natural Language Processing". In: *Proceedings of the 2019 Conference of the North American Chapter of the Association for Computational Linguistics: Tutorials*. Ed. by Anoop Sarkar and Michael Strube. Minneapolis, Minnesota: Association for Computational Lin guistics, June 2019, pp. 15–18. DOI: https://doi.org/10.18653/v1/N19-5004. URL: https://www.aclanthology.org/N19-5004/.
373. Koustav Rudra, Pawan Goyal, Niloy Ganguly, Prasenjit Mitra, and Muhammad Imran. "Identifying Sub-events and Summarizing Disaster-Related Information from Microblogs". In: *The 41st International ACM SIGIR Conference on Research & Development in Information Retrieval, SIGIR 2018, Ann Arbor, MI, USA, July 08–12, 2018*. Ed. by Kevyn Collins-Thompson, Qiaozhu Mei, Brian D. Davison, Yiqun Liu, and Emine Yilmaz. ACM, 2018, pp. 265–274. DOI: https://doi.org/10.1145/3209978.3210030. URL: https://doi.org/10.1145/3209978.3210030.

374. Carl Sabottke, Octavian Suciu, and Tudor Dumitras. "Vulnerability Disclosure in the Age of Social Media: Exploiting Twitter for Predicting Real-World Exploits". In: *24th USENIX Security Symposium, USENIX Security 15, Washington, D.C., USA, August 12–14, 2015*. Ed. by Jaeyeon Jung and Thorsten Holz. USENIX Association, 2015, pp. 1041–1056. URL: https://www.usenix.org/conference/usenixsecurity15/technical-sessions/presentation/sabottke.
375. Gözde Gül Sahin and Mark Steedman. "Data Augmentation via Dependency Tree Morphing for Low-Resource Languages". In: *Proceedings of the 2018 Conference on Empirical Methods in Natural Language Processing, Brussels, Belgium, October 31—November 4, 2018*. Ed. by Ellen Riloff, David Chiang, Julia Hockenmaier, and Jun'ichi Tsujii. Association for Computational Linguistics, 2018, pp. 5004–5009. DOI: https://doi.org/10.18653/V1/D18-1545. URL: https://doi.org/10.18653/v1/d18-1545.
376. Tatsuhiro Sakai, Keiichi Tamura, and Hajime Kitakami. "Emergency Situation Awareness During Natural Disasters Using Density-Based Adaptive Spatiotemporal Clustering". In: *Database Systems for Advanced Applications—DASFAA 2015 International Workshops, SeCoP, BDMS, and Posters, Hanoi, Vietnam, April 20–23, 2015, Revised Selected Papers. Ed.* by An Liu, Yoshiharu Ishikawa, Tieyun Qian, Sarana Nutanong, and Muhammad Aamir Cheema. Vol. 9052. Lecture Notes in Computer Science. Springer, 2015, pp. 155–169. DOI: https://doi.org/10.1007/978-3-319-22324-7_13. URL: https://doi.org/10.1007/978-3-319-22324-7_13.
377. Aleieldin Salem, Sebastian Banescu, and Alexander Pretschner. "Maat: Automatically Analyzing VirusTotal for Accurate Labeling and Effective Malware Detection". *In: ACM Trans. Priv. Secur*. 24.4 (2021), 25:1–25:35. DOI: https://doi.org/10.1145/3465361. URL: https://doi.org/10.1145/3465361.
378. David Samuel. "BERTs are Generative In-Context Learners". In: *Advances in Neural Information Processing Systems 38: Annual Conference on Neural Information Processing Systems 2024, NeurIPS 2024, Vancouver, BC, Canada, December 10–15, 2024*. Ed. by Amir Globersons, Lester Mackey, Danielle Belgrave, Angela Fan, Ulrich Paquet, Jakub M. Tomczak, and Cheng Zhang. 2024. URL: http://papers.nips.cc/paper_files/paper/2024/hash/04ea184dfb5f1babb78c093e850a83f9-Abstract-Conference.html.
379. Victor Sanh, Lysandre Debut, Julien Chaumond, and Thomas Wolf. "DistilBERT, a distilled version of BERT: smaller, faster, cheaper and lighter". In: *CoRR* abs/1910.01108 (2019). arXiv:1910.01108. URL: http://arxiv.org/abs/1910.01108.
380. Iqbal H. Sarker, A. S. M. Kayes, Shahriar Badsha, Hamed AlQahtani, Paul A. Watters, and Alex Ng. "Cybersecurity data science: an overview from machine learning perspective". In: *J. Big Data* 7.1 (2020), p. 41. DOI: https://doi.org/10.1186/S40537-020-00318-5. URL: https://doi.org/10.1186/s40537-020-00318-5.
381. Quentin Le Sceller, ElMouatez Billah Karbab, Mourad Debbabi, and Farkhund Iqbal. "SONAR: Automatic Detection of Cyber Security Events over the Twitter Stream". In: *Proceedings of the 12th International Conference on Availability, Reliability and Security, Reggio Calabria, Italy, August 29—September 01, 2017*. ACM, 2017, 23:1–23:11. DOI: https://doi.org/10.1145/3098954.3098992. URL: https://doi.org/10.1145/3098954.3098992.
382. Timo Schick, Jane Dwivedi-Yu, Roberto Dessì, Roberta Raileanu, Maria Lomeli, Eric Hambro, Luke Zettlemoyer, Nicola Cancedda, and Thomas Scialom. "Toolformer: Language Models Can Teach Themselves to Use Tools". In: *Advances in Neural*

Information Processing Systems 36: Annual Conference on Neural Information Processing Systems 2023, NeurIPS 2023, New Orleans, LA, USA, December 10–16, 2023. Ed. by Alice Oh, Tristan Naumann, Amir Globerson, Kate Saenko, Moritz Hardt, and Sergey Levine. 2023. URL: http://papers.nips.cc/paper_files/paper/2023/hash/d842425e4bf79ba039352da0f658a906-Abstract-Conference.html.

383. Timo Schick and Hinrich Schütze. "Exploiting Cloze-Questions for Few-Shot Text Classification and Natural Language Inference". In: *Proceedings of the 16th Conference of the European Chapter of the Association for Computational Linguistics: Main Volume, EACL 2021, Online, April 19–23, 2021*. Ed. by Paola Merlo, Jörg Tiedemann, and Reut Tsarfaty. Association for Computational Linguistics, 2021, pp. 255–269. DOI: https://doi.org/10.18653/V1/2021.EACL-MAIN.20. URL: https://doi.org/10.18653/v1/2021.eacl-main.20.
384. Timo Schick and Hinrich Schütze. "It's Not Just Size That Matters: Small Language Models Are Also Few-Shot Learners". In: *Proceedings of the 2021 Conference of the North American Chapter of the Association for Computational Linguistics: Human Language Technologies, NAACL-HLT 2021, Online, June 6–11, 2021*. Ed. by Kristina Toutanova, Anna Rumshisky, Luke Zettle moyer, Dilek Hakkani-Tür, Iz Beltagy, Steven Bethard, Ryan Cotterell, Tanmoy Chakraborty, and Yichao Zhou. Association for Computational Linguistics, 2021, pp. 2339–2352. DOI: https://doi.org/10.18653/V1/2021.NAACL-MAIN.185. URL: https://doi.org/10.18653/v1/2021.naaclmain.185.
385. Tobias Schnabel, Igor Labutov, David M. Mimno, and Thorsten Joachims. "Evaluation methods for unsupervised word embeddings". In: *Proceedings of the 2015 Conference on Empirical Methods in Natural Language Processing, EMNLP 2015, Lisbon, Portugal, September 17–21, 2015*. Ed. by Lluís Màrquez, Chris Callison-Burch, Jian Su, Daniele Pighin, and Yuval Marton. The Association for Computational Linguistics, 2015, pp. 298–307. DOI: https://doi.org/10.18653/V1/D15-1036. URL: https://doi.org/10.18653/v1/d15-1036.
386. Christopher Schröder, Lydia Müller, Andreas Niekler, and Martin Potthast. "Small-Text: Active Learning for Text Classification in Python". In: *Proceedings of the 17th Conference of the European Chapter of the Association for Computational Linguistics: System Demonstrations*. Dubrovnik, Croatia: Association for Computational Linguistics, May 2023, pp. 84–95. URL: https://www.aclanthology.org/2023.eacl-demo.11.
387. Christopher Schröder, Andreas Niekler, and Martin Potthast. "Revisiting Uncertainty-based Query Strategies for Active Learning with Transformers". In: *Findings of the Association for Computational Linguistics: ACL 2022, Dublin, Ireland, May 22–27, 2022*. Ed. by Smaranda Muresan, Preslav Nakov, and Aline Villavicencio. Association for Computational Linguistics, 2022, pp. 2194–2203. DOI: https://doi.org/10.18653/V1/2022.FINDINGS-ACL.172. URL: https://doi.org/10.18653/v1/2022.findingsacl.172.
388. Axel Schulz, Christian Guckelsberger, and Frederik Janssen. "Semantic Abstraction for generalization of tweet classification: An evaluation of incident-related tweets". In: *Semantic Web 8.3* (2017), pp. 353–372. DOI: https://doi.org/10.3233/SW-150188. URL: https://doi.org/10.3233/SW150188.
389. Eli Schwartz, Leonid Karlinsky, Joseph Shtok, Sivan Harary, Mattias Marder, Abhishek Kumar, Rogé Schmidt Feris, Raja Giryes, and Alexander M. Bronstein. "Delta-encoder: an effective sample synthesis method for few-shot object recognition". In: *Advances in Neural Information Processing Systems 31: Annual Conference on Neural Information Processing Systems 2018, NeurIPS 2018, December 3–8, 2018, Montréal, Canada*. Ed.

by Samy Bengio, Hanna M. Wallach, Hugo Larochelle, Kristen Grauman, Nicolo Cesa-Bianchi, and Roman Garnett. 2018, pp. 2850–2860. URL: https://www.proceedings.neurips.cc/paper/2018/hash/1714726c817af50457d810aae9d27a2e-Abstract.html.

390. Ravid Schwartz-Ziv and Naftali Tishby. "Opening the black box of deep neural networks via information". In: (2017).
391. Rico Sennrich, Barry Haddow, and Alexandra Birch. "Improving Neural Machine Translation Models with Monolingual Data". In: *Proceedings of the 54th Annual Meeting of the Association for Computational Linguistics, ACL 2016, August 7–12, 2016, Berlin, Germany, Volume 1: Long Papers*. The Association for Computer Linguistics, 2016. DOI: https://doi.org/10.18653/V1/P16-1009. URL: https://doi.org/10.18653/v1/p16-1009.
392. Seungmin Seo, Donghyun Kim, Youbin Ahn, and Kyong-Ho Lee. "Active Learning on Pre-trained Language Model with Task-Independent Triplet Loss". In: *Thirty-Sixth AAAI Conference on Artificial Intelligence, AAAI 2022, Thirty-Fourth Conference on Innovative Applications of Artificial Intelligence, IAAI 2022, The Twelveth Symposium on Educational Advances in Artificial Intelligence, EAAI 2022 Virtual Event, February 22—March 1, 2022*. AAAI Press, 2022, pp. 11276–11284. DOI: https://doi.org/10.1609/AAAI.V36I10.21378. URL: https://doi.org/10.1609/aaai.v36i10.21378.
393. Burr Settles. *Active Learning*. Synthesis Lectures on Artificial Intelligence and Machine Learning. Morgan & Claypool Publishers, 2012. ISBN: 978-3-031-00432-2. DOI: https://doi.org/10.2200/S00429ED1V01Y201207AIM018. URL: https://doi.org/10.2200/S00429ED1V01Y201207AIM018.
394. Burr Settles and Mark Craven. "An Analysis of Active Learning Strategies for Sequence Labeling Tasks". In: *2008 Conference on Empirical Methods in Natural Language Processing, EMNLP 2008, Proceedings of the Conference, 25–27 October 2008, Honolulu, Hawaii, USA, A meeting of SIGDAT, a Special Interest Group of the ACL*. ACL, 2008, pp. 1070–1079. URL: https://www.aclanthology.org/D08-1112/.
395. Ali Shafahi, Mahyar Najibi, Amin Ghiasi, Zheng Xu, John P. Dickerson, Christoph Studer, Larry S. Davis, Gavin Taylor, and Tom Goldstein. "Adversarial training for free!" In: *Advances in Neural Information Processing Systems 32: Annual Conference on Neural Information Processing Systems 2019, NeurIPS 2019, December 8-14, 2019, Vancouver, BC, Canada*. Ed. by Hanna M. Wallach, Hugo Larochelle, Alina Beygelzimer, Florence d'Alché-Buc, Emily B. Fox, and Roman Garnett. 2019, pp. 3353–3364. URL: https://www.proceedings.neurips.cc/paper/2019/hash/7503cfacd12053d309b6bed5c89de212-Abstract.html.
396. Ronghua Shang, Songling Zhu, Jinhong Ren, Hangcheng Liu, and Licheng Jiao. "Evolutionary neural architecture search based on evaluation correction and functional units". In: *Knowl. Based Syst.* 251 (2022), p. 109206. DOI: https://doi.org/10.1016/J.KNOSYS.2022.109206.DOI: https://doi.org/10.1016/j.knosys.2022.109206.
397. Artem Shelmanov, Dmitry Puzyrev, Lyubov Kupriyanova, Denis Belyakov, Daniil Larionov, Nikita Khromov, Olga Kozlova, Ekaterina Artemova, Dmitry V. Dylov, and Alexander Panchenko. "Active Learning for Sequence Tagging with Deep Pre-trained Models and Bayesian Uncertainty Estimates". In: *Proceedings of the 16th Conference of the European Chapter of the Association for Computational Linguistics: Main Volume, EACL 2021, Online, April 19–23, 2021*. Ed. by Paola Merlo, Jorg Tiedemann, and Reut Tsarfaty. Association for Computational Linguistics, 2021, pp. 1698–1712. DOI: https://doi.org/10.18653/V1/2021.EACL-MAIN.145. URL: https://doi.org/10.18653/v1/2021.eacl-main.145.

398. Dinghan Shen, Mingzhi Zheng, Yelong Shen, Yanru Qu, and Weizhu Chen. "A Simple but Tough-to-Beat Data Augmentation Approach for Natural Language Understanding and Generation". In: *CoRR* abs/2009.13818 (2020). arXiv:2009.13818. URL: https://www.arxiv.org/abs/2009.13818.
399. Haoyue Shi, Karen Livescu, and Kevin Gimpel. "Substructure Substitution: Structured Data Augmentation for NLP". In: *Findings of the Association for Computational Linguistics: ACL/IJCNLP 2021, Online Event, August 1–6, 2021*. Ed. by Chengqing Zong, Fei Xia, Wenjie Li, and Roberto Navigli. Vol. ACL/IJCNLP 2021. Findings of ACL. Association for Computational Linguistics, 2021, pp. 3494–3508. DOI: https://doi.org/10.18653/V1/2021.FINDINGS-ACL.307. URL: https://doi.org/10.18653/v1/2021.findings-acl.307.
400. Oleh Shliazhko, Alena Fenogenova, Maria Tikhonova, Anastasia Kozlova, Vladislav Mikhailov, and Tatiana Shavrina. "mGPT: Few-Shot Learners Go Multilingual". In: *Trans. Assoc. Comput. Linguistics* 12 (2024), pp. 58–79. DOI: https://doi.org/10.1162/TACL_A_00633. URL: https://doi.org/10.1162/tacl_a_00633.
401. Connor Shorten and Taghi M. Khoshgoftaar. "A survey on Image Data Augmentation for Deep Learning". In: *J. Big Data* 6 (2019), p. 60. DOI: https://doi.org/10.1186/S40537-019-0197-0. URL: https://doi.org/10.1186/s40537-019-0197-0.
402. Connor Shorten, Taghi M. Khoshgoftaar, and Borko Furht. "Text Data Augmentation for Deep Learning". In: *J. Big Data* 8.1 (2021), p. 101. DOI: https://doi.org/10.1186/S40537-021-00492-0. URL: https://doi.org/10.1186/s40537-021-00492-0.
403. Kai Shu, Deepak Mahudeswaran, Suhang Wang, Dongwon Lee, and Huan Liu. "FakeNewsNet: A Data Repository with News Content, Social Context and Dynamic Information for Studying Fake News on Social Media". In: *CoRR* abs/1809.01286 (2018). arXiv:1809.01286. URL: http://arxiv.org/abs/1809.01286.
404. Kai Shu, Amy Sliva, Suhang Wang, Jiliang Tang, and Huan Liu. "Fake News Detection on Social Media: A Data Mining Perspective". In: *SIGKDD Explor.* 19.1 (2017), pp. 22–36. DOI: https://doi.org/10.1145/3137597.3137600. URL: https://doi.org/10.1145/3137597.3137600.
405. Samuel Henrique Silva and Peyman Najafirad. "Opportunities and Challenges in Deep Learning Adversarial Robustness: A Survey". In: *CoRR* abs/2007.00753 (2020). arXiv:2007.00753. URL: https://www.arxiv.org/abs/2007.00753.
406. Patrice Y. Simard, David Steinkraus, and John C. Platt. "Best Practices for Convolutional Neural Networks Applied to Visual Document Analysis". In: *7th International Conference on Document Analysis and Recognition (ICDAR 2003), 2-Volume Set, 3–6 August 2003, Edinburgh, Scotland, UK*. IEEE Computer Society, 2003, pp. 958–962. DOI: https://doi.org/10.1109/ICDAR.2003.1227801. URL: https://doi.org/10.1109/ICDAR.2003. 1227801.
407. Dan Siroker and Steve Miller. *Topical Clustering, Summarization, and Visualization*. 2008.
408. Leslie N. Smith. "A disciplined approach to neural network hyper-parameters: Part 1—learning rate, batch size, momentum, and weight decay". In: *CoRR* abs/1803.09820 (2018). arXiv:1803.09820. URL: http://arxiv.org/abs/1803.09820.
409. Richard Socher, Alex Perelygin, Jean Wu, Jason Chuang, Christopher D. Manning, Andrew Y. Ng, and Christopher Potts. "Recursive Deep Models for Semantic Compositionality Over a Sentiment Treebank". In: *Proceedings of the 2013 Conference on Empirical Methods in Natural Language Processing, EMNLP 2013, 18–21 October*

2013, Grand Hyatt Seattle, Seattle, Washington, USA, A meeting of SIGDAT, a Special Interest Group of the ACL. ACL, 2013, pp. 1631–1642. URL: https://www.aclanthology.org/D13-1170/.

410. Richard Socher, Alex Perelygin, Jean Wu, Jason Chuang, Christopher D. Manning, Andrew Y. Ng, and Christopher Potts. "Recursive Deep Models for Semantic Compositionality Over a Sentiment Treebank". In: *Proceedings of the 2013 Conference on Empirical Methods in Natural Language Processing, EMNLP 2013, 18–21 October 2013, Grand Hyatt Seattle, Seattle, Washington, USA, A meeting of SIGDAT, a Special Interest Group of the ACL*. ACL, 2013, pp. 1631–1642. URL: https://www.aclanthology.org/D13-1170/.
411. Robert Soden and Leysia Palen. "Informating Crisis: Expanding Critical Perspectives in Crisis Informatics". In: *Proc. ACM Hum. Comput. Interact.* 2.CSCW (2018), 162:1–162:22. DOI: https://doi.org/10.1145/3274431. URL: https://doi.org/10.1145/3274431.
412. Irene Solaiman, Miles Brundage, Jack Clark, Amanda Askell, Ariel Herbert-Voss, Jeff Wu, Alec Radford, and Jasmine Wang. "Release Strategies and the Social Impacts of Language Models". In: *CoRR* abs/1908.09203 (2019). arXiv:1908.09203. URL: http://arxiv.org/abs/1908.09203.
413. Bhavna Soman. Death to the IOC: *What's Next in Threat Intelligence*. https://www.blackhat.com/us-19/briefings/schedule/#death-to-the-ioc-whats-next-in-threat-intelligence-15392. Online; accessed 28-December-2020. 2019.
414. Junshuai Song, Jiangshan Zhang, Jifeng Zhu, Mengyun Tang, and Yong Yang. "TRAttack": "Text Rewriting Attack Against Text Retrieval". In: *Proceedings of the 7th Workshop on Representation Learning for NLP, RepL4NLP at ACL 2022, Dublin, Ireland, May 26, 2022*. Ed. by Spandana Gella, He He, Bodhisattwa Prasad Majumder, Burcu Can, Eleonora Giunchiglia, Samuel Cahyawijaya, Sewon Min, Maximilian Mozes, Xiang Lorraine Li, Isabelle Augenstein, Anna Rogers, Kyunghyun Cho, Edward Grefenstette, Laura Rimell, and Chris Dyer. Association for Computational Linguistics, 2022, pp. 191–203. DOI: https://doi.org/10.18653/V1/2022.REPL4NLP-1.20. URL: https://doi.org/10.18653/v1/2022.repl4nlp-1.20.
415. Fabio Souza, Rodrigo Frassetto Nogueira, and Roberto de Alencar Lotufo. "Portuguese Named Entity Recognition using BERT-CRF". In: *CoRR* abs/1909.10649 (2019). arXiv:1909.10649. URL: http://arxiv.org/abs/1909.10649.
416. T Spielhofer, R Greenlaw, D Markham, and A Hahne. "Data mining Twitter during the UK floods: Investigating the potential use of social media in emergency management". In: *2016 3rd International Conference on Information and Communication Technologies for Disaster Management (ICT-DM)* (2016), pp. 1–6. DOI: https://doi.org/10.1109/ICT-DM.2016.7857213.
417. Nitish Srivastava, Geoffrey E. Hinton, Alex Krizhevsky, Ilya Sutskever, and Ruslan Salakhutdinov. "Dropout: a simple way to prevent neural networks from overfitting". In: *J. Mach. Learn. Res.* 15.1 (2014), pp. 1929–1958. DOI: https://doi.org/10.5555/2627435.2670313. URL: https://doi.org/10.5555/2627435.2670313.
418. Johan Frederik Steffensen. *Interpolation*. Courier Corporation, 2006.
419. Stefan Stieglitz, Linh Dang-Xuan, Axel Bruns, and Christoph Neuberger. "Social Media Analytics—An Interdisciplinary Approach and Its Implications for Information Systems". In: *Bus. Inf. Syst. Eng.* 6.2 (2014), pp. 89–96. DOI: https://doi.org/10.1007/S12599-014-0315-7. URL: https://doi.org/10.1007/s12599-014-0315-7.

420. Stefan Stieglitz, Milad Mirbabaie, Jennifer Fromm, and Stefanie Melzer. "The Adoption of social media analytics for crisis management—Challenges and Opportunities". In: *26th European Conference on Information Systems: Beyond Digitization—Facets of Socio-Technical Change, ECIS 2018, Portsmouth, UK, June 23–28, 2018*. Ed. by Peter M. Bednar, Ulrich Frank, and Karlheinz Kautz. 2018, p. 4. URL: https://www.aisel.aisnet.org/ecis2018_rp/4.
421. Stefan Stieglitz, Milad Mirbabaie, Björn Ross, and Christoph Neuberger. "Social media analytics—Challenges in topic discovery, data collection, and data preparation". In: *Int. J. Inf. Manag.* 39 (2018), pp. 156–168. DOI: https://doi.org/10.1016/J.IJINFOMGT.2017.12.002. URL: https://doi.org/10.1016/j.ijinfomgt.2017.12.002.
422. Emma Strubell, Ananya Ganesh, and Andrew McCallum. "Energy and Policy Considerations for Deep Learning in NLP". In: *Proceedings of the 57th Conference of the Association for Computational Linguistics, ACL 2019, Florence, Italy, July 28- August 2, 2019, Volume 1: Long Papers*. Ed. by Anna Korhonen, David R. Traum, and Lluís Marquez. Association for Computational Linguistics, 2019, pp. 3645–3650. DOI: https://doi.org/10.18653/V1/P19-1355. URL: https://doi.org/10.18653/v1/p19-1355.
423. Lixin Su, Jiafeng Guo, Ruqing Zhang, Yixing Fan, Yanyan Lan, and Xueqi Cheng. "Continual Domain Adaptation for Machine Reading Comprehension". In: *CIKM '20: The 29th ACM International Conference on Information and Knowledge Management, Virtual Event, Ireland, October 19–23, 2020*. Ed. by Mathieu d'Aquin, Stefan Dietze, Claudia Hauff, Edward Curry, and Philippe Cudre-Mauroux. ACM, 2020, pp. 1395–1404. DOI: https://doi.org/10.1145/3340531.3412047. URL: https://doi.org/10.1145/3340531.3412047.
424. Sandhya Sukhabogi and M. Anusha. "A Theoretical review on the importance of Threat Intelligence Sharing & The challenges intricated". en. In: *Turkish Journal of Computer and Mathematics Education (TURCOMAT)* 12.3 (Apr. 2021), pp. 3950–3956. ISSN: 1309–4653. DOI: https://doi.org/10.17762/turcomat.v12i3.1684. URL: https://www.turcomat.org/index.php/turkbilmat/article/view/1684 (visited on 11/22/2023).
425. Chen Sun, Abhinav Shrivastava, Saurabh Singh, and Abhinav Gupta. "Revisiting Unreasonable Effectiveness of Data in Deep Learning Era". In: *IEEE International Conference on Computer Vision, ICCV 2017, Venice, Italy, October 22–29, 2017*. IEEE Computer Society, 2017, pp. 843–852. DOI: https://doi.org/10.1109/ICCV.2017.97. URL: https://doi.org/10.1109/ICCV.2017.97.
426. Chi Sun, Xipeng Qiu, Yige Xu, and Xuanjing Huang. "How to Fine-Tune BERT for Text Classification?" In: *Chinese Computational Linguistics—18th China National Conference, CCL 2019, Kunming, China, October 18–20, 2019, Proceedings*. Ed. by Maosong Sun, Xuanjing Huang, Heng Ji, Zhiyuan Liu, and Yang Liu. Vol. 11856. Lecture Notes in Computer Science. Springer, 2019, pp. 194–206. DOI: https://doi.org/10.1007/978-3-030-32381-3_16. URL: https://doi.org/10.1007/978-3-030-32381-3_16.
427. Lichao Sun, Congying Xia, Wenpeng Yin, Tingting Liang, Philip S. Yu, and Lifang He. "Mixup-Transfomer: Dynamic Data Augmentation for NLP Tasks". In: *CoRR* abs/2010.02394 (2020). arXiv:2010.02394. URL: https://www.arxiv.org/abs/2010.02394.
428. Xiao Sun and Jiajin He. "A novel approach to generate a large scale of supervised data for short text sentiment analysis". In: *Multim. Tools Appl.* 79.9-10 (2020), pp. 5439–5459. DOI: https://doi.org/10.1007/S11042-018-5748-4. URL: https://doi.org/10.1007/s11042-018-5748-4.

429. Rich Sutton. *The Bitter Lesson*. 2019. URL: https://www.cs.utexas.edu/~eunsol/courses/data/bitter_lesson.pdf.
430. Christian Szegedy, Wojciech Zaremba, Ilya Sutskever, Joan Bruna, Dumitru Erhan, Ian J. Goodfellow, and Rob Fergus. "Intriguing properties of neural networks". In: *2nd International Conference on Learning Representations, ICLR 2014, Banff, AB, Canada, April 14–16, 2014, Conference Track Proceedings*. Ed. by Yoshua Bengio and Yann LeCun. 2014. URL: http://arxiv.org/abs/1312.6199.
431. Derek Tam, Rakesh R. Menon, Mohit Bansal, Shashank Srivastava, and Colin Raffel. "Improving and Simplifying Pattern Exploiting Training". In: *Proceedings of the 2021 Conference on Empirical Methods in Natural Lan guage Processing, EMNLP 2021, Virtual Event / Punta Cana, Dominican Republic, 7–11 November, 2021*. Ed. by Marie-Francine Moens, Xuanjing Huang, Lucia Specia, and Scott Wen-tau Yih. Association for Computational Linguistics, 2021, pp. 4980–4991. DOI: https://doi.org/10.18653/V1/2021.EMNLP-MAIN.407. URL: https://doi.org/10.18653/v1/2021.emnlp-main.407.
432. Edson C. Tandoc, Zheng Wei Lim, and Richard Ling. "Defining Fake news : A typology of scholarly definitions". In: *Digital Journalism* 6.2 (2018), pp. 137–153. ISSN: 2167082X.
433. Rohan Taori, Ishaan Gulrajani, Yann Dubois, Xuechen Li, Percy Liang, and Tatsunori B Hashimoto. Alpaca: *A Strong, Replicable Instruction- Following Model*. en. 2023.
434. Luke Taylor and Geoff Nitschke. "Improving Deep Learning with Generic Data Augmentation". In: *IEEE Symposium Series on Computational Intelligence, SSCI 2018, Bangalore, India, November 18–21, 2018*. IEEE, 2018, pp. 1542–1547. DOI: https://doi.org/10.1109/SSCI.2018.8628742. URL: https://doi.org/10.1109/SSCI.2018.8628742.
435. Wilson L. Taylor. "Cloze Procedure : A New Tool for Measuring Readability". en. In: *Journalism Quarterly* 30.4 (Sept. 1953), pp. 415–433. ISSN: 0022–5533. DOI: https://doi.org/10.1177/107769905303000401. URL: http://journals.sagepub.com/doi/10.1177/107769905303000401 (visited on 07/05/2022).
436. Ugur Tekin and Ercan Nurcan Yilmaz. "Obtaining Cyber Threat Intelligence Data From Twitter With Deep Learning Methods". en. In: *2021 5th International Symposium on Multidisciplinary Studies and Innovative Technologies (ISMSIT)*. Ankara, Turkey: IEEE, Oct. 2021, pp. 82–86. ISBN: 978-1-66544-930-4. DOI: https://doi.org/10.1109/ISMSIT52890.2021.9604715. URL: https://www.ieeexplore.ieee.org/document/9604715/ (vis ited on 12/20/2023).
437. Ian Tenney, Patrick Xia, Berlin Chen, Alex Wang, Adam Poliak, R. Thomas McCoy, Najoung Kim, Benjamin Van Durme, Samuel R. Bowman, Dipanjan Das, and Ellie Pavlick. "What do you learn from context? Probing for sentence structure in contextualized word representations". In: *7th International Conference on Learning Representations, ICLR 2019, New Orleans, LA, USA, May 6–9, 2019*. OpenReview.net, 2019. URL: https://www.openreview.net/forum?id=SJzSgnRcKX.
438. James Thorne and Andreas Vlachos. "Adversarial attacks against Fact Extraction and VERification". In: *CoRR* abs/1903.05543 (2019). arXiv:1903.05543. URL: http://arxiv.org/abs/1903.05543.
439. Lin Tian, Xiuzhen Zhang, and Min Peng. "FakeFinder: Twitter Fake News Detection on Mobile". In: *Companion of The 2020 Web Conference 2020, Taipei, Taiwan, April 20–24, 2020*. Ed. by Amal El Fallah Seghrouchni, Gita Sukthankar, Tie-Yan Liu, and Maarten van Steen. ACM / IW3C2, 2020, pp. 79–80. DOI: https://doi.org/10.1145/3366424.3382706. URL: https://doi.org/10.1145/3366424.3382706.

440. Lin Tian, Xiuzhen Zhang, Yan Wang, and Huan Liu. "Early Detection of Rumours on Twitter via Stance Transfer Learning". In: *Advances in Information Retrieval—42nd European Conference on IR Research, ECIR 2020, Lisbon, Portugal, April 14–17, 2020, Proceedings, Part I*. Ed. by Joemon M. Jose, Emine Yilmaz, João Magalhães, Pablo Castells, Nicola Ferro, Mário J. Silva, and Flávio Martins. Vol. 12035. Lecture Notes in Computer Science. Springer, 2020, pp. 575–588. DOI: https://doi.org/10.1007/978-3-030-45439-5_38. URL: https://doi.org/10.1007/978-3-030-45439-5_38.
441. Naftali Tishby and Noga Zaslavsky. "Deep learning and the information bottleneck principle". In: *2015 IEEE Information Theory Workshop, ITW 2015, Jerusalem, Israel, April 26—May 1, 2015*. IEEE, 2015, pp. 1–5. DOI: https://doi.org/10.1109/ITW.2015.7133169. URL: https://doi.org/10.1109/ITW.2015.7133169.
442. Josh Tobin, Rachel Fong, Alex Ray, Jonas Schneider, Wojciech Zaremba, and Pieter Abbeel. "Domain randomization for transferring deep neural networks from simulation to the real world". In: *2017 IEEE/RSJ International Conference on Intelligent Robots and Systems, IROS 2017, Vancouver, BC, Canada, September 24–28, 2017*. IEEE, 2017, pp. 23–30. DOI: https://doi.org/10.1109/IROS.2017.8202133. URL: https://doi.org/10.1109/IROS.2017.8202133.
443. Simon Tong and Daphne Koller. "Support Vector Machine Active Learning with Applications to Text Classification". In: *J. Mach. Learn. Res.* 2 (2001), pp. 45–66. URL: https://www.jmlr.org/papers/v2/tong01a.html.
444. Lisa Torrey and Jude Shavlik. "Transfer learning". In: *Handbook of research on machine learning applications and trends: algorithms, methods, and techniques*. IGI global, 2010, pp. 242–264.
445. Wiem Tounsi and Helmi Rais. "A survey on technical threat intelligence in the age of sophisticated cyber attacks". In: *Comput. Secur.* 72 (2018), pp. 212–233. DOI: https://doi.org/10.1016/J.COSE.2017.09.001. URL: https://doi.org/10.1016/j.cose.2017.09.001.
446. Hugo Touvron, Thibaut Lavril, Gautier Izacard, Xavier Martinet, Marie-Anne Lachaux, Timothée Lacroix, Baptiste Roziére, Naman Goyal, Eric Hambro, Faisal Azhar, Aurélien Rodriguez, Armand Joulin, Edouard Grave, and Guillaume Lample. "LLaMA: Open and Efficient Foundation Language Models". In: *CoRR* abs/2302.13971 (2023). DOI: https://doi.org/10.48550/ARXIV.2302.13971. arXiv:2302.13971. URL: https://doi.org/10.48550/arXiv.2302.13971.
447. Marcos V. Treviso, Ji-Ung Lee, Tianchu Ji, Betty van Aken, Qingqing Cao, Manuel R. Ciosici, Michael Hassid, Kenneth Heafield, Sara Hooker, Colin Raffel, Pedro Henrique Martins, André F. T. Martins, Jessica Zosa Forde, Peter A. Milder, Edwin Simpson, Noam Slonim, Jesse Dodge, Emma Strubell, Niranjan Balasubramanian, Leon Derczynski, Iryna Gurevych, and Roy Schwartz. "Efficient Methods for Natural Language Processing: A Survey". In: Trans. Assoc. Comput. Linguistics 11 (2023), pp. 826–860. DOI:'https://doi.org/10.1162/TACL_A_00577. URL: https://doi.org/10.1162/tacl_a_00577.
448. Akim Tsvigun, Artem Shelmanov, Gleb Kuzmin, Leonid Sanochkin, Daniil Larionov, Gleb Gusev, Manvel Avetisian, and Leonid Zhukov. "wards Computationally Feasible Deep Active Learning". In: *Findings of the Association for Computational Linguistics: NAACL 2022, Seattle, WA, United States, July 10–15, 2022*. Ed. by Marine Carpuat, Marie-Catherine de Marneffe, and Iván Vladimir Meza Ruíz. Association for Computational Linguistics, 2022, pp. 1198–1218. DOI: https://doi.org/10.18653/V1/2022.FINDINGS-NAACL.90. URL: https://doi.org/10.18653/v1/2022.findings-naacl.90.

449. Lewis Tunstall, Nils Reimers, Unso Eun Seo Jo, Luke Bates, Daniel Korat, Moshe Wasserblat, and Oren Pereg. "Efficient Few-Shot Learning Without Prompts". In: *CoRR* abs/2209.11055 (2022). DOI: https://doi.org/10.48550/ARXIV.2209.11055. arXiv:2209.11055. URL: https://doi.org/10.48550/arXiv.2209.11055.
450. Aleksandra Urman and Mykola Makhortykh. *The Silence of the LLMs: Cross-Lingual Analysis of Political Bias and False Information Prevalence in ChatGPT, Google Bard, and Bing Chat*. en. 2023. URL: https://doi.org/10.31219/osf.io/q9v8f.
451. Clara Vania, Yova Kementchedjhieva, Anders Søgaard, and Adam Lopez. "A systematic comparison of methods for low-resource dependency parsing on genuinely low-resource languages". In: *Proceedings of the 2019 Conference on Empirical Methods in Natural Language Processing and the 9th International Joint Conference on Natural Language Processing, EMNLPIJCNLP 2019, Hong Kong, China, November 3–7, 2019*. Ed. by Kentaro Inui, Jing Jiang, Vincent Ng, and Xiaojun Wan. Association for Computational Linguistics, 2019, pp. 1105–1116. DOI: https://doi.org/10.18653/V1/D19-1102. URL: https://doi.org/10.18653/v1/D19-1102.
452. Joaquin Vanschoren. "Meta-Learning: A Survey". In: *CoRR* (2018). arXiv:1810.03548. URL: http://arxiv.org/abs/1810.03548.
453. Ashish Vaswani, Noam Shazeer, Niki Parmar, Jakob Uszkoreit, Llion Jones, Aidan N. Gomez, Lukasz Kaiser, and Illia Polosukhin. "Attention is All you Need". In: *Advances in Neural Information Processing Systems 30: Annual Conference on Neural Information Processing Systems 2017, December 4–9, 2017, Long Beach, CA, USA*. Ed. by Isabelle Guyon, Ulrike von Luxburg, Samy Bengio, Hanna M. Wallach, Rob Fergus, S. V. N. Vishwanathan, and Roman Garnett. 2017, pp. 5998–6008. URL: https://www.proceedings.neurips.cc/paper/2017/hash/3f5ee243547dee91fbd053c1c4a845aa-Abstract.html.
454. Deepika Verma, Duncan Jansen, Kerstin Bach, Mannes Poel, Paul Jarle Mork, and Wendy Oude Nijeweme d'Hollosy. "Exploratory application of machine learning methods on patient reported data in the development of supervised models for predicting outcomes". In: *BMC Medical Informatics Decis*. Mak. 22.1 (2022), p. 227. DOI: https://doi.org/10.1186/S12911-022-01973-9. URL: https://doi.org/10.1186/s12911-022-01973-9.
455. Sudha Verma, Sarah Vieweg, William J. Corvey, Leysia Palen, James H. Martin, Martha Palmer, Aaron Schram, and Kenneth Mark Anderson. "Natural Language Processing to the Rescue? Extracting "Situational Awareness" Tweets During Mass Emergency". In: *Proceedings of the Fifth International Conference on Weblogs and Social Media, Barcelona, Catalonia, Spain, July 17–21, 2011*. Ed. by Lada A. Adamic, Ricardo Baeza-Yates, and Scott Counts. The AAAI Press, 2011. URL: http://www.aaai.org/ocs/index.php/ICWSM/ICWSM11/paper/view/2834.
456. Vikas Verma, Alex Lamb, Christopher Beckham, Amir Najafi, Ioannis Mitliagkas, David Lopez-Paz, and Yoshua Bengio. "Manifold Mixup: Better Representations by Interpolating Hidden States". In: *Proceedings of the 36th International Conference on Machine Learning, ICML 2019, 9–15 June 2019, Long Beach, California, USA*. Ed. by Kamalika Chaudhuri and Ruslan Salakhutdinov. Vol. 97. Proceedings of Machine Learning Research. PMLR, 2019, pp. 6438–6447. URL: http://proceedings.mlr.press/v97/verma19a.html.
457. Sarah Elizabeth Vieweg. "Situational Awareness in Mass Emergency: A Behavioral and Linguistic Analysis of Microblogged Communications". In: (2012), pp. 1–300.

458. Marco Viviani and Gabriella Pasi. "Credibility in social media: Opinions, news, and health information-a survey: Credibility in social media". In: *Wiley Interdisciplinary Reviews: Data Mining and Knowledge Discovery* 7.5 (2017), e1209. ISSN: 19424787.
459. Thomas D. Wagner, Khaled Mahbub, Esther Palomar, and Ali E. Abdallah. "Cyber threat intelligence sharing: Survey and research directions". In: *Comput. Secur.* 87 (2019). DOI: https://doi.org/10.1016/J.COSE.2019.101589. URL: https://doi.org/10.1016/j.cose.2019.101589.
460. Thomas Walshe, Sae Young Moon, Chunyang Xiao, Yawwani Gunawar-dana, and Fran Silavong. "Automatic Labelling with Open-source LLMs using Dynamic Label Schema Integration". In: *CoRR* abs/2501.12332 (2025). DOI: https://doi.org/10.48550/ARXIV.2501.12332. arXiv:2501.12332. URL: https://doi.org/10.48550/arXiv.2501.12332.
461. Juncheng Wan, Jian Yang, Shuming Ma, Dongdong Zhang, Weinan Zhang, Yong Yu, and Zhoujun Li. "PAEG: Phrase-level Adversarial Example Generation for Neural Machine Translation". In: *Proceedings of the 29th International Conference on Computational Linguistics, COLING 2022, Gyeongju, Republic of Korea, October 12–17, 2022*. Ed. by Nicoletta Calzolari, Chu-Ren Huang, Hansaem Kim, James Pustejovsky, Leo Wanner, Key-Sun Choi, Pum-Mo Ryu, Hsin-Hsi Chen, Lucia Donatelli, Heng Ji, Sadao Kurohashi, Patrizia Paggio, Nianwen Xue, Seokhwan Kim, Younggyun Hahm, Zhong He, Tony Kyungil Lee, Enrico Santus, Francis Bond, and Seung-Hoon Na. International Committee on Computational Linguistics, 2022, pp. 5085–5097. URL: https://www.aclanthology.org/2022.coling-1.451.
462. Zhaohong Wan, Xiaojun Wan, and Wenguang Wang. "Improving Grammatical Error Correction with Data Augmentation by Editing Latent Representation". In: *Proceedings of the 28th International Conference on Computational Linguistics, COLING 2020, Barcelona, Spain (Online), December 8–13, 2020*. Ed. by Donia Scott, Núria Bel, and Chengqing Zong. International Committee on Computational Linguistics, 2020, pp. 2202–2212. DOI: https://doi.org/10.18653/V1/2020.COLING-MAIN.200. URL: https://doi.org/10.18653/v1/2020.coling-main.200.
463. Alex Wang, Yada Pruksachatkun, Nikita Nangia, Amanpreet Singh, Julian Michael, Felix Hill, Omer Levy, and Samuel R. Bowman. "SuperGLUE: A Stickier Benchmark for General-Purpose Language Understanding Systems". In: *Advances in Neural Information Processing Systems 32: Annual Conference on Neural Information Processing Systems 2019, NeurIPS 2019, December 8–14, 2019, Vancouver, BC, Canada*. Ed. by Hanna M. Wallach, Hugo Larochelle, Alina Beygelzimer, Florence d'Alché-Buc, Emily B. Fox, and Roman Garnett. 2019, pp. 3261–3275. URL: https://www.proceedings.neurips.cc/paper/2019/hash/4496bf24afe7fab6f046bf4923da8de6-Abstract.html.
464. Alex Wang, Amanpreet Singh, Julian Michael, Felix Hill, Omer Levy, and Samuel R. Bowman. "GLUE: A Multi-Task Benchmark and Analysis Platform for Natural Language Understanding". In: *Proceedings of the Workshop: Analyzing and Interpreting Neural Networks for NLP, BlackboxNLP at EMNLP 2018, Brussels, Belgium, November 1, 2018*. Ed. by Tal Linzen, Grzegorz Chrupala, and Afra Alishahi. Association for Computational Linguistics, 2018, pp. 353–355. DOI: https://doi.org/10.18653/V1/W18-5446. URL: https://doi.org/10.18653/v1/w18-5446.
465. Bing Wang, Liang Ding, Qihuang Zhong, Ximing Li, and Dacheng Tao. "A Contrastive Cross-Channel Data Augmentation Framework for Aspect-Based Sentiment Analysis". In: *Proceedings of the 29th International Conference on Computational Linguistics*. Ed. by Nicoletta Calzolari, Chu-Ren Huang, Hansaem Kim, James Pustejovsky, Leo

Wanner, Key-Sun Choi, Pum-Mo Ryu, Hsin-Hsi Chen, Lucia Donatelli, Heng Ji, Sadao Kurohashi, Patrizia Paggio, Nianwen Xue, Seokhwan Kim, Younggyun Hahm, Zhong He, Tony Kyungil Lee, Enrico Santus, Francis Bond, and Seung-Hoon Na. Gyeongju, Republic of Korea: International Committee on Computational Linguistics, Oct. 2022, pp. 6691–6704. URL: https://www.aclanthology.org/2022.coling-1.581/.

466. Boxin Wang, Chejian Xu, Shuohang Wang, Zhe Gan, Yu Cheng, Jianfeng Gao, Ahmed Hassan Awadallah, and Bo Li. "Adversarial GLUE: A Multi-Task Benchmark for Robustness Evaluation of Language Models". In: *Proceedings of the Neural Information Processing Systems Track on Datasets and Benchmarks 1, NeurIPS Datasets and Benchmarks 2021, December 2021, virtual*. Ed. by Joaquin Vanschoren and Sai-Kit Yeung. 2021. URL: https://www.datasets-benchmarks-proceedings.neurips.cc/paper/2021/hash/335f5352088d7d9bf74191e006d8e24c-Abstractround2.html.

467. Congcong Wang and David Lillis. "Classification for Crisis-Related Tweets Leveraging Word Embeddings and Data Augmentation". In: *Proceedings of the Twenty-Eighth Text REtrieval Conference, TREC 2019, Gaithersburg, Maryland, USA, November 13–15, 2019. Ed.* by Ellen M. Voorhees and Angela Ellis. Vol. 1250. NIST Special Publication. National Institute of Standards and Technology (NIST), 2019. URL: https://www.trec.nist.gov/pubs/trec28/papers/CS-UCD.IS.pdf.

468. Dilin Wang, ChengYue Gong, and Qiang Liu. "Improving Neural Language Modeling via Adversarial Training". In: *Proceedings of the 36th International Conference on Machine Learning, ICML 2019, 9–15 June 2019, Long Beach, California, USA*. Ed. by Kamalika Chaudhuri and Ruslan Salakhut dinov. Vol. 97. Proceedings of Machine Learning Research. PMLR, 2019, pp. 6555–6565. URL: http://proceedings.mlr.press/v97/wang19f.html.

469. Fali Wang, Zhiwei Zhang, Xianren Zhang, Zongyu Wu, Tzuhao Mo, Qiuhao Lu, Wanjing Wang, Rui Li, Junjie Xu, Xianfeng Tang, Qi He, Yao Ma, Ming Huang, and Suhang Wang. "A Comprehensive Survey of Small Language Models in the Era of Large Language Models: Techniques, Enhancements, Applications, Collaboration with LLMs, and Trustworthiness". In: *CoRR* abs/2411.03350 (2024). DOI: https://doi.org/10.48550/ARXIV.2411.03350. arXiv:2411.03350. URL: https://doi.org/10.48550/arXiv.2411.03350.

470. Han Wang, Ming Shan Hee, Md. Rabiul Awal, Kenny Tsu Wei Choo, and Roy Ka-Wei Lee. "Evaluating GPT-3 Generated Explanations for Hateful Content Moderation". In: *Proceedings of the Thirty-Second International Joint Conference on Artificial Intelligence, IJCAI 2023, 19th–25th August 2023, Macao, SAR, China*. ijcai.org, 2023, pp. 6255–6263. DOI: https://doi.org/10.24963/IJCAI.2023/694. URL: https://doi.org/10.24963/ijcai.2023/694.

471. Haoyu Wang, Guozheng Ma, Cong Yu, Ning Gui, Linrui Zhang, Zhiqi Huang, Suwei Ma, Yongzhe Chang, Sen Zhang, Li Shen, Xueqian Wang, Peilin Zhao, and Dacheng Tao. "Are Large Language Models Really Robust to Word-Level Perturbations?" In: *CoRR* abs/2309.11166 (2023). DOI: https://doi.org/10.48550/ARXIV.2309.11166. arXiv:2309.11166. URL: https://doi.org/10.48550/arXiv.2309.11166.

472. Wenqi Wang, Lina Wang, Run Wang, Aoshuang Ye, and Jianpeng Ke. "Better constraints of imperceptibility, better adversarial examples in the text". In: *Int. J. Intell. Syst.* 37.6 (2022), pp. 3440–3459. DOI: https://doi.org/10.1002/INT.22696. URL: https://doi.org/10.1002/int.22696.

473. William Yang Wang and Diyi Yang. "That's So Annoying!!!: A Lexical and Frame-Semantic Embedding Based Data Augmentation Approach to Automatic Categorization of Annoying Behaviors using #petpeeve Tweets". In: *Proceedings of the 2015 Conference on Empirical Methods in Natural Language Processing, EMNLP 2015, Lisbon, Portugal, September 17–21, 2015*. Ed. by Lluís Márquez, Chris Callison-Burch, Jian Su, Daniele Pighin, and Yuval Marton. The Association for Computational Linguistics, 2015, pp. 2557–2563. DOI: https://doi.org/10.18653/V1/D15-1306. URL: https://doi.org/10.18653/v1/d15-1306.

474. Xingkai Wang, Yiqiang Sheng, Haojiang Deng, and Zhenyu Zhao. "Charcnn-svm for chinese text datasets sentiment classification with data augmentation". In: International Journal of Innovative Computing, Information and Control (2019). ISSN: 13494198. DOI: https://doi.org/10.24507/ijicic.15.01.227.

475. Xinru Wang, Hannah Kim, Sajjadur Rahman, Kushan Mitra, and Zhengjie Miao. "Human-LLM Collaborative Annotation Through Effective Verification of LLM Labels". In: *Proceedings of the 2024 CHI Conference on Human Factors in Computing Systems*. CHI '24. Hon olulu, HI, USA: Association for Computing Machinery, 2024. ISBN: 9798400703300. DOI: https://doi.org/10.1145/3613904.3641960. URL: https://doi.org/10.1145/3613904.3641960.

476. Xinyi Wang, Hieu Pham, Zihang Dai, and Graham Neubig. "SwitchOut: an Efficient Data Augmentation Algorithm for Neural Machine Translation". In: *Proceedings of the 2018 Conference on Empirical Methods in Natural Language Processing, Brussels, Belgium, October 31—November 4, 2018*. Ed. by Ellen Riloff, David Chiang, Julia Hockenmaier, and Jun'ichi Tsujii. Association for Computational Linguistics, 2018, pp. 856–861. DOI: https://doi.org/10.18653/V1/D18-1100. URL: https://doi.org/10.18653/v1/d18-1100.

477. Xuezhi Wang, Haohan Wang, and Diyi Yang. "Measure and Improve Robustness in NLP Models: A Survey". In: *Proceedings of the 2022 Conference of the North American Chapter of the Association for Computational Linguistics: Human Language Technologies, NAACL 2022, Seattle, WA, United States, July 10–15, 2022*. Ed. by Marine Carpuat, Marie-Catherine de Marneffe, and Iván Vladimir Meza Ruíz. Association for Computational Linguistics, 2022, pp. 4569–4586. DOI: https://doi.org/10.18653/V1/2022.NAACL-MAIN.339. URL: https://doi.org/10.18653/v1/2022.naaclmain.339.

478. Yufei Wang, Wanjun Zhong, Liangyou Li, Fei Mi, Xingshan Zeng, Wenyong Huang, Lifeng Shang, Xin Jiang, and Qun Liu. "Aligning Large Language Models with Human: A Survey". In: *CoRR* abs/2307.12966 (2023). DOI: https://doi.org/10.48550/ARXIV.2307.12966. arXiv:2307.12966. URL: https://doi.org/10.48550/arXiv.2307.12966.

479. Jason Wei, Xuezhi Wang, Dale Schuurmans, Maarten Bosma, Brian Ichter, Fei Xia, Ed H. Chi, Quoc V. Le, and Denny Zhou. "Chain-of Thought Prompting Elicits Reasoning in Large Language Models". In: *Advances in Neural Information Processing Systems 35: Annual Conference on Neural Information Processing Systems 2022, NeurIPS 2022, New Orleans, LA, USA, November 28—December 9, 2022*. Ed. by Sanmi Koyejo, S. Mohamed, A. Agarwal, Danielle Belgrave, K. Cho, and A. Oh. 2022. URL: http://papers.nips.cc/paper_files/paper/2022/hash/9d5609613524ecf4f15af0f7b31abca4-Abstract-Conference.html.

480. Jason Wei, Xuezhi Wang, Dale Schuurmans, Maarten Bosma, Brian Ichter, Fei Xia, Ed H. Chi, Quoc V. Le, and Denny Zhou. "Chain-of-Thought Prompting Elicits Reasoning in Large Language Models". In: *Advances in Neural Informa-*

tion Processing Systems 35: Annual Conference on Neural Information Processing Systems 2022, NeurIPS 2022, New Orleans, LA, USA, November 28—December 9, 2022. Ed. by Sanmi Koyejo, S. Mohamed, A. Agarwal, Danielle Belgrave, K. Cho, and A. Oh. 2022. URL: http://papers.nips.cc/paper_files/paper/2022/hash/9d5609613524ecf4f15af0f7b31abca4-AbstractConference.html.

481. Jason W. Wei and Kai Zou. "EDA: Easy Data Augmentation Techniques for Boosting Performance on Text Classification Tasks". In: *Proceedings of the 2019 Conference on Empirical Methods in Natural Language Processing and the 9th International Joint Conference on Natural Language Processing, EMNLP-IJCNLP 2019, Hong Kong, China, November 3–7, 2019*. Ed. by Kentaro Inui, Jing Jiang, Vincent Ng, and Xiaojun Wan. Association for Computational Linguistics, 2019, pp. 6381–6387. DOI: https://doi.org/10.18653/V1/D19-1670. URL: https://doi.org/10.18653/v1/D19-1670.
482. Qingsong Wen, Liang Sun, Fan Yang, Xiaomin Song, Jingkun Gao, Xue Wang, and Huan Xu. "Time Series Data Augmentation for Deep Learning: A Survey". In: *Proceedings of the Thirtieth International Joint Conference on Artificial Intelligence, IJCAI 2021, Virtual Event / Montreal, Canada, 19–27 August 2021*. Ed. by Zhi-Hua Zhou. ijcai.org, 2021, pp. 4653–4660. DOI: https://doi.org/10.24963/IJCAI.2021/631. URL: https://doi.org/10.24963/ijcai.2021/631.
483. Moira J West-Brown, Don Stikvoort, Klaus-Peter Kossakowski, Georgia Killcrece, Robin Ruefle, and Mark Zajicek. *Handbook for Computer Security Incident Response Teams (CSIRTs)*. 2003.
484. Remco H.S. Westerink and Henk P.M. Vijverberg. "Universal Sentence Encoder". In: *Toxicology and Applied Pharmacology* (2018). ISSN: 0041008X. DOI: https://doi.org/10.1006/taap.2002.9482.
485. Suzanne Widup, Dave Hylender, Gabriel Bassett, Philippe Langlois, and Alex Pinto. "2020 Verizon Data Breach Investigations Report". In: *Verizon* (May 2020). DOI: https://doi.org/10.13140/RG.2.2.21300.48008.
486. Mark A. Williams, Sumi Dey, Roberto Camacho Barranco, Sheikh Motahar Naim, Mahmud Shahriar Hossain, and Monika Akbar. "Analyzing Evolving Trends of Vulnerabilities in National Vulnerability Database". In: *IEEE International Conference on Big Data (IEEE BigData 2018), Seattle, WA, USA, December 10–13, 2018*. Ed. by Naoki Abe, Huan Liu, Calton Pu, Xiaohua Hu, Nesreen K. Ahmed, Mu Qiao, Yang Song, Donald Kossmann, Bing Liu, Kisung Lee, Jiliang Tang, Jingrui He, and Jeffrey S. Saltz. IEEE, 2018, pp. 3011–3020. DOI: https://doi.org/10.1109/BIGDATA.2018.8622299. URL: https://doi.org/10.1109/BigData.2018.8622299.
487. Ian H. Witten, Eibe Frank, Mark A. Hall, and Christopher J. Pal. *Data Mining: Practical Machine Learning Tools and Techniques*. 2016. ISBN: 978-0-12-804291-5. DOI: https://doi.org/10.1016/c2009-0-19715-5.
488. Sebastien C. Wong, Adam Gatt, Victor Stamatescu, and Mark D. Mc-Donnell. "Understanding Data Augmentation for Classification: When to Warp?" In: *2016 International Conference on Digital Image Computing: Techniques and Applications, DICTA 2016, Gold Coast, Australia, Novem ber 30—December 2, 2016*. IEEE, 2016, pp. 1–6. DOI: https://doi.org/10.1109/DICTA.2016.7797091. URL: https://doi.org/10.1109/DICTA.2016.7797091.
489. Max Woolf. *GitHub—gpt-2-simple: Python package to easily retrain OpenAI's GPT-2 text-generating model on new texts*. 2019. URL: https://www.github.com/minimaxir/gpt-2-simple.

490. Lianwei Wu, Yuan Rao, Hualei Yu, Yiming Wang, and Ambreen Nazir. "False Information Detection on Social Media via a Hybrid Deep Model". In: *Social Informatics—10th International Conference, SocInfo 2018, St. Petersburg, Russia, September 25–28, 2018, Proceedings, Part II*. Ed. by Steffen Staab, Olessia Koltsova, and Dmitry I. Ignatov. Vol. 11186. Lecture Notes in Computer Science. Springer, 2018, pp. 323–333. DOI: https://doi.org/10.1007/978-3-030-01159-8_31. URL: https://doi.org/10.1007/978-3-030-01159-8_31.
491. Xing Wu, Shangwen Lv, Liangjun Zang, Jizhong Han, and Songlin Hu. "Conditional BERT Contextual Augmentation". In: *Computational Science—ICCS 2019—19th International Conference, Faro, Portugal, June 12–14, 2019, Proceedings, Part IV*. Ed. by João M. F. Rodrigues, Pedro J. S. Cardoso, Jânio M. Monteiro, Roberto Lam, Valeria V. Krzhizhanovskaya, Michael Harold Lees, Jack J. Dongarra, and Peter M. A. Sloot. Vol. 11539. Lecture Notes in Computer Science. Springer, 2019, pp. 84–95. DOI: https://doi.org/10.1007/978-3-030-22747-0_7. URL: https://doi.org/10.1007/978-3-030-22747-0_7.
492. Yonghui Wu, Mike Schuster, Zhifeng Chen, Quoc V. Le, Mohammad Norouzi, Wolfgang Macherey, Maxim Krikun, Yuan Cao, Qin Gao, Klaus Macherey, Jeff Klingner, Apurva Shah, Melvin Johnson, Xiaobing Liu, Lukasz Kaiser, Stephan Gouws, Yoshikiyo Kato, Taku Kudo, Hideto Kazawa, Keith Stevens, George Kurian, Nishant Patil, Wei Wang, Cliff Young, Jason Smith, Jason Riesa, Alex Rudnick, Oriol Vinyals, Greg Corrado, Macduff Hughes, and Jeffrey Dean. "Google's Neural Machine Translation System: Bridging the Gap between Human and Machine Translation". In: *CoRR* abs/1609.08144 (2016). arXiv:1609.08144. URL: http://arxiv.org/abs/1609.08144.
493. Zhuofeng Wu, Sinong Wang, Jiatao Gu, Madian Khabsa, Fei Sun, and Hao Ma. "CLEAR: Contrastive Learning for Sentence Representation". In: *CoRR* abs/2012.15466 (2020). arXiv:2012.15466. URL: https://www.arxiv.org/abs/2012.15466.
494. Xin Xia, Xiaohu Yang, Chao Wu, Shanping Li, and Linfeng Bao. "Information Credibility on Twitter in Emergency Situation". In: *Intelligence and Security Informatics—Pacific Asia Workshop, PAISI 2012, Kuala Lumpur, Malaysia, May 29, 2012. Proceedings*. Ed. by Michael Chau, G. Alan Wang, Wei Thoo Yue, and Hsinchun Chen. Vol. 7299. Lecture Notes in Computer Science. Springer, 2012, pp. 45–59. DOI: https://doi.org/10.1007/978-3-642-30428-6_4. URL: https://doi.org/10.1007/978-3-642-30428-6_4.
495. Guang Xiang, Jason I. Hong, Carolyn P. Rosé, and Lorrie Faith Cranor. "CANTINA+: A Feature-Rich Machine Learning Framework for Detecting Phishing Web Sites". In: *ACM Trans. Inf. Syst. Secur.* 14.2 (2011), 21:1–21:28. DOI: https://doi.org/10.1145/2019599.2019606. URL: https://doi.org/10.1145/2019599.2019606.
496. Rong Xiang, Emmanuele Chersoni, Qin Lu, Chu-Ren Huang, Wenjie Li, and Yunfei Long. "Lexical data augmentation for sentiment analysis". In: *J. Assoc. Inf. Sci. Technol.* 72.11 (2021), pp. 1432–1447. DOI: https://doi.org/10.1002/ASI.24493. URL: https://doi.org/10.1002/asi.24493.
497. Guangxuan Xiao, Ji Lin, Mickaël Seznec, Hao Wu, Julien Demouth, and Song Han. "SmoothQuant: Accurate and Efficient Post-Training Quantization for Large Language Models". In: *International Conference on Machine Learning, ICML 2023, 23–29 July 2023, Honolulu, Hawaii, USA*. Ed. by Andreas Krause, Emma Brunskill, Kyunghyun Cho, Barbara Engelhardt, Sivan Sabato, and Jonathan Scarlett. Vol. 202. Proceedings

of Machine Learning Research. PMLR, 2023, pp. 38087–38099. URL: https://www.proceedings.mlr.press/v202/xiao23c.html.

498. Qizhe Xie, Zihang Dai, Eduard H. Hovy, Thang Luong, and Quoc Le. "Unsupervised Data Augmentation for Consistency Training". In: *Advances in Neural Information Processing Systems 33: Annual Conference on Neural Information Processing Systems 2020, NeurIPS 2020, December 6–12, 2020, virtual*. Ed. by Hugo Larochelle, Marc'Aurelio Ran-zato, Raia Hadsell, Maria-Florina Balcan, and Hsuan-Tien Lin. 2020. URL: https://www.proceedings.neurips.cc/paper/2020/hash/44feb0096faa8326192570788b38c1d1-Abstract.html.

499. Ziang Xie, Sida I. Wang, Jiwei Li, Daniel Lévy, Aiming Nie, Dan Jurafsky, and Andrew Y. Ng. "Data Noising as Smoothing in Neural Network Language Models". In: *5th International Conference on Learning Representations, ICLR 2017, Toulon, France, April 24–26, 2017, Conference Track Proceedings*. OpenReview.net, 2017. URL: https://www.openreview.net/forum?id=H1VyHY9gg.

500. Mingyuan Xin and Yong Wang. "Research on Feature Selection of Intrusion Detection Based on Deep Learning". In: *16th International Wireless Communications and Mobile Computing Conference, IWCMC 2020, Limassol, Cyprus, June 15–19, 2020*. IEEE, 2020, pp. 1431–1434. DOI: https://doi.org/10.1109/IWCMC48107.2020.9148217. URL: https://doi.org/10.1109/IWCMC48107.2020.9148217.

501. Yang Xin, Lingshuang Kong, Zhi Liu, Yuling Chen, Yan-Miao Li, Hongliang Zhu, Mingcheng Gao, Haixia Hou, and Chunhua Wang. "Machine Learning and Deep Learning Methods for Cybersecurity". In: *IEEE Access* 6 (2018), pp. 35365–35381. DOI: https://doi.org/10.1109/ACCESS.2018.2836950. URL: https://doi.org/10.1109/ACCESS.2018.2836950.

502. Mengyang Xu, Qihuang Zhong, and Juhua Liu. "LLM-as-an-Augmentor: Improving the Data Augmenta-tion for Aspect-Based Sentiment Analysis with Large Language Models". In: *ICIC2024* (2024).

503. Rui Xu and Donald C. Wunsch II. "Survey of clustering algorithms". In: *IEEE Trans. Neural Networks* 16.3 (2005), pp. 645–678. DOI: https://doi.org/10.1109/TNN.2005.845141. URL: https://doi.org/10.1109/TNN.2005.845141.

504. Yan Xu, Ran Jia, Lili Mou, Ge Li, Yunchuan Chen, Yangyang Lu, and Zhi Jin. "Improved relation classification by deep recurrent neural networks with data augmentation". In: *COLING 2016, 26th International Conference on Computational Linguistics, Proceedings of the Conference: Technical Papers, December 11–16, 2016, Osaka, Japan*. Ed. by Nicoletta Calzolari, Yuji Matsumoto, and Rashmi Prasad. ACL, 2016, pp. 1461–1470. URL: https://www.aclanthology.org/C16-1138/.

505. Yuanmeng Yan, Rumei Li, Sirui Wang, Fuzheng Zhang, Wei Wu, and Weiran Xu. "ConSERT: A Contrastive Framework for Self-Supervised Sentence Representation Transfer". In: *Proceedings of the 59th Annual Meeting of the Association for Computational Linguistics and the 11th International Joint Conference on Natural Language Processing, ACL/IJCNLP 2021, (Volume 1: Long Papers), Virtual Event, August 1–6, 2021*. Ed. by Chengqing Zong, Fei Xia, Wenjie Li, and Roberto Navigli. Association for Computational Linguistics, 2021, pp. 5065–5075. DOI: https://doi.org/10.18653/V1/2021.ACL-LONG.393. URL: https://doi.org/10.18653/v1/2021.acllong.393.

506. Aimin Yang, Chaomeng Lu, Jie Li, Xiangdong Huang, Tianhao Ji, Xi-chang Li, and Yichao Sheng. "Application of meta-learning in cyberspace security: a survey". In: *Digit.*

Commun. Networks 9.1 (2023), pp. 67–78. DOI: https://doi.org/10.1016/J.DCAN.2022.03.007. URL: https://doi.org/10.1016/j.dcan.2022.03.007.

507. Peng Yang, Guangzhen Zhao, and Peng Zeng. "Phishing Website Detection Based on Multidimensional Features Driven by Deep Learning". In: *IEEE Access* 7 (2019), pp. 15196–15209. DOI: https://doi.org/10.1109/ACCESS.2019.2892066. URL: https://doi.org/10.1109/ACCESS.2019.2892066.
508. Yiben Yang, Chaitanya Malaviya, Jared Fernandez, Swabha Swayamdipta, Ronan Le Bras, Ji-Ping Wang, Chandra Bhagavatula, Yejin Choi, and Doug Downey. "Generative data augmentation for commonsense reasoning". In: *arXiv preprint* arXiv:2004.11546 (2020).
509. Muchao Ye, Chenglin Miao, Ting Wang, and Fenglong Ma. "TextHoaxer: Budgeted Hard-Label Adversarial Attacks on Text". In: *Thirty-Sixth AAAI Conference on Artificial Intelligence, AAAI 2022, Thirty-Fourth Conference on Innovative Applications of Artificial Intelligence, IAAI 2022, The Twel veth Symposium on Educational Advances in Artificial Intelligence, EAAI 2022 Virtual Event, February 22—March 1, 2022*. AAAI Press, 2022, pp. 3877–3884. DOI: https://doi.org/10.1609/AAAI.V36I4.20303. URL: https://doi.org/10.1609/aaai.v36i4.20303.
510. Jiao Yin, MingJian Tang, Jinli Cao, and Hua Wang. "Apply transfer learning to cybersecurity: Predicting exploitability of vulnerabilities by description". In: *Knowl. Based Syst.* 210 (2020), p. 106529. DOI: https://doi.org/10.1016/J.KNOSYS.2020.106529. URL: https://doi.org/10.1016/j.knosys.2020.106529.
511. Jie Yin, Andrew Lampert, Mark A. Cameron, Bella Robinson, and Robert Power. "Using Social Media to Enhance Emergency Situation Awareness". In: *IEEE Intell. Syst.* 27.6 (2012), pp. 52–59. DOI: https://doi.org/10.1109/MIS.2012.6. URL: https://doi.org/10.1109/MIS.2012.6.
512. Kang Min Yoo, Dongju Park, Jaewook Kang, Sang-Woo Lee, and Woo-Myoung Park. "GPT3Mix: Leveraging Large-scale Language Models for Text Augmentation". In: *Findings of the Association for Computational Linguistics: EMNLP 2021, Virtual Event / Punta Cana, Dominican Republic, 16–20 November, 2021*. Ed. by Marie-Francine Moens, Xuanjing Huang, Lucia Specia, and Scott Wen-tau Yih. Association for Computational Linguistics, 2021, pp. 2225–2239. DOI: https://doi.org/10.18653/V1/2021.FINDINGS-EMNLP.192. URL: https://doi.org/10.18653/v1/2021.findings-emnlp.192.
513. Yang You, Jing Li, Jonathan Hseu, Xiaodan Song, James Demmel, and Cho-Jui Hsieh. "Reducing BERT Pre-Training Time from 3 Days to 76 Minutes". In: *CoRR* abs/1904.00962 (2019). arXiv:1904.00962. URL: http://arxiv.org/abs/1904.00962.
514. Adams Wei Yu, David Dohan, Minh-Thang Luong, Rui Zhao, Kai Chen, Mohammad Norouzi, and Quoc V. Le. "QANet: Combining Local Convolution with Global Self-Attention for Reading Comprehension". In: *6th International Conference on Learning Representations, ICLR 2018, Vancouver, BC, Canada, April 30–May 3, 2018, Conference Track Proceedings*. OpenReview.net, 2018. URL: https://www.openreview.net/forum?id=B14TlG-RW.
515. Lantao Yu, Weinan Zhang, Jun Wang, and Yong Yu. "SeqGAN: Sequence Generative Adversarial Nets with Policy Gradient". In: *Proceedings of the Thirty-First AAAI Conference on Artificial Intelligence, February 4–9, 2017, San Francisco, California, USA*. Ed. by Satinder Singh and Shaul Markovitch. AAAI Press, 2017, pp. 2852–2858. DOI: https://doi.org/10.1609/AAAI.V31I1.10804. URL: https://doi.org/10.1609/aaai.v31i1.10804.

516. Zhongjie Yu and Sebastian Raschka. "Looking Back to Lower-Level Information in Few-Shot Learning". In: *Inf.* 11.7 (2020), p. 345. DOI: https://doi.org/10.3390/INFO11070345. URL: https://doi.org/10.3390/info11070345.
517. Michelle Yuan, Hsuan-Tien Lin, and Jordan L. Boyd-Graber. "Cold-start Active Learning through Self-supervised Language Modeling". In: *Proceedings of the 2020 Conference on Empirical Methods in Natural Language Processing, EMNLP 2020, Online, November 16–20, 2020*. Ed. by Bonnie Webber, Trevor Cohn, Yulan He, and Yang Liu. Association for Computational Linguistics, 2020, pp. 7935–7948. DOI: https://doi.org/10.18653/V1/2020.EMNLP-MAIN.637. URL: https://doi.org/10.18653/v1/2020.emnlp-main.637.
518. Matthew D. Zeiler and Rob Fergus. "Stochastic Pooling for Regularization of Deep Convolutional Neural Networks". In: *1st International Conference on Learning Representations, ICLR 2013, Scottsdale, Arizona, USA, May 2–4, 2013, Conference Track Proceedings*. Ed. by Yoshua Bengio and Yann LeCun. 2013. URL: http://arxiv.org/abs/1301.3557.
519. Junhai Zhai, Jiaxing Qi, and Sufang Zhang. "Imbalanced data classification based on diverse sample generation and classifier fusion". In: *Int. J. Mach. Learn. Cybern.* 13.3 (2022), pp. 735–750. DOI: https://doi.org/10.1007/S13042-021-01321-9. URL: https://doi.org/10.1007/s13042-021-01321-9.
520. Dinghuai Zhang, Tianyuan Zhang, Yiping Lu, Zhanxing Zhu, and Bin Dong. "You Only Propagate Once: Accelerating Adversarial Training via Maximal Principle". In: *Advances in Neural Information Processing Systems 32: Annual Conference on Neural Information Processing Systems 2019, NeurIPS 2019, December 8–14, 2019, Vancouver, BC, Canada*. Ed. by Hanna M. Wallach, Hugo Larochelle, Alina Beygelzimer, Florence d'Alche-Buc, Emily B. Fox, and Roman Garnett. 2019, pp. 227–238. URL: https://www.proceedings.neurips.cc/paper/2019/hash/812b4ba287f5ee0bc9d43bbf5bbe87fb-Abstract.html.
521. Hongyi Zhang, Moustapha Cissé, Yann N. Dauphin, and David Lopez-Paz. "mixup: Beyond Empirical Risk Minimization". In: *6th International Conference on Learning Representations, ICLR 2018, Vancouver, BC, Canada, April 30–May 3, 2018, Conference Track Proceedings*. OpenReview.net, 2018. URL: https://www.openreview.net/forum?id=r1Ddp1-Rb.
522. Ningyu Zhang, Luoqiu Li, Xiang Chen, Shumin Deng, Zhen Bi, Chuanqi Tan, Fei Huang, and Huajun Chen. "Differentiable Prompt Makes Pre-trained Language Models Better Few-shot Learners". In: *The Tenth International Conference on Learning Representations, ICLR 2022, Virtual Event, April 25–29, 2022*. OpenReview.net, 2022. URL: https://www.openreview.net/forum?id=ek9a0qIafW.
523. Ningyu Zhang, Luoqiu Li, Xiang Chen, Shumin Deng, Zhen Bi, Chuanqi Tan, Fei Huang, and Huajun Chen. "Differentiable Prompt Makes Pre-trained Language Models Better Few-shot Learners". In: *The Tenth International Conference on Learning Representations, ICLR 2022, Virtual Event, April 25–29, 2022*. OpenReview.net, 2022. URL: https://www.openreview.net/forum?id=ek9a0qIafW.
524. Xiang Zhang and Yann LeCun. "Text Understanding from Scratch". In: *CoRR* abs/1502.01710 (2015). arXiv:1502.01710. URL: http://arxiv.org/abs/1502.01710.
525. Xiang Zhang, Junbo Jake Zhao, and Yann LeCun. "Character-level Convolutional Networks for Text Classification". In: *Advances in Neural Information Processing Systems 28: Annual Conference on Neural Information Processing Sys-*

tems 2015, December 7–12, 2015, Montreal, Quebec, Canada. Ed. by Corinna Cortes, Neil D. Lawrence, Daniel D. Lee, Masashi Sugiyama, and Roman Garnett. 2015, pp. 649–657. URL: https://www.proceedings.neurips.cc/paper/2015/hash/250cf8b51c773f3f8dc8b4be867a9a02-Abstract.html.

526. Xiang Zhang, Junbo Jake Zhao, and Yann LeCun. "Character-level Convolutional Networks for Text Classification". In: *Advances in Neural Information Processing Systems 28: Annual Conference on Neural Information Processing Systems 2015, December 7–12, 2015, Montreal, Quebec, Canada*. Ed. by Corinna Cortes, Neil D. Lawrence, Daniel D. Lee, Masashi Sugiyama, and Roman Garnett. 2015, pp. 649–657. URL: https://www.proceedings.neurips.cc/paper/2015/hash/250cf8b51c773f3f8dc8b4be867a9a02-Abstract.html.
527. Yiming Zhang, Shi Feng, and Chenhao Tan. "Active Example Selection for In-Context Learning". In: *Proceedings of the 2022 Conference on Empirical Methods in Natural Language Processing, EMNLP 2022, Abu Dhabi, United Arab Emirates, December 7–11, 2022*. Ed. by Yoav Goldberg, Zornitsa Kozareva, and Yue Zhang. Association for Computational Linguistics, 2022, pp. 9134–9148. DOI: https://doi.org/10.18653/V1/2022.EMNLP-MAIN.622. URL: https://doi.org/10.18653/v1/2022.emnlp-main.622.
528. Yue Zhang, Yafu Li, Leyang Cui, Deng Cai, Lemao Liu, Tingchen Fu, Xinting Huang, Enbo Zhao, Yu Zhang, Yulong Chen, Longyue Wang, Anh Tuan Luu, Wei Bi, Freda Shi, and Shuming Shi. "Siren's Song in the AI Ocean: A Survey on Hallucination in Large Language Models". In: *CoRR* abs/2309.01219 (2023). DOI: https://doi.org/10.48550/ARXIV.2309.01219. arXiv:2309.01219. URL: https://doi.org/10.48550/arXiv.2309.01219.
529. Zhisong Zhang, Emma Strubell, and Eduard H. Hovy. "A Survey of Active Learning for Natural Language Processing". In: *Proceedings of the 2022 Conference on Empirical Methods in Natural Language Processing, EMNLP 2022, Abu Dhabi, United Arab Emirates, December 7–11, 2022*. Ed. by Yoav Goldberg, Zornitsa Kozareva, and Yue Zhang. Association for Computational Linguistics, 2022, pp. 6166–6190. DOI: https://doi.org/10.18653/V1/2022.EMNLP-MAIN.414. URL: https://doi.org/10.18653/v1/2022.emnlp-main.414.
530. Zhuosheng Zhang, Aston Zhang, Mu Li, and Alex Smola. "Automatic Chain of Thought Prompting in Large Language Models". In: *The Eleventh International Conference on Learning Representations, ICLR 2023, Kigali, Rwanda, May 1–5, 2023*. OpenReview.net, 2023. URL: https://www.openreview.net/pdf?id=5NTt8GFjUHkr.
531. Zhuosheng Zhang, Aston Zhang, Mu Li, and Alex Smola. "Automatic Chain of Thought Prompting in Large Language Models". In: *The Eleventh International Conference on Learning Representations, ICLR 2023, Kigali, Rwanda, May 1–5, 2023*. OpenReview.net, 2023. URL: https://www.openreview.net/pdf?id=5NTt8GFjUHkr.
532. Wayne Xin Zhao, Kun Zhou, Junyi Li, Tianyi Tang, Xiaolei Wang, Yupeng Hou, Yingqian Min, Beichen Zhang, Junjie Zhang, Zican Dong, Yifan Du, Chen Yang, Yushuo Chen, Zhipeng Chen, Jinhao Jiang, Ruiyang Ren, Yifan Li, Xinyu Tang, Zikang Liu, Peiyu Liu, Jian-Yun Nie, and Ji-Rong Wen. "A Survey of Large Language Models". In: *CoRR* abs/2303.18223 (2023). DOI: https://doi.org/10.48550/ARXIV.2303.18223. arXiv:2303.18223. URL: https://doi.org/10.48550/arXiv.2303.18223.
533. Wei Zhao, Haiyun Peng, Steffen Eger, Erik Cambria, and Min Yang. "Towards Scalable and Reliable Capsule Networks for Challenging NLP Applications". In: *Proceedings of the 57th Conference of the Association for Computational Linguistics, ACL 2019,*

Florence, Italy, July 28- August 2, 2019, Volume 1: Long Papers. Ed. by Anna Korhonen, David R. Traum, and Lluís Màrquez. Association for Computational Linguistics, 2019, pp. 1549–1559. DOI: https://doi.org/10.18653/V1/P19-1150. URL: https://doi.org/10.18653/v1/p19-1150.

534. Haoqi Zheng, Qihuang Zhong, Liang Ding, Zhiliang Tian, Xin Niu, Changjian Wang, Dongsheng Li, and Dacheng Tao. "Self-Evolution Learning for Mixup: Enhance Data Augmentation on Few-Shot Text Classification Tasks". In: *Proceedings of the 2023 Conference on Empirical Methods in Natural Language Processing*. Ed. by Houda Bouamor, Juan Pino, and Kalika Bali. Singapore: Association for Computational Linguistics, Dec. 2023, pp. 8964–8974. DOI: https://doi.org/10.18653/v1/2023.emnlp-main.555. URL: https://www.aclanthology.org/2023.emnlp-main.555/.
535. Zhun Zhong, Liang Zheng, Guoliang Kang, Shaozi Li, and Yi Yang. "Random Erasing Data Augmentation". In: *The Thirty-Fourth AAAI Conference on Artificial Intelligence, AAAI 2020, The Thirty-Second Innovative Applications of Artificial Intelligence Conference, IAAI 2020, The Tenth AAAI Symposium on Educational Advances in Artificial Intelligence, EAAI 2020, New York, NY, USA, February 7–12, 2020*. AAAI Press, 2020, pp. 13001–13008. DOI: https://doi.org/10.1609/AAAI.V34I07.7000. URL: https://doi.org/10.1609/aaai.v34i07.7000.
536. Kun Zhou, Yutao Zhu, Zhipeng Chen, Wentong Chen, Wayne Xin Zhao, Xu Chen, Yankai Lin, Ji-Rong Wen, and Jiawei Han. "Don't Make Your LLM an Evaluation Benchmark Cheater". In: *CoRR* abs/2311.01964 (2023). DOI: https://doi.org/10.48550/ARXIV.2311.01964. arXiv:2311.01964. URL: https://doi.org/10.48550/arXiv.2311.01964.
537. Xiaokang Zhou, Wei Liang, Shohei Shimizu, Jianhua Ma, and Qun Jin. "Siamese Neural Network Based Few-Shot Learning for Anomaly Detection in Industrial Cyber-Physical Systems". In: *IEEE Trans. Ind. Informatics* 17.8 (2021), pp. 5790–5798. DOI: https://doi.org/10.1109/TII.2020.3047675. URL: https://doi.org/10.1109/TII.2020.3047675.
538. Chen Zhu, Yu Cheng, Zhe Gan, Siqi Sun, Tom Goldstein, and Jingjing Liu. "FreeLB: Enhanced Adversarial Training for Natural Language Understanding". In: *8th International Conference on Learning Representations, ICLR 2020, Addis Ababa, Ethiopia, April 26–30, 2020*. OpenReview.net, 2020. URL: https://www.openreview.net/forum?id=BygzbyHFvB.
539. Xinyue Zhu, Yifan Liu, Jiahong Li, Tao Wan, and Zengchang Qin. "Emotion Classification with Data Augmentation Using Generative Adversarial Networks". In: *Advances in Knowledge Discovery and Data Mining—22nd Pacific-Asia Conference, PAKDD 2018, Melbourne, VIC, Australia, June 3–6, 2018, Proceedings, Part III*. Ed. by Dinh Q. Phung, Vincent S. Tseng, Geoffrey I. Webb, Bao Ho, Mohadeseh Ganji, and Lida Rashidi. Vol. 10939. Lecture Notes in Computer Science. Springer, 2018, pp. 349–360. DOI: https://doi.org/10.1007/978-3-319-93040-4_28. URL: https://doi.org/10.1007/978-3-319-93040-4_28.
540. Arkaitz Zubiaga, Geraldine Wong Sak Hoi, Maria Liakata, Rob Procter, and Peter Tolmie. "Analysing How People Orient to and Spread Rumours in Social Media by Looking at Conversational Threads". In: *CoRR* abs/1511.07487 (2015). arXiv:1511.07487. URL: http://arxiv.org/abs/1511.07487.

The manufacturer's authorised representative in the EU is Springer Nature Customer Service Centre GmbH, Europaplatz 3, 69115 Heidelberg, Germany. If you have any concerns regarding our products, please contact ProductSafety@springernature.com

Printed and bound by CPI Group (UK) Ltd, Croydon, CR0 4YY

30/09/2026

02225566-0005